# Global Energy Demand and 2-degree Target, Report 2014

Valentin Crastan

# Global Energy Demand and 2-degree Target, Report 2014

Springer

Valentin Crastan
Evilard
Switzerland

ISBN 978-3-319-12309-7 ISBN 978-3-319-12310-3 (eBook)
DOI 10.1007/978-3-319-12310-3

Library of Congress Control Number: 2014953231

Springer Cham Heidelberg New York Dordrecht London



Printed on acid-free paper

Springer is part of Springer Science+Business Media (www.springer.com)

# Preface

Climate science requires global warming to be limited to 2 °C, or even less if possible, relative to the pre-industrial era, as reported both in the latest IPCC climate report of the United Nations and in other studies. The reports confirm to us once again that the increased greenhouse effect is man-made and is about 75 % due to the unchecked burning of fossil fuels.

The goal can only be achieved with rapid and radical measures in the field of energy conversion and use, which will have economic consequences that encounter political resistance. That is why there is a temptation to give up this goal in favour of easier 2.5- or 3-degree targets or to suggest that we should simply adapt.

But adaptation will be necessary in any case, even if we achieve the 2-degree target, and is neither easier nor cheaper than reducing $CO_2$ emissions. The reports from climate science make it clear that with a greater level of warming, adaptation is the more expensive option, to say nothing of the destruction of many ecosystems and the associated human suffering.

Compliance with the 2-degree climate target is a challenge, but not impossible. The longer we delay taking action, the more expensive and difficult it gets. We can't say better than Thomas F. Stocker: "Both delay and insufficient mitigation efforts close the door on limiting global mean warming permanently. This constitutes more than a climate change commitment: It is the fast and irreversible shrinking, and eventual disappearance, of the mitigation options with every year of increasing greenhouse gas emissions" (The Closing Door of Climate Targets, Science 339, January 2013).

The present report takes a neutral and independent point of view in attempting to show pragmatic but concrete ways to keep open the mitigation effort, overall and for all key countries.

The feasibility of the scenarios requires detail discussion and testing, both technically and politically, and the distribution of the reductions in emissions can then be adjusted accordingly. But the overall objective cannot be abandoned. If someone generates more emissions, somebody else must compensate by producing less. This is a question of overall economic optimization that must be pursued by the efforts of local politics and achieved with the support of the global market economy (higher $CO_2$ price). To this end, promotion by international forums and purposeful international cooperation, both bilateral and multilateral, are imperative.

Evilard, April 2014 Valentin Crastan

# Contents

# Chapter 1
# Introduction

The fifth IPCC report on climate change [1–3] essentially confirmed the statements of the fourth report of 2007. Warnings about man-made global warming are even stronger. More insistently than before, the need is emphasized to reduce $CO_2$ emissions quickly, to prevent the mean temperature rise of the earth from exceeding 2 °C (the 2-degree target).

A report by the Oeschger Centre, Bern, calls for a stricter reduction of $CO_2$ emissions to apply more limiting ocean acidification (corals, calcareous shells of marine animals), carbon loss on arable land and rising of sea levels [4].

Figure 1.1 shows the annual global emission profile (Gt $CO_2$/a) from 1970 to 2011 (only fossil-fuel emissions) according to the IEA statistics [5] and some of the possible scenarios for further development up to 2100. The corresponding cumulative values in Gt of carbon from 1870 to 2100 are also shown. For the period 1870–1970, about 100 GtC were taken into account [1, 4]. For each cumulative value, there is (with 66 % probability) a certain temperature increase.

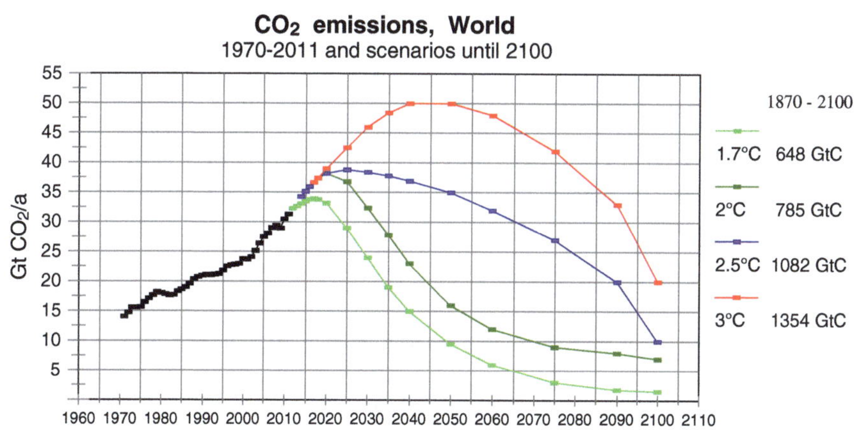


**Fig. 1.1** Worldwide $CO_2$ emissions 1970 to 2011 [5] and possible scenarios until 2100



V. Crastan, *Global Energy Demand and 2-degree Target, Report 2014*,
DOI 10.1007/978-3-319-12310-3_1

For the 2 °C limit, which is the main concern of the present study, four different scenario-variants are shown in Fig. 1.2, all with the same value of 16 Gt $CO_2$ in 2050 (only fossil fuels). Variant *b* corresponds to the 2-degree variant of Fig. 1.1.

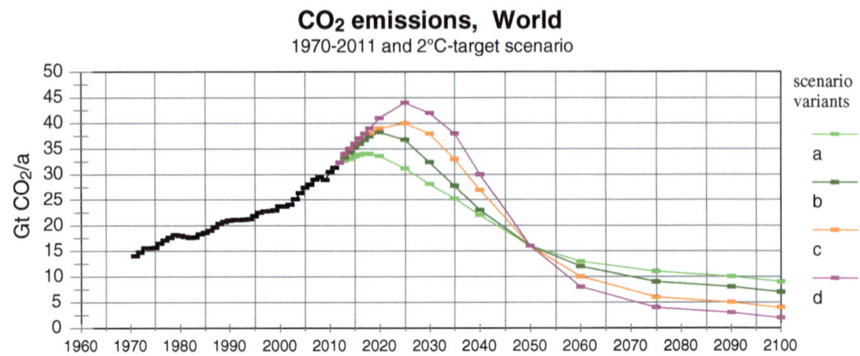


**Fig. 1.2** Four variants of the 2 °C scenario

The graphs of cumulative emissions from 1870 to 2100 for the scenarios of Figs. 1.1 and 1.2 are shown in Fig. 1.3 (assuming 100 GtC from 1870 to 1970).

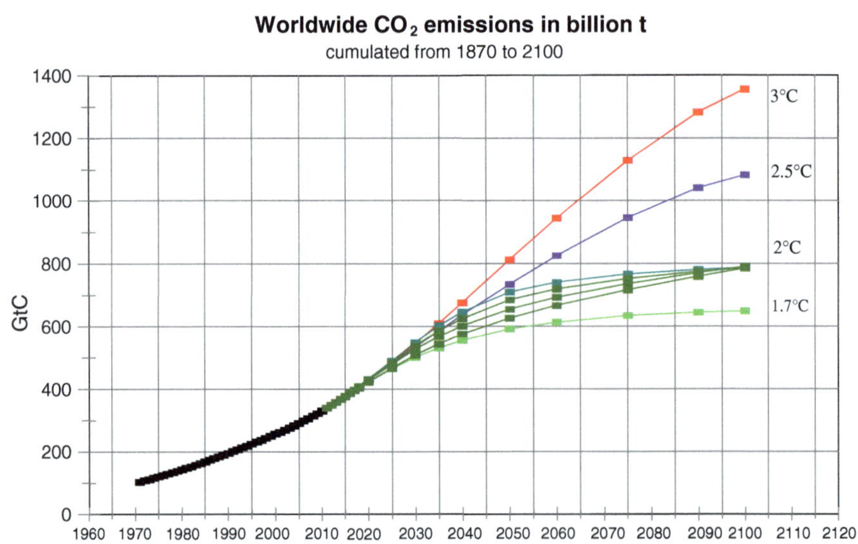


**Fig. 1.3** Cumulated carbon emissions (only fossil fuels) from 1870 to 2011 and scenarios until 2100

The 2-degree target can only be achieved, with 66 % probability, when the total energy-related $CO_2$ emissions from 1870 to 2100 do not exceed 800 GtC (about 2,900 Gt $CO_2$).

Of the four 2 °C variants of Fig. 1.2, only the two stricter variants *a* and *b* are shown in Fig. 1.4, whose feasibility is analyzed in more detail below. Variant *b* is, compared with *a*, more generous in the period up to 2030 but requires from that date stricter reduction targets, not only until 2050 but also after that date, to meet the required cumulative emissions limit (identical for both variants) in 2100.

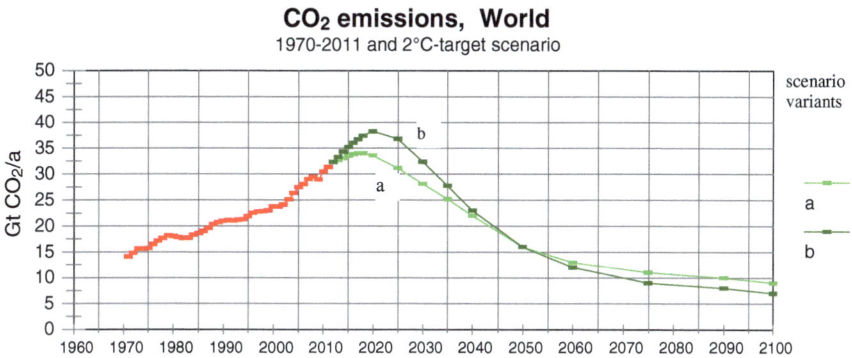


**Fig. 1.4** Worldwide annual emissions from 1971 to 2011 (only fossil fuels) and necessary trend for the 2-degree-target until 2100 for both variants *a* and *b*

The variant *a* has the advantage that with a stricter trend from 2030 (approximately corresponding to that of variant *b*), the cumulative emissions would be further reduced and keep the temperature rise below 2 °C, according to the requirements in [4].

More pessimistic is the last IEA Outlook 2013 [6] which assumes around 38 Gt $CO_2$ in 2035, rather corresponding to variant *d* of Fig. 1.2 or the 2.5 °C target of Fig. 1.1. To achieve the 2-degree target, the emissions would then have to go down rapidly from this date, which in practice requires very wide deployment of Carbon Capture and Storage (CCS), and from 2050 also nuclear fusion. CCS enables the $CO_2$ from combustion to be kept out of the atmosphere by capturing and storing it. It is not yet entirely clear that this technique will turn out to be environmentally acceptable. In some studies, the possibility of "negative emissions" (or BECCS, bioenergy with CCS) is also considered, with the combustion of fast-growing plants, absorbing $CO_2$ from the atmosphere, and subsequent CCS [3, 7].

We investigate below the conditions that **the energy system of all countries has to fulfill in order to comply with Fig. 1.4**. This scenario is, at present, in our opinion, a feasible, although difficult, option. According to this scenario, the $CO_2$ emissions of just 32 Gt in 2011 [5] must be reduced to 28 Gt in 2030 (variant *a*) or limited to 32.5 Gt (variant *b*). In order to meet the 2-degree target, the consumption of fossil fuels must therefore be significantly and rapidly reduced and replaced by low-$CO_2$ energy sources, whereby a deployment of CCS, within limits, will be unavoidable.

The only alternative would be to adapt to higher temperatures, with serious, and in some cases dramatic, consequences, expressed clearly by climate science in the

IPCC report [2]. We are not considering here other possibilities of geo-engineering, which still rather lie in the realm of science fiction.

**In the following report**, more concretely, the temporal evolution of the main parameters is given from 1970 to 2011 for all regions of the world and all G-20 countries, starting from the basic data (population, gross domestic product at purchasing power parity (GDP PPP), gross inland consumption (gross energy) and $CO_2$ emissions). The parameters are then extrapolated to 2030, taking into account current trends, local factors and the requirements of the 2-degree climate target. The database consists of the worldwide available statistics of the International Energy Agency (IEA) [5, 9, 11, 12], with the exception of GDP (PPP), taken over by the International Monetary Fund (IMF), with the advantage of corresponding predictions up to 2018 [8] or recently up to 2019 [13].

An important basis is the structure of the current energy consumption and energy flows of all regions of the world and all G-20 countries, which is **analyzed in the appendix in detail** and reproduced as clearly as possible (last available worldwide IEA data are from 2011 or recently up to 2012).

Following (essentially) the IEA, the world is divided as follows:

| | |
|---|---|
| OECD-34 | Consisting of EU-15, USA, Japan, the rest of the OECD (17 countries) |
| Non-OECD | Eurasia+ (including non-OECD Europe) |
| | China |
| | India |
| | Middle East |
| | Rest of Asia/Oceania (without China, India and OECD members) |
| | Non-OECD America (Central and South America without Chile) |
| | Africa |

Of great importance are the G-20 countries, which in 2011 together comprise 65 % of the world population and 83 % of the gross domestic product at purchasing power parity (GDP PPP). They use 79 % of the world's gross energy and are responsible for 84 % of global $CO_2$ emissions. Given their weight, economically and politically, they are the first areas to address for measures to contain $CO_2$ emissions. These 20 will therefore be analyzed together and individually.

With regard to Europe, it should be noted that 21 members of the EU-27 are also members of the OECD (Germany, France, United Kingdom, Ireland, Italy, Austria, Spain, Portugal, Greece, Netherlands, Belgium, Luxembourg, Denmark, Sweden, Finland, Poland, Czech Republic, Slovakia, Hungary, Slovenia, Estonia). A further 6 States are allocated to Eurasia+ (Latvia, Lithuania, Cyprus, Malta, Romania, Bulgaria). The above-mentioned EU-15 includes the Western European EU members (being also OECD members). Norway, Iceland and Switzerland are OECD members but not in the EU.

**Indicators**

The main characteristics are [10]:

- The **energy intensity**, in kWh/$ (measure of energy efficiency of the region or country),
- **$CO_2$ intensity** (of the energy used), in g $CO_2$/kWh, depending on the energy-mix (fossil, nuclear, renewable),
- The resulting **indicator of $CO_2$ sustainability**, defined as the product of these two variables (and thus expressed in g $CO_2$/$)

For 2011, the following figures are valid worldwide:
Energy intensity: 1.96 kWh/$,
$CO_2$ intensity of energy: 219 g $CO_2$/kWh,
Corresponding indicator of $CO_2$ sustainability: 429 g $CO_2$/$.

For the calculation, a GDP (PPP) in international dollars of 2007, according to IMF, has been used (value of $2007: 0.94175 relative to $2005 or 1.085 relative to $2011, [8]).

According to the development shown in Fig. 1.4, necessary for compliance with the 2-degree target, these indicators would have the following values in 2030, with reasonable estimation of the development of the world's energy needs (see Sect. 2.2) and of the GDP (taking into account, and extrapolating from, the forecasts for 2018 published by the IMF), for example for scenario variant *a*:
Energy intensity: 1.35 kWh/$ (−1.94 %/year).
$CO_2$ intensity of energy: 166 g $CO_2$/kWh, (−1.45 %/year).
Indicator of the $CO_2$ sustainability: 225 g $CO_2$/$, (−3.4 %/year).
The percentages state the necessary average rate of reduction from 2011 to 2030.

These figures assume an energy demand, slightly higher than that of the 450-scenario of the IEA (Sect. 2.2). This includes a great improvement in energy efficiency, but, only moderate use of Carbon Capture and Storage (CCS). Increased energy consumption would result in a greater energy intensity, that in order to meet the climate target would have to be compensated by a greater decrease in $CO_2$ intensity of energy, not only by means of renewable energy sources or nuclear energy but also with wider use of CCS. For variant *b* there are lower (and therefore better affordable) rates of decrease until 2030, but correspondingly greater rates of decrease from that date.

With the problem how the overall necessary rates of decrease could be achieved, comes the important question **how to distribute the efforts among the individual world regions or countries**. An attempt is made to give an answer based on the emissions relative to economic performance. Crucial to the implementation will finally be economic considerations, which can be effectively influenced by local politics but also by international forums and bilateral negotiations.

# Chapter 2
# Summarizing Preview

## 2.1 Global Distribution of $CO_2$ Emissions

Comparing, worldwide for 2011, the $CO_2$ emissions caused by burning fossil fuels, in all regions of the world, in the G-20 countries as a whole and in some important members of the G-20 group, the results are shown in Fig. 2.1. In 2011, the G-20 countries were responsible for a total of 84 %, China and the OECD together for 67 %, China and the U.S. together for 44 % of global $CO_2$ emissions. Meeting the 2-degree climate target is therefore realistic only if these countries and groups of countries participate actively and purposefully.

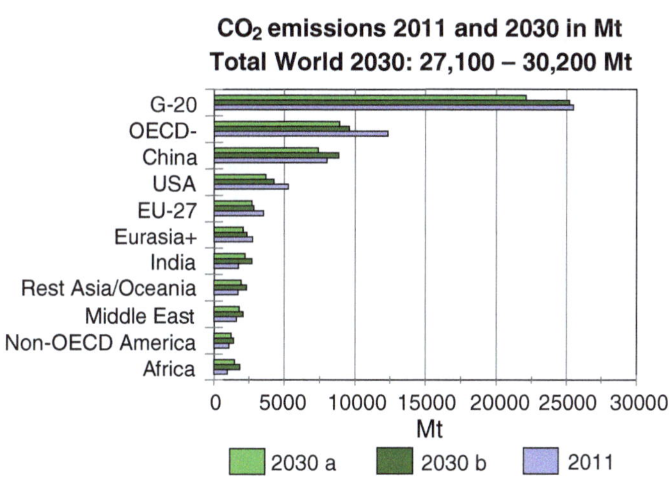


**Fig. 2.1** Comparison of $CO_2$ emissions from the regions of the world and some countries in 2011 (excluding marine and aviation bunkers, see Appendix) and necessary reduction by 2030 (variants *a* and *b*) to achieve the 2 °C climate target



V. Crastan, *Global Energy Demand and 2-degree Target, Report 2014*,
DOI 10.1007/978-3-319-12310-3_2

The same graph shows for 2030 a scenario resulting from the present study, that includes the reduction in total emissions (variants a and b) required by Fig. 1.4. The main burden of the reduction must be carried by the industrialized countries (main actors: the U.S., Japan and EU-27). However, the success can only be guaranteed with a contribution from China in particular, but also of Eurasia (primarily Russia). More details in the following chapters.

Currently, the total emissions of $CO_2$ are still rising. The **trends** of the indicators already mentioned (energy intensity of GDP and $CO_2$ intensity of energy), which are defined and analyzed in more detail below, are very important.

## 2.2 Energy Intensity

Energy consumption is rising as a consequence of the world-wide increase in gross domestic product (GDP). This grows with the increase in world population and, above all, because of the legitimate increase in prosperity in the emerging and developing countries. The ratio energy/GDP is the ***energy intensity*** $\varepsilon$, that can be quantified, for example in **kWh/$**. To achieve the goals of climate protection requires the conversion and use of energy to be as efficient as possible, in order to have the lowest possible energy intensity. Hence, the concept of energy efficiency is also used: it is the reciprocal of energy intensity.

To compare the energy intensity of the economies of various countries, a measurement unit must be established that enables the economic capacity of a country to be defined in an objective and fair way. The absolute value of the GDP in $ at the currency's rate of exchange does not seem to be particularly suitable for that because the level of prices and therefore the buying power, that is the real indicator of capacity and competitiveness, can be very different. Therefore international organisations (World Bank, IMF [8, 13]) determine the GDP of all countries at purchasing power parity, the GDP (PPP). There have also been attempts to define other quantities. However, despite its shortcomings as a welfare indicator, the GDP (PPP) is the only statistically available characteristic that makes reasonable world-wide comparisons possible.

Extrapolation of the energy intensity to the year 2030 or 2050 requires scenarios that represent the development of world-wide energy demand corresponding to the growth of population and the GDP. Scenarios of this type have been developed for 2030 and 2035 by the International Energy Agency (IEA). They are shown in Fig. 2.2, together with the energy needs of the year 2009 [9]. The first two scenarios (*Reference and Alternative Scenario* for 2030) were presented in 2004. The *reference scenario* leads to an energy consumption of 22 TWa (1 TWa = 753 Mtoe), which would correspond to an average growth rate of primary energy demand of 1.8 %/year from 2009. The *alternative scenario* that meets the requirements of climate protection somewhat better, leads, for 2030, thanks to improved efficiency, to an energy consumption of 19.7 TWa (average growth rate 1.3 %/year). The *450 Scenario* of 2009 best meets the requirements of climate protection and for 2030 expects an energy demand of 18.5 TWa (corresponds to about 14,000 Mtoe). It requires a large decrease in both energy intensity and $CO_2$ intensity of energy.

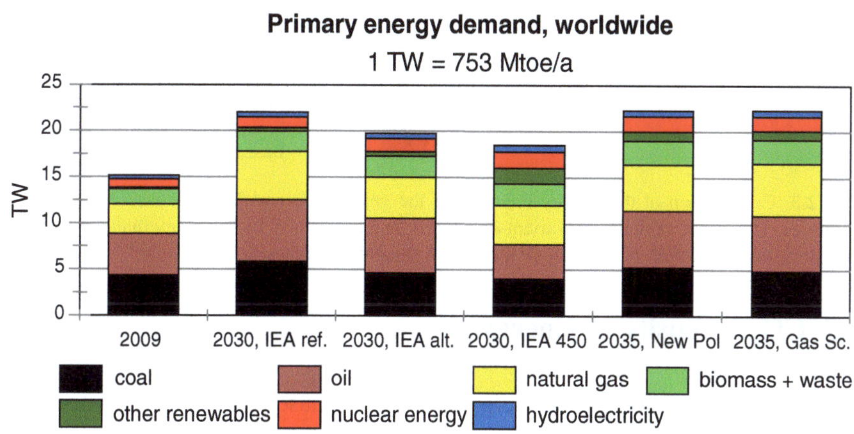


Fig. 2.2 IEA global demand scenarios for primary energy (2030, 2035)

With an energy demand of 19.3 TWa, slightly more than in the 450 Scenario, and a GDP (PPP) of $125,000 billion in 2030, the energy intensity (for variant *a*) would amount to 1.35 kWh/$, as already noted in the introduction (see also Fig. 2.3), i.e. a reduction of 31 % compared to 2011 (reduction rate 1.94 %/year).

The scenarios presented in 2011 for 2035 (New Policies Scenario and Gas Scenario, with about 22.5 TWa (growth rate of about 1.5 %/a), allow a greater share of fossil fuels and thus a slightly greater energy intensity. To achieve the 2-degree target, this would have to be compensated by a greater reduction in the $CO_2$ intensity of energy. These scenarios take account of the needs of climate protection only if it is assumed that a significant portion of the consumption of fossil fuels, thanks to CCS, does not contribute to $CO_2$ enrichment of the atmosphere.

The global comparison of the energy intensity (gross energy per $) is given in Fig. 2.3 for 2000 and 2011. Progress from 2000 to 2011 is considerable, but more effort is essential, especially in regions of the world with poor energy efficiency (Eurasia, China, Africa, Middle East).

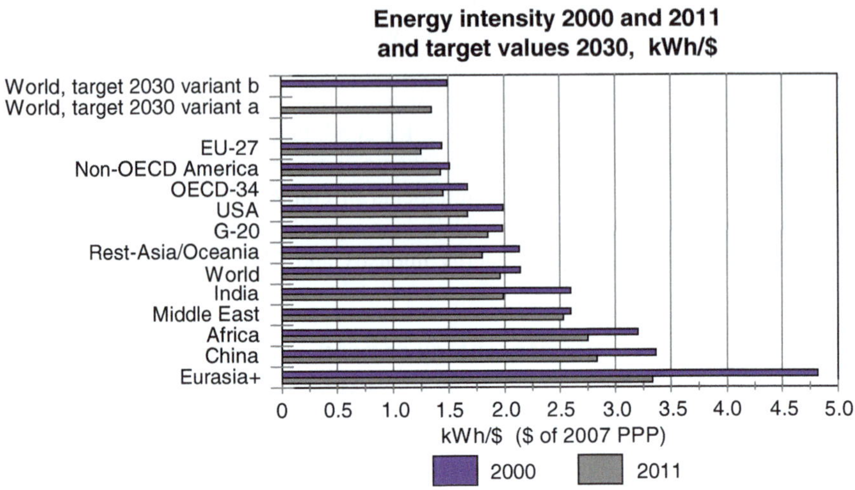


**Fig. 2.3** Comparison of the energy intensity of the years 2000 and 2011 for world regions and countries. The target value 2030 (variant *a* of Fig. 1.2) requires 4 % more energy than the IEA 450 Scenario, that of variant *b*, about 4 % more than the New Policies Scenario

## 2.3 $CO_2$ Intensity of Energy

$CO_2$ emissions from the use of energy are rising because, particularly (but not only) in the emerging nations, there is increasing use of energy sources (oil, coal, gas) that emit large amounts of $CO_2$. This is expressed in the ***$CO_2$ intensity k*** of the consumed energy in units of **g $CO_2$/kWh**. It is dependent on the type of energy used. The $CO_2$ released in burning coal gives a value of around $k$ = 350 g $CO_2$/kWh, burning oil gives about 260 g $CO_2$/kWh and burning natural gas is about 200 g $CO_2$/kWh. The use of low-$CO_2$ energy sources (water power, solar radiation, wind energy, solar heating, geothermal energy, biomass energy, sea currents and wave power and also nuclear energy) reduces the resulting value of $CO_2$ intensity.

Figure 2.4 compares the $CO_2$ intensity of energy for 2000 and 2011 worldwide. The progress in this decade is not only inadequate, but $CO_2$ intensity has actually increased slightly worldwide. Slight but not sufficient progress is only recorded in Europe and America (North and South America). Eurasia is almost stationary and Asia and the Middle East generally have a clear deterioration of the specific emissions. The target value for 2030 can only be achieved if not only the replacement of coal by gas but also the use of renewable energy, nuclear energy and CCS quickly receives a higher priority in emerging markets.

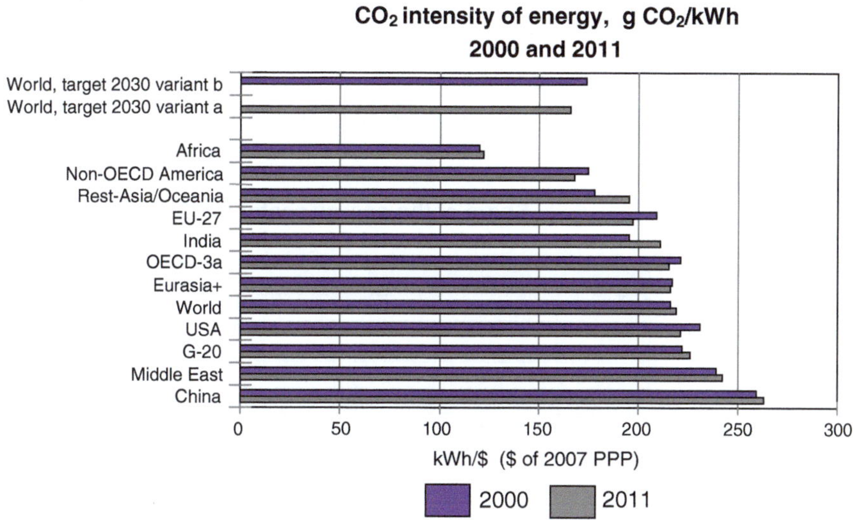


**Fig. 2.4** Worldwide comparison 2000 and 2011 of the $CO_2$ intensity of energy and target values 2030 (variant *a* and *b*)

## 2.4 Energy Sector

The representation of the energy flows in the appendix shows clearly that worldwide more than half of the $CO_2$ emissions (only fossil fuels) come from the energy sector that serves mainly for the production of electricity. The worldwide level of electrification corresponds to a large extent to the level of economic development. In addition, $CO_2$-reducing actions in the area of heating and mobility have the effect of increasing the demand for electricity (heat pumps, hybrid and electric cars).

Figure 2.5 shows the share of $CO_2$ emissions coming from the energy sector (electricity + district heating + losses in the energy sector; see also the appendix for exact definition). The type of electricity production and efficiency of the energy sector play the leading role. The need to catch up is greatest in China, India and Eurasia. The low share of Central and South America (non-OECD America) can be explained with the prominence of water power in this continent.

It is therefore of primary importance to generate electricity with the least possible production of $CO_2$. The following are possible ways, ***which should all be used*** (weighted differently depending on the country), to achieve this [10]:

(a) Large reduction of losses in the energy sector by significantly increasing the **efficiency of thermal power plants** (cogeneration, combined-cycle generation).

(b) **$CO_2$ capture and storage** (**CCS**) in coal, oil and gas-fired power plants; with a significant limitation: the technology is not yet mature, is probably expensive and is not yet fully tested for environmental sustainability. But without CCS the aforementioned objectives are unlikely to be reached.

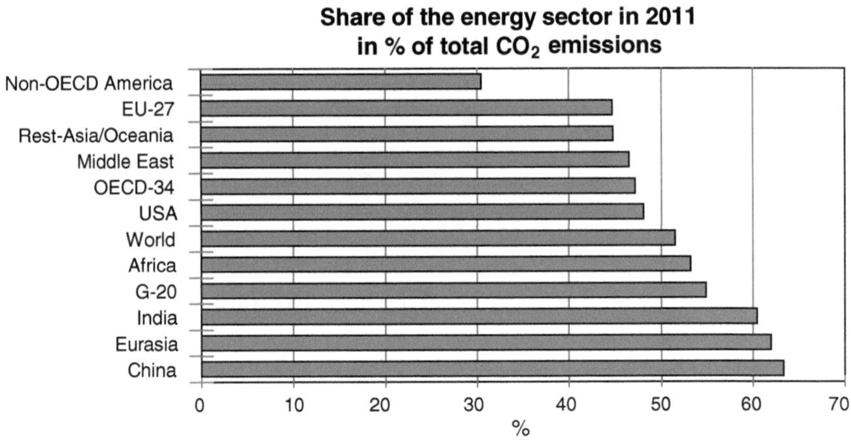


**Fig. 2.5** Share of the energy sector in % of total $CO_2$ emissions in 2011

(c) Use of **natural gas** instead of coal and oil: $CO_2$ emissions, compared to coal, are reduced to about 55 % (compared to oil, to about 75 %), with limitations: world-wide gas reserves are not unlimited and political dependencies. Fracking significantly increase the reserves, but environmental risks have not yet been sufficiently clarified.

(d) Use of **nuclear energy**: the power stations are nearly free of $CO_2$ emissions; restrictions: reserves of uranium, used in 3rd-generation reactors, are also limited. Use of reactors of the 4th generation is possible, but requires careful technical and political consideration. Nuclear fission has met increasing resistance after Fukushima. Nuclear fusion will only be available in the second half of the century.

(e) Use of all opportunities to generate electricity from **water power**: Restrictions: the potential for this is limited, resistances from nature conservation and ecology.

(f) Use of **wind energy**: the technology is mature and, where the wind conditions are favourable, cost-efficient. With the use of off-shore installations, the potential is considerable. Limitations: power grids and energy storage require significant adjustment.

(g) Use of **geothermal energy**: Limitations: geothermal power stations are especially suitable for sites with geothermal anomalies, but there they have a large potential.

(h) Use of **biomass and waste**: Restrictions: the capability of biomass is limited. Biomass should therefore be used primarily, and provided its use is ecologically acceptable, for heat, combined heat and power and motor fuels. But the production of bio-fuels is often far from $CO_2$-neutral.

(i) Use of solar thermal energy and photovoltaic: **Solar thermal power stations** are suitable for countries with a low proportion of diffuse light, and there they have a large potential. **Photovoltaic** generation is currently hindered by its high cost, but as its potential is practically unlimited and has widespread availability, its further development must be pursued with determination, and as long as economically worthwhile including reasonable feed-in tariffs. Here too, adjustments in the transmission and distribution grid as well as in energy storage (partly through load control) are necessary.

## 2.5 Indicator of $CO_2$ Sustainability

The upward trend in $CO_2$ emissions can only be broken by influencing both of its causes. The important number is the product of the two factors ***energy intensity*** $\varepsilon$ and ***$CO_2$ intensity of energy k****, used as* ***indicator*** $\eta$ ***of $CO_2$ sustainability.***

The absolute values and the yearly changes are given by [10]:

$$\eta\left[\frac{\text{gCO}_2}{\$}\right] = k\left[\frac{\text{g CO}_2}{\text{kWh}}\right] \times \varepsilon\left[\frac{\text{kWh}}{\$}\right]$$

$$\Delta\eta\ [\%/\text{a}] = \Delta k\ [\%/\text{a}] + \Delta\varepsilon\ [\%/\text{a}]$$

With a reasonable estimate of their gross domestic products for 2030 (GDP PPP), taking into account the estimate for 2018 of the International Monetary Fund, the result is, with the assumptions of Fig. 2.1 (variants *a* and *b*), the scenarios for the decrease of the indicator shown in Fig. 2.6. All of them, but especially China, Eurasia (crucially influenced by Russia) and the Middle East have a very strong need to catch up. The 2-degree target can only be achieved if in these regions of the world also, both the energy intensity and $CO_2$ intensity of energy are reduced in accordance with the following chapters (see Sect. 12.1 for China, Chap. 7 + Sect. 12.4 for Eurasia/Russia and Chap. 6 for the Middle East).

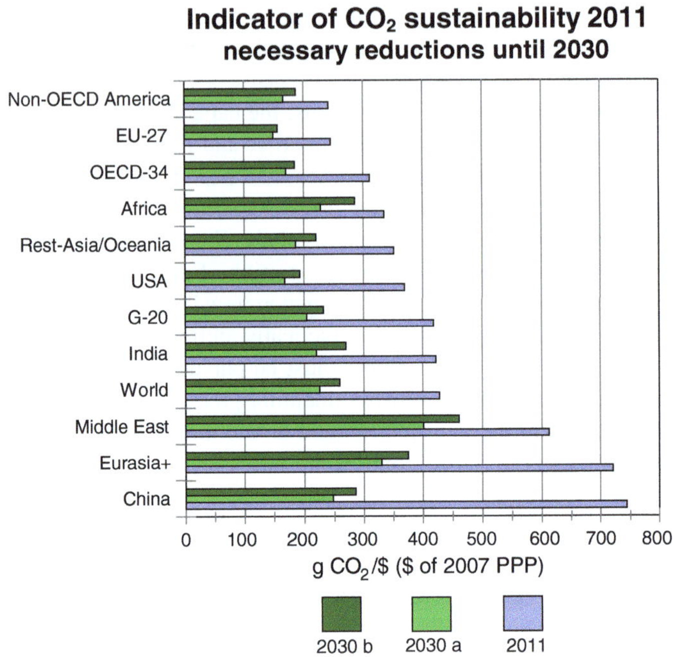


**Fig. 2.6** Sustainability indicator of world regions and important countries for 2011 and necessary development to 2030 for the 2 °C target, variants *a* and *b*

The resulting per-head indicators: **e** for energy (from **ε**), and **α** for $CO_2$ emissions (from **η**) are also of interest:

$$e\left[\frac{\text{kW}}{\text{capita}}\right] = \varepsilon\left[\frac{\text{kWa}}{10{,}000\ \$}\right] \times y\left[\frac{10{,}000\ \$}{\text{a, capita}}\right]$$

$$\alpha\left[\frac{\text{t}\,\text{CO}_2}{\text{a, capita}}\right] = \eta\left[\frac{\text{t}\,\text{CO}_2}{10{,}000\ \$}\right] \times y\left[\frac{10{,}000\ \$}{\text{a, capita}}\right]$$

where y is the PPP-corrected GDP (PPP = purchasing power parity) per capita.

(1 kWa/10,000 $ = 0.876 kWh $, 1 t $CO_2$/10,000 $ = 100 g $CO_2$/$).

The indicator **α** (t $CO_2$ per capita per year, with a worldwide value in 2011 of about 4.5 t/capita, and target 2030 just 3.4 t/capita) makes sense in the long term as a global target, as the population growth is most likely to be predictable. It is, however, less suitable for current comparisons between countries, and therefore as short-term basis for negotiation, because of the greatly differing levels of development of the regions of the world.

The indicator **e** (kW per capita, with a worldwide value in 2011 and target for 2030 both, about 2.3 kW/capita) is also not suitable as a comparison value for the same reason. It can be determined from energy intensity and GDP per capita. It is primarily of interest in connection with studies on the feasibility of a 2,000-W society.

# Chapter 3
# World

## 3.1 Population and Gross Domestic Product 2011

The world is divided (according essentially to the IEA) into the following zones: OECD-34 (split into EU-15, USA, Japan and the other 17 OECD countries), Eurasia+ (with Non-OECD-Europe), Middle East, China, India, Rest of Asia/ Oceania (excluding OECD members), Central + South America (without Chile) and Africa. The percentage shares of population are shown in Fig. 3.1. The GDP (PPP) of the world is $73,000 billion ($ of 2007), but there is a wide gap, with a factor of about 5.4 between the GDP per capita in OECD countries and the rest of the world (Fig. 3.2). China, with 19 % of the population, has a GDP (PPP) of $10,750 billion/a, or almost 15 % of the world GDP. India produces, with 18 % of the population, a GDP (PPP) of $4,080 billion/a, or about 6.8 % of world GDP.

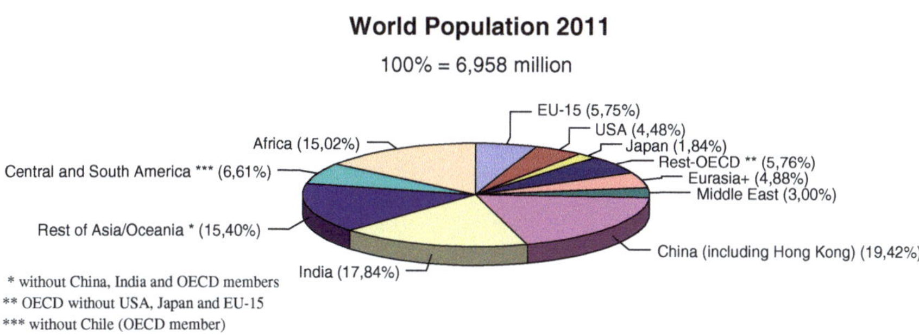


**Fig. 3.1** Percent breakdown of the world population



V. Crastan, *Global Energy Demand and 2-degree Target, Report 2014*,
DOI 10.1007/978-3-319-12310-3_3

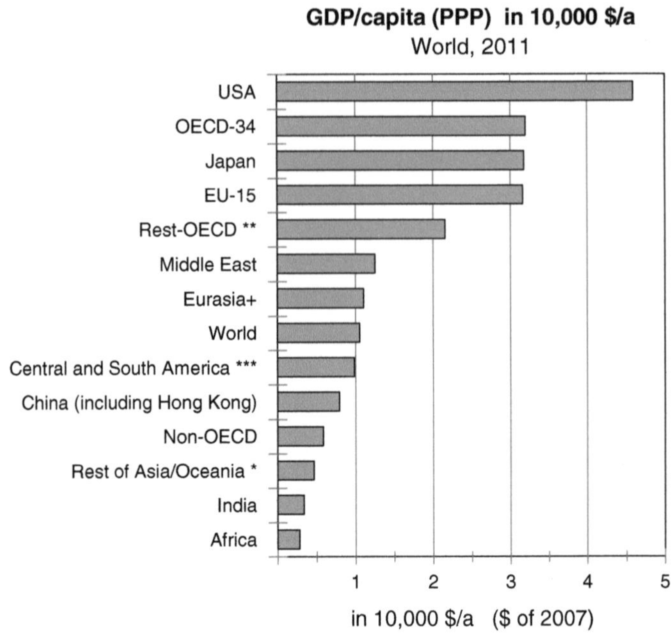


**Fig. 3.2** GDP (PPP) per capita of world zones and countries

The world average GDP (PPP) per capita is \$10,500 (Fig. 3.2) and varies between \$20,000–\$45,000/a in the industrialized world and between \$2,800–\$13,000/a in the remaining parts of the world. The per-capita GDP (PPP) of China is \$8,000/a and that of India is \$3,300/a.

## 3.2 Energy Intensity 2011

The world's mean gross energy intensity is 1.95 kWh/$. Clear differences can be observed all the way through the OECD and non-OECD countries. For example the efficiency of energy use in the USA is 1.67 kWh/$, those of EU-15 and Japan are significantly lower at around 1.20 kWh/$. On the other hand, Central + South America have an energy intensity (1.43 kWh/$), close to the OECD average (1.45 kWh/$). Particularly blatant is the energy waste in the Middle East (2.5 kWh/$), in China (2.8 kWh/$), and especially in Eurasia, with 3.3 kWh/$, which might be primarily due to the low-cost availability of the fossil energy carriers. However, the progress made from 2000 to 2011 is encouraging (Fig. 3.3). The energy intensity has decreased everywhere. The 2-degree targets for 2018 and 2030 (variants *a* and *b*, see also Sect. 3.5) are also plotted. The target for 2018 is based on the GDP (PPP) forecast by the IMF [8].

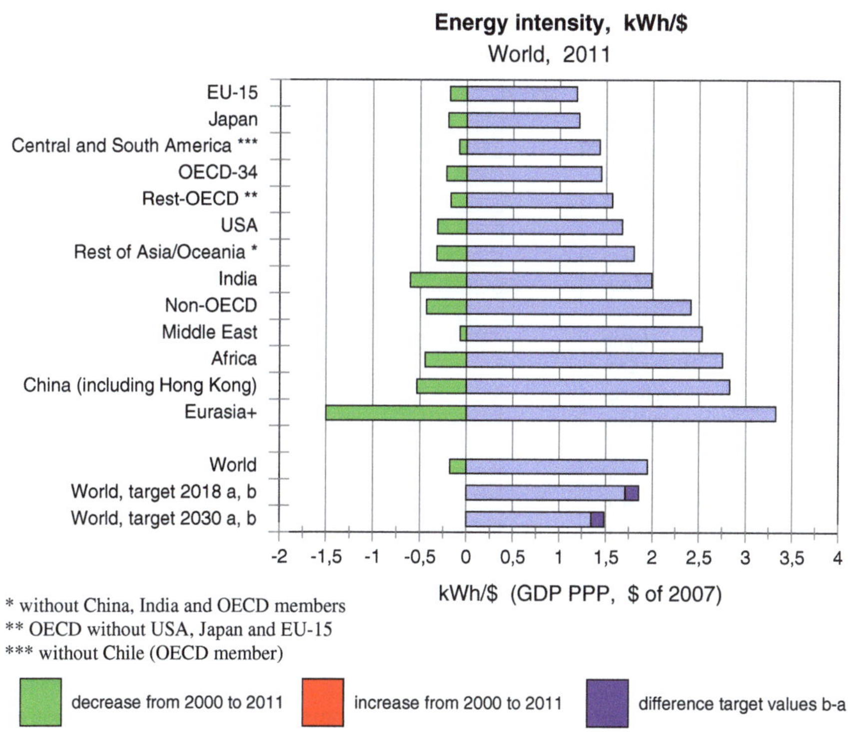


**Fig. 3.3** Energy intensity of the world zones and countries, and progress from 2000 to 2011

## 3.3 $CO_2$ Intensity of Energy 2011

The global $CO_2$ intensity is 219 g $CO_2$/kWh and is thus slightly higher than the value for the more advanced OECD countries. Exceptions are Africa (122 g $CO_2$/kWh) with an energy economy still partly based on biomass, and Central and South America (168 g $CO_2$/kWh) thanks to the high proportion of hydropower in its electricity production. Close to the world average is Eurasia (Russia), the latter because of its energy industry that is largely based on natural gas. Slightly above the world average are the USA (221 g $CO_2$/kWh) and considerably higher are the Middle East and China (the latter with 263 g $CO_2$/kWh). The main reason is the generation of electricity mainly from oil or coal. The Fukushima accident (March 2011) pushed Japan several steps backwards. In the last decade, unfortunately, a slight overall increase can be seen. A significant trend reversal in the decade 2020–2030 is essential to achieve the two-degree target (see target values) (Fig. 3.4).

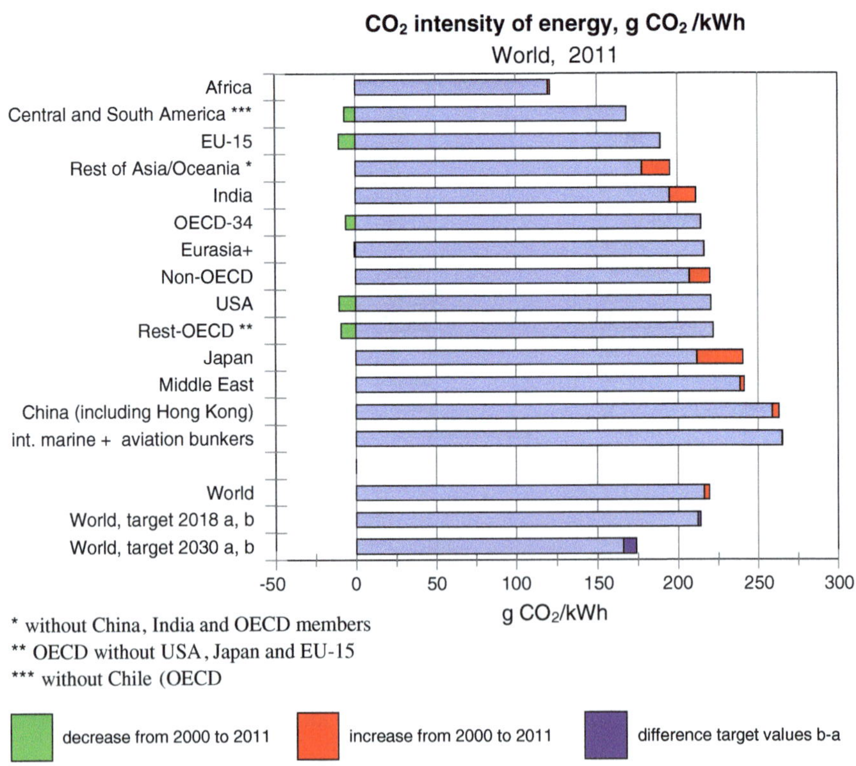


**Fig. 3.4** $CO_2$ intensity of world zones and countries, and changes from 2000 to 2011

## 3.4 Indicator of $CO_2$ Sustainability 2011

In climate protection matters, the measure for the sustainability of energy consumption is the $CO_2$ indicator obtained as the product of energy intensity and $CO_2$ intensity of energy. The worldwide average of the $CO_2$ indicator for 2011 is 427 g $CO_2$/\$. The countries below 300 g $CO_2$/\$ are Central and South America, the EU-15, and Japan. The USA with 369 g $CO_2$/\$ is somewhat more sustainable than the world average. The greatest need to catch up is in China and Eurasia (≥700 g $CO_2$/\$, despite remarkable progress) and the Middle East (600 g $CO_2$/\$). The targeted 2-degree values require increased efforts but do not seem to be unachievable provided that efforts are made in all the world's regions in good time for the decade 2020–2030 (also see the following sections) (Fig. 3.5).

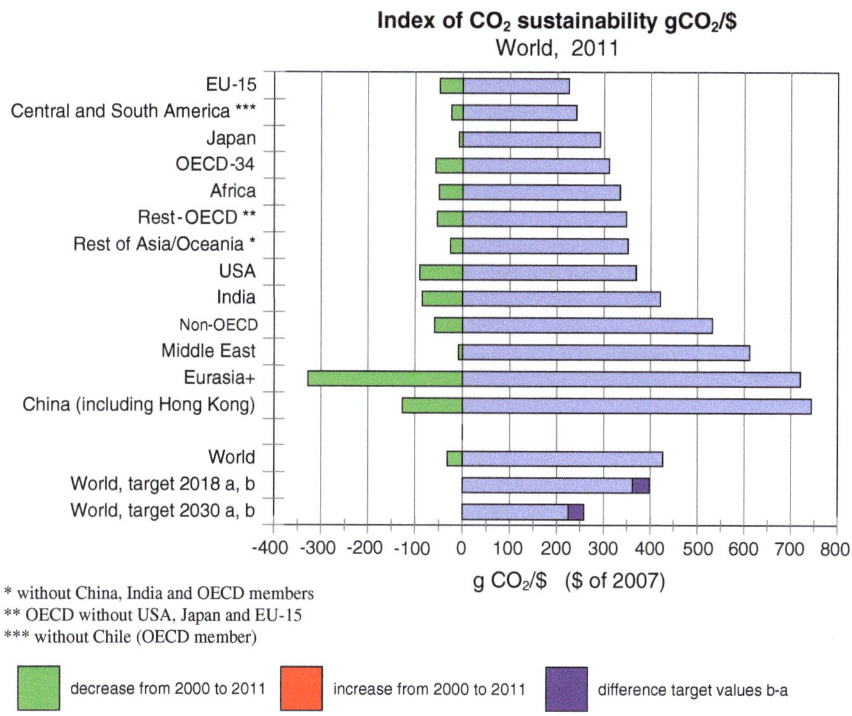


**Fig. 3.5** $CO_2$ sustainability indicator of world zones and countries, and progress from 2000 to 2011

The change in the $CO_2$ sustainability index is shown in more detail in Fig. 3.6 for the period 2000–2008 and for 2008–2011. In the last 3 years because of the economic crisis, there was rather a slowdown in the positive development. Eurasia (significantly influenced by Russia) has made great progress, especially up to 2008, but has stagnated since then. The progress of the U.S. and China from 2008, gives rise to some hope.

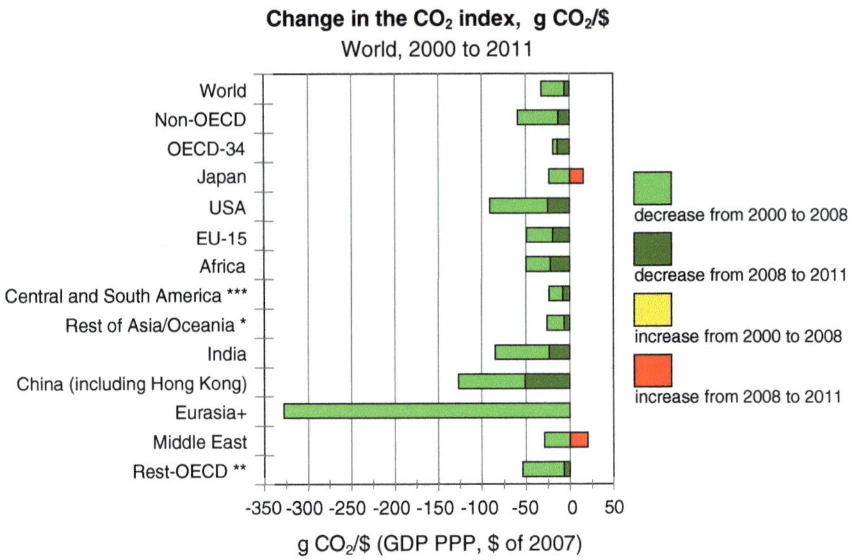


**Fig. 3.6** Change in the $CO_2$ index of world zones and countries from 2000 to 2008 and 2008 to 2011

## 3.5 Actual Indicators and $CO_2$ Emissions from 1980 to 2011 and Values Required from 2011 to meet the 2 °C Target

### 3.5.1 *Energy Intensity, $CO_2$ Intensity and $CO_2$ Sustainability up to 2030*

In order to meet the 2-degree target, $CO_2$ emissions in 2030 must not exceed 28,200 Mt worldwide for the stricter variant *a* (with marine and aviation bunkers). At an assumed GDP (PPP) of 125,000 billion ($ of 2007), this would require an indicator of $CO_2$ sustainability of $\eta = 225$ g $CO_2$/$. For variant *b* the limit is at 32,400 Mt, which, with the same GDP, gives an indicator value of 260 g $CO_2$/$.

The actual figure from 1980 to 2011 and the graph of the indicators (energy intensity, $CO_2$ intensity of energy and resulting $CO_2$ indicator of sustainability) up to 2030, necessary for the 2-degree variants *a* and *b*, are shown in Fig. 3.7. The mildest variant *d*, from Fig. 1.2 is also shown (could be called the "emergency" variant) with the values 42,000 Mt $CO_2$ and 336 g $CO_2$/$.

The values for 2018 correspond to the data of the IMF (International Monetary Fund), with a GDP (PPP) forecast level of $94,000 billion ($ of 2007).

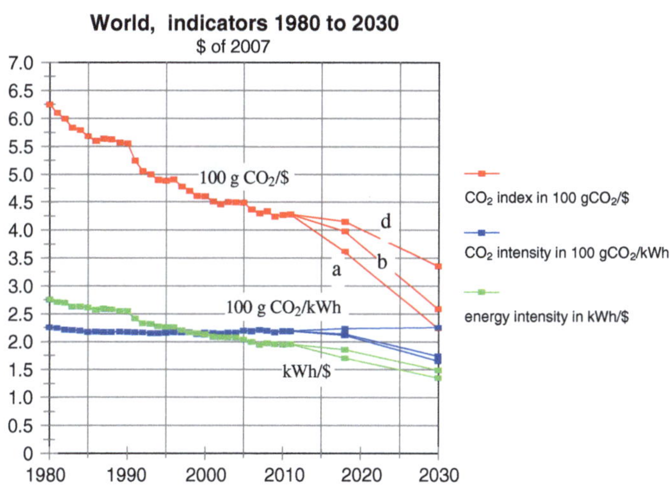


**Fig. 3.7** World-wide indicators from 1980 to 2011 and climate protection scenario up to 2030

With this scenario the gross energy consumption in 2030 would be 14,550 Mtoe, or 19.2 TWa for variant *a*, which is slightly above the IEA 450 scenario of Fig. 1.3 and 16,000 Mtoe (21 TWa) for the variants b and d which are more like the IEA New Policies or Gas Scenario.

## 3.5.2 Indicators and $CO_2$ Emissions up to 2050

Figure 3.8 clearly shows, for variant *a*, and Fig. 3.9 for variant *b*, the required tightening up of the global indicator trend up to 2050.

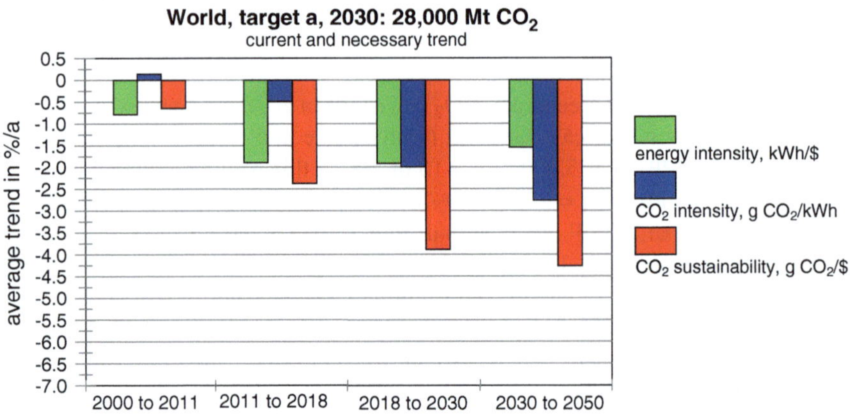


**Fig. 3.8** Trend from 2000 to 2011 and necessary change in the trend from 2011 to achieve the 2-degree target, variant *a*

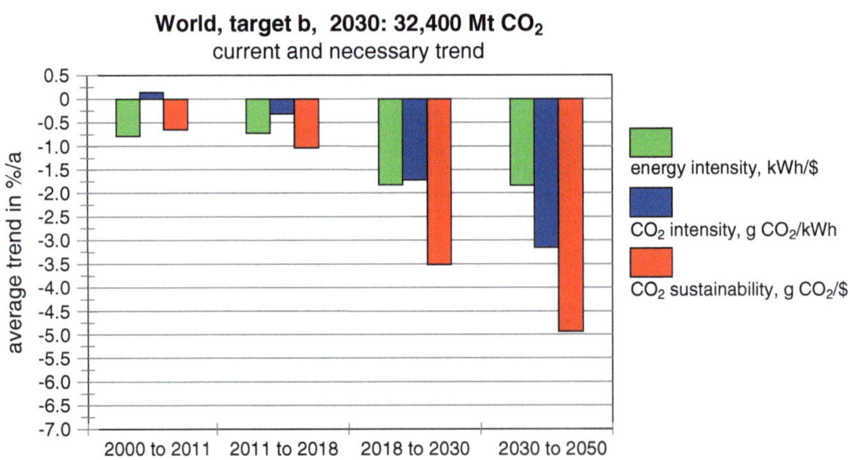


**Fig. 3.9** Trend from 2000 to 2011 and necessary change in the trend from 2011 to achieve the 2-degree target, variant *b*

The variant *a* requires a rapid reduction in *global energy intensity*, which could, however, start somewhat later, in 2018, for variant *b*. A trend reversal of *global $CO_2$ intensity of energy* is necessary for both variants, slightly greater with variant *a*, and in both cases the trend towards reduction must be considerably tightened up from 2018.

Figure 3.10 gives the actual $CO_2$ emissions from 1970 to 2011, and the necessary values up to 2050 required by the 2-degree targets variants *a* and *b* (looked at in more detail in the following sections). The emissions profile for the emergency variant *d* is also plotted.

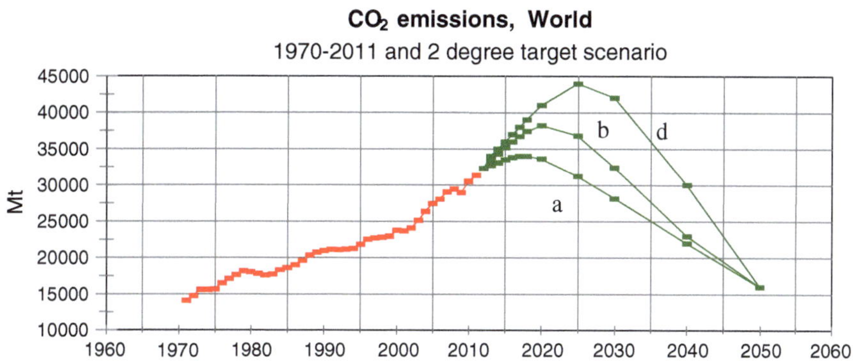


**Fig. 3.10** World-wide $CO_2$ emissions 1970–2011 and 2-degree climate protection scenario up to 2050 variants *a*, *b* and *d* (with marine and aviation bunkers)

Figure 3.11 shows for variant *d*, that should be avoided if possible, the trend in the global indicators up to 2050 required by the 2-degree target. Variant *d* is less stringent up to 2030, but then becomes much stricter (at least in respect of $CO_2$ intensity: reduction by nearly 4.5 %/a).

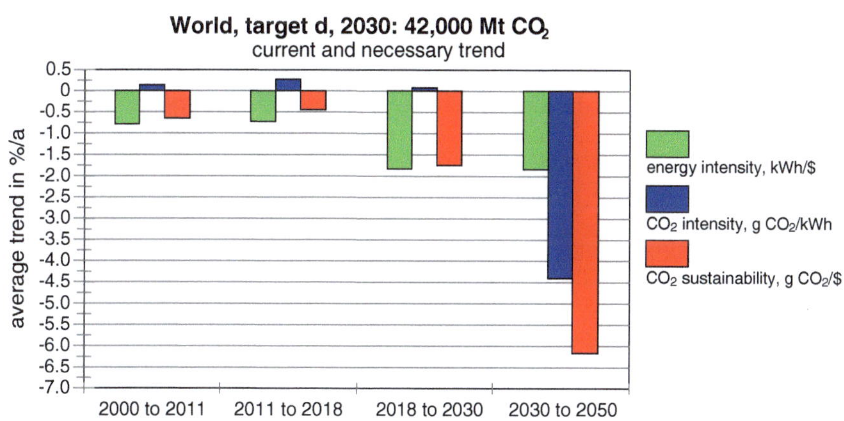


**Fig. 3.11** Trend from 2000 to 2011 and necessary change in the trend from 2011 to achieve the 2-degree target, variant *d*

The changes in the different regions of the world required for meeting the 2-degree target are estimated and presented in the following chapters for variants *a* and *b*. We restrict ourselves to these two variants as they are the ones to aim at in the first place.

# Chapter 4
# OECD-34

## 4.1 Population and Gross Domestic Product 2011

The 34 countries of the Organization for Economic Cooperation and Development (OECD) had in 2011 a population of 1,241 million (Fig. 4.1), which is about 18 % of the world population. They generate a GDP (PPP) of $39,700 billion ($ of 2007), representing 54 % of the world GDP, after correction for purchasing power. The main players are the EU-15, the USA and Japan, which together have 68 % of the population and generate 78 % of OECD GDP.

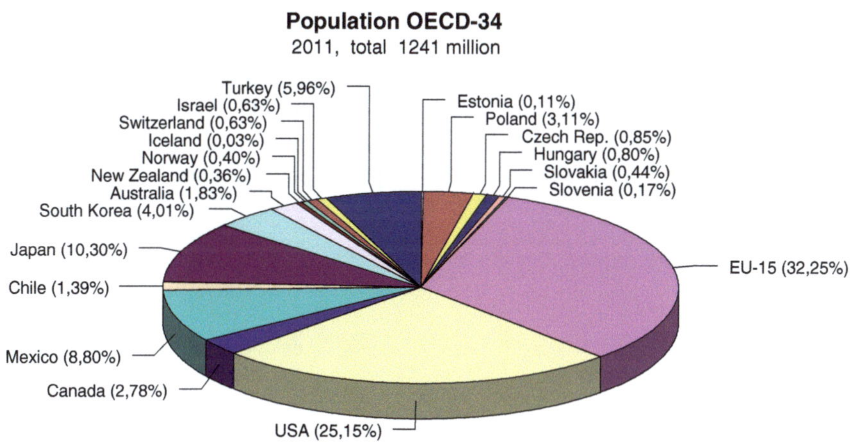


**Fig. 4.1** Percent partition of the population of the OECD-34



V. Crastan, *Global Energy Demand and 2-degree Target, Report 2014*,
DOI 10.1007/978-3-319-12310-3_4

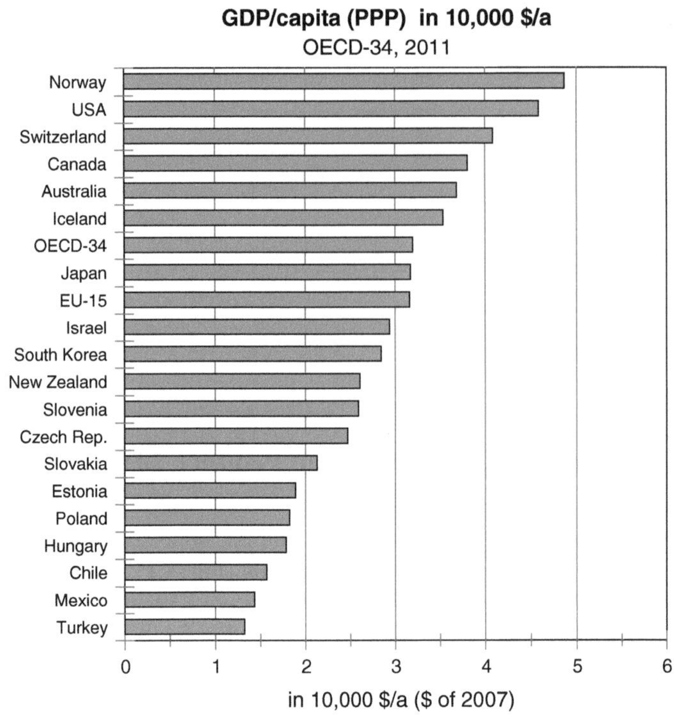


**Fig. 4.2** GDP (PPP) per capita of the OECD-34 countries

The GDP (PPP) per capita (Fig. 4.2) has an average of $31,200 and ranges from almost $49,000 in Norway to $13,300 in Turkey. After Norway, the leaders are the USA and Switzerland. In addition to Turkey, the purchasing power is also relatively low in Mexico, Chile and the east European countries (Estonia, Poland, Slovakia, Hungary), all between $14,000 and $21,000 per capita and year.

## 4.2 Energy Intensity 2011

The average energy intensity (gross energy consumption per \$, only energy use) in OECD countries is 1.45 kWh/\$, which is significantly lower than the world mean value. The very good figure for Switzerland (0.90 kWh/\$) and the good ones for Japan and the EU-15 (<1.25 kWh/\$) are offset by the poorer efficiencies of the USA and South Korea (1.67 and 1.81 kWh/\$), and Canada (2.0 kWh/\$). The extremely large value for Iceland (5.5 kWh/\$) is a consequence of the climate, but also of the great availability of $CO_2$-free energy sources in the form of water power and geothermal power and thus climatically irrelevant. The target values required for 2018 and 2030 are given for the variants *a* and *b*. Details of the EU-15 countries are considered in Sect. 4.5 (Fig. 4.3).

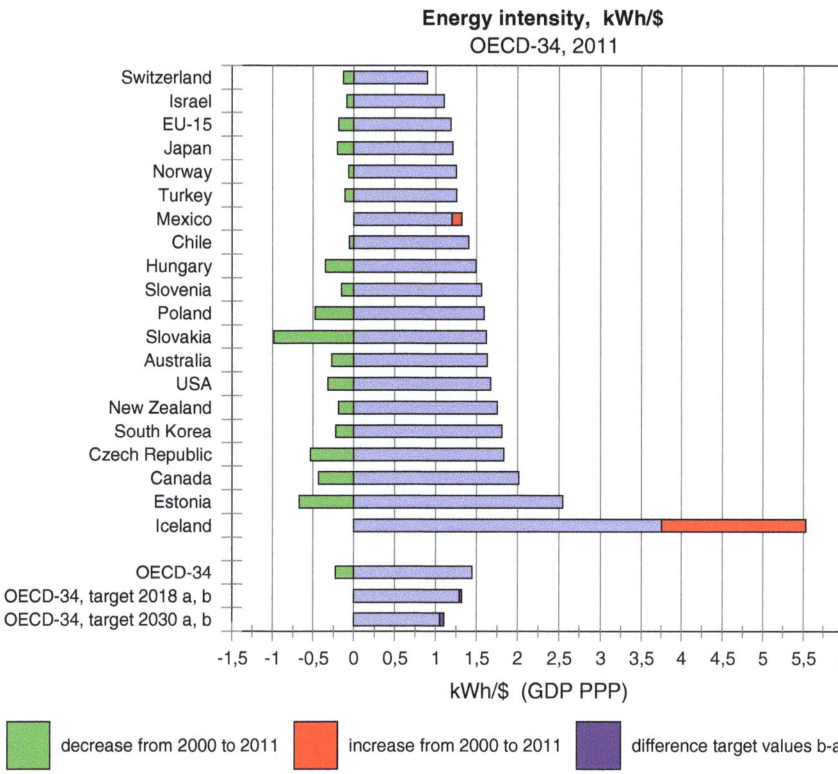


**Fig. 4.3** Energy intensity of the OECD-34 countries and changes from 2000 to 2011

## 4.3 $CO_2$ Intensity of Energy 2011

Iceland is the leader with a very low value of $CO_2$ intensity (30 g $CO_2$/kWh) thanks to water power and geothermal energy. Norway, Switzerland and New Zealand also do well in the international comparison (≤150 g $CO_2$/kWh) on account of their electricity generation purely from water power in one case and mainly water power and nuclear or geothermal energy in the others. The mean value of $CO_2$ intensity of the OECD in 2011 was 215 g $CO_2$/kWh. A little above this is the value for the United States (221 g $CO_2$/kWh) and there is a great need for improvement by Estonia, Israel, Poland and Australia (all over 260 g $CO_2$/kWh). The main reason is the high proportion of coal-fired power plants in electricity production in these countries. The increase in Japan is a consequence of the Fukushima accident (March 2011). The target value for 2030 is based on a share of 8,900 Mt (variant *a*) and 9,600 Mt (variant *b*) of global $CO_2$ emissions (without marine and aviation bunkers) (Fig. 4.4).

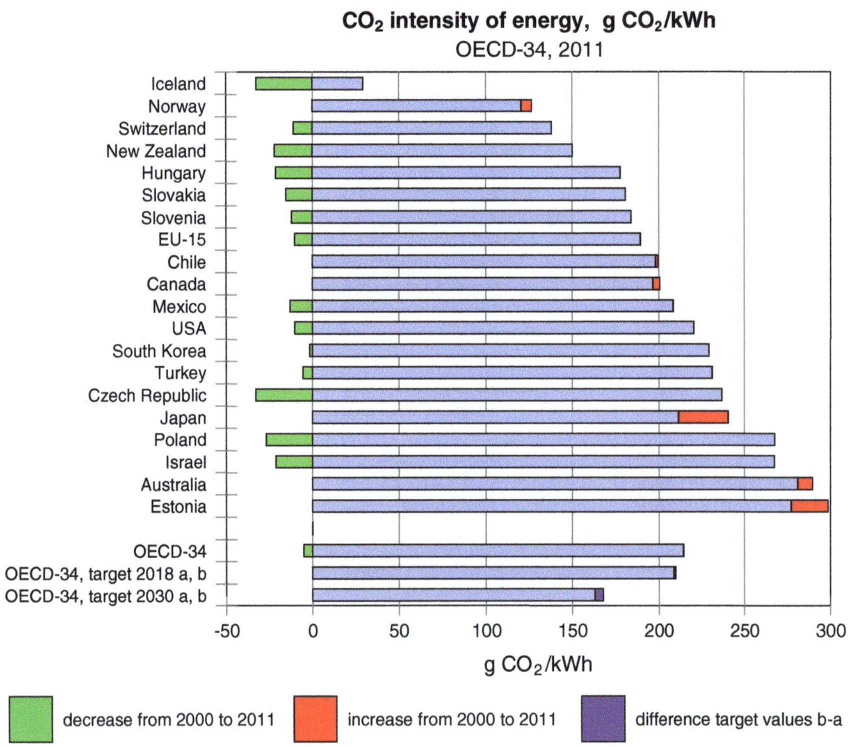


**Fig. 4.4** $CO_2$ intensity of the OECD-34 countries and changes from 2000 to 2011

## 4.4 Indicator of $CO_2$ Sustainability 2011

The sustainability of the energy consumption in terms of climate change can be characterised by the product of energy intensity and $CO_2$ intensity. The $CO_2$ indicator calculated this way is 311 g $CO_2$/\$ for the OECD. Values below 200 g $CO_2$/\$ apply to Switzerland (the leader with 124 g $CO_2$/\$), Norway, Iceland and countries of the EU-15 such as Sweden and France. The reasons have already been mentioned (electricity wholly or mainly from hydroelectricity, nuclear or geothermal energy). At the other end of the scale are countries that generate electricity mainly from coal: USA (369 g $CO_2$/\$) as well as Poland, the Czech Republic and Australia (all exceeding 400 g $CO_2$/\$), Estonia (>700 g $CO_2$/\$) or have poor energy efficiency (Canada 404 g $CO_2$/\$, partly due to climate) (Fig. 4.5).

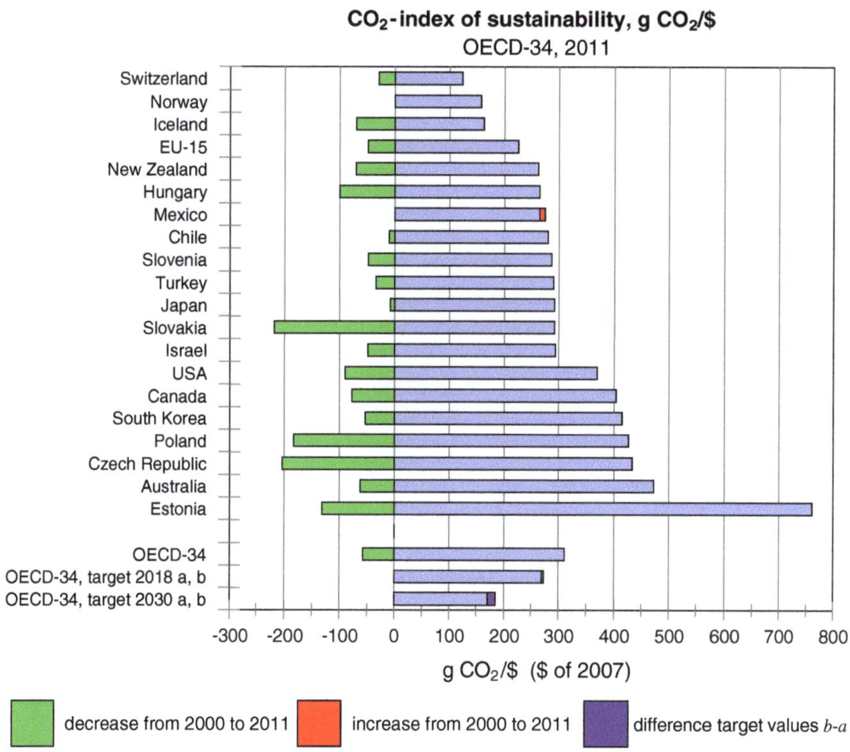


**Fig. 4.5** Indicator of $CO_2$ sustainability of the OECD countries and progress from 2000 to 2011

The change in the $CO_2$ sustainability index is shown in more detail in Fig. 4.6 for the period 2000–2008 and for 2008–2011. In the last three years (partly because of the economic crisis), there was rather a slowdown in the positive development or in some countries even a step backwards.

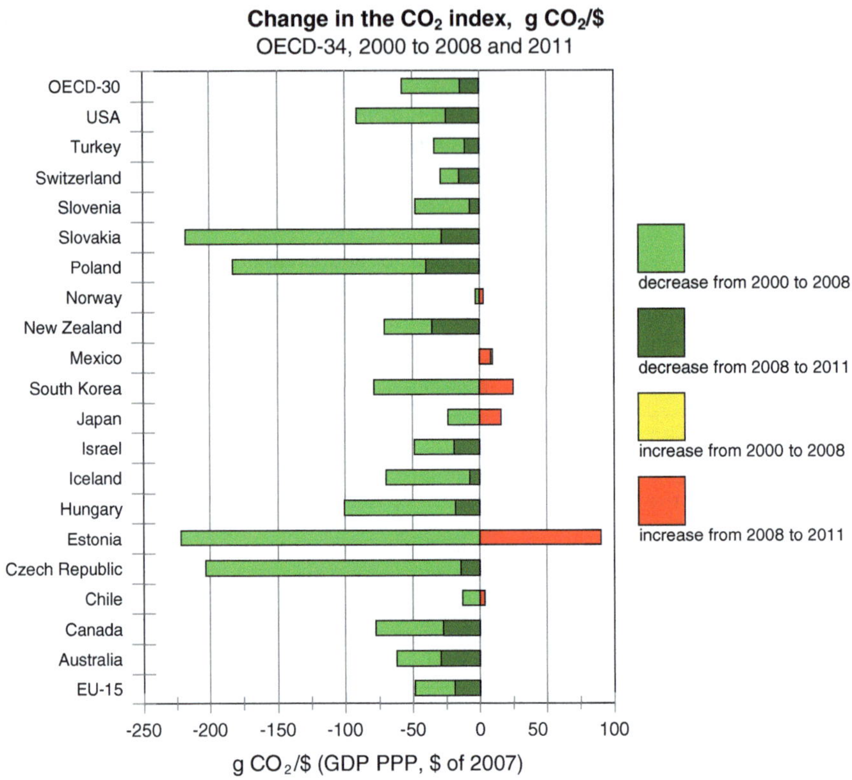


**Fig. 4.6** Change in the $CO_2$ index from 2000 to 2008 and 2008 to 2011 for the OECD-34 countries

## 4.5 Actual Indicators and $CO_2$ Emissions from 1980 to 2011 and Values Required from 2011 to meet the 2 °C Target

### *4.5.1 Energy Intensity, $CO_2$ Intensity and $CO_2$ Sustainability up to 2030*

In order to meet the 2-degree target, the $CO_2$ emissions share of the OECD in 2030 was set at 8,900 Mt (variant *a*) and 9,600 Mt (variant *b*). With an assumed GDP (PPP) of 52,000 billion ($ of 2007), this would result in a required indicator of $CO_2$ sustainability of $\eta = 171–185$ g $CO_2$/$. The actual figures from 1980 to 2011 and the graph of the indicators up to 2030 required for meeting the 2-degree target are shown in Fig. 4.7. The values for 2018 correspond to the data of the IMF (International Monetary Fund), with a GDP (PPP) forecast level of $43,800 billion ($ of 2007).

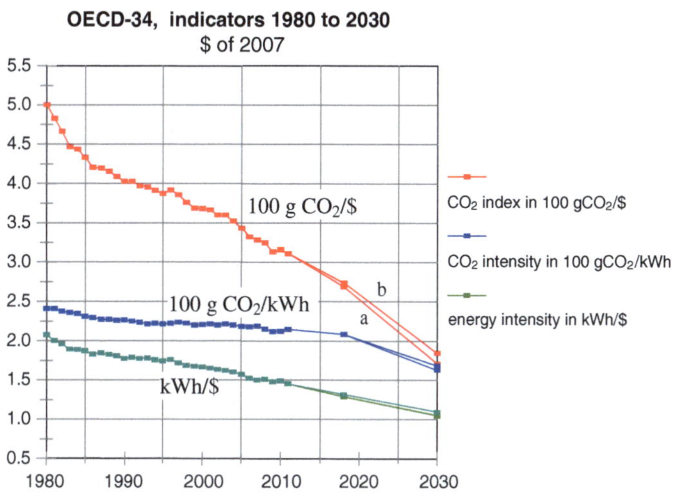


**Fig. 4.7** Indicators of the OECD-34, 1980–2011 and climate protection scenario up to 2030

The gross energy demand for 2030 (only energy use and without marine and aviation bunkers) is, in this scenario, 4,700–4,900 Mtoe. The global per capita indicators for energy and $CO_2$ emissions would then be $e = 4.6–4.9$ kW/capita (decrease of 12–8 % relative to 2011), and $\alpha = 6.6–7.2$ t $CO_2$/capita (−33 to −28 % relative to 2011).

The change in the trend of the indicators to be aimed at to meet the 2-degree target is illustrated in Fig. 4.8 for variant *a*. The trend in energy intensity must be slightly improved and the $CO_2$ intensity trend must be considerably improved, particularly after 2018.

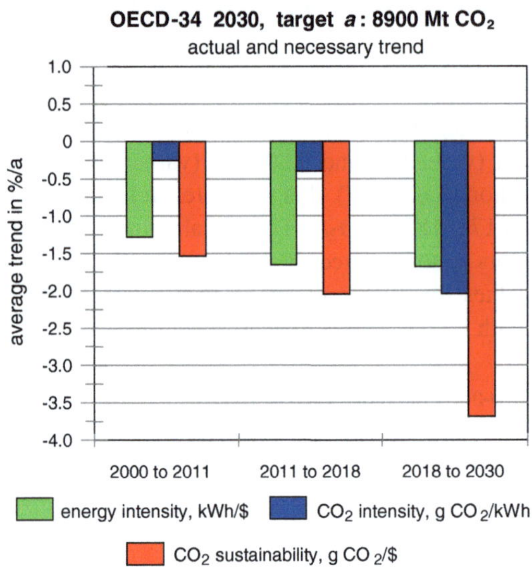


**Fig. 4.8** Trend from 2000 to 2011 and necessary change in the trend from 2011 to achieve the 2-degree target, variant *a*

With variant *b* (Fig. 4.9) the required change in the trend is similar, but a little smaller.

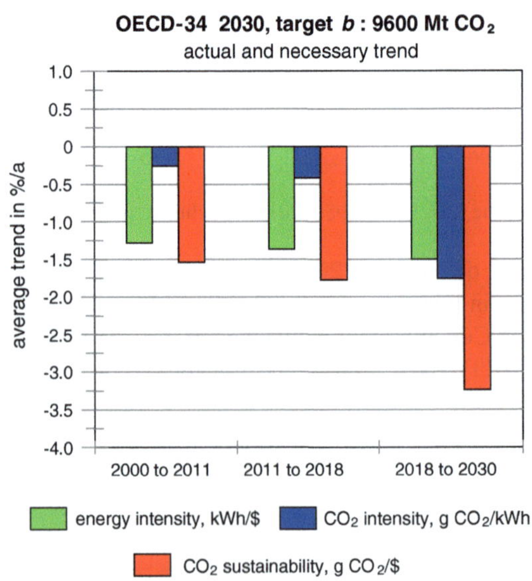


**Fig. 4.9** Trend from 2000 to 2011 and necessary change in the trend from 2011 to achieve the 2-degree target, variant *b*

### 4.5.2 $CO_2$ Emissions until 2050

Figure 4.10 shows the actual $CO_2$ emissions from 1980 to 2011, and those corresponding to Fig. 4.7 and permissible for meeting the 2-degree climate target up to 2050, for variants *a* and *b*.

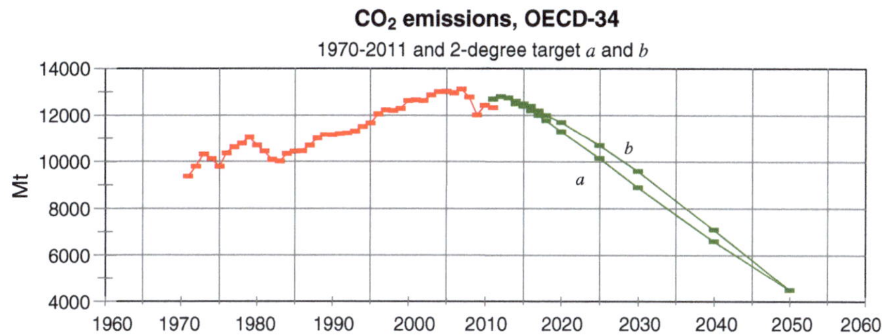


**Fig. 4.10** $CO_2$ emissions of the OECD-34 countries, 1970–2011 and 2-degrees scenario until 2050

### 4.5.3 Per Capita Indicators up to 2030

Figure 4.11 shows the per capita indicators of the OECD-34 from 1980 to 2011 as well as the profile up to 2030 resulting from the above considerations to meet the 2-degree target, for both variants *a* and *b*. The GDP (PPP) data up to 2018 correspond to the statistics and forecasts of the IMF.

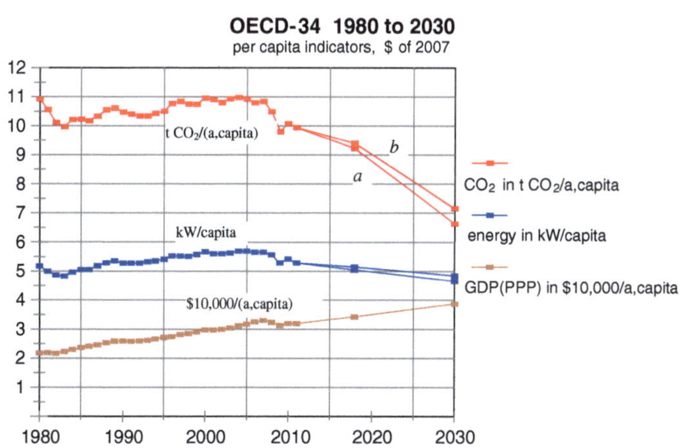


**Fig. 4.11** Per capita indicators of the OECD-34 from 1980 to 2011 and 2-degrees-scenario until 2030

Figure 4.12 shows the per capita emissions of 2011 of the resident population, according to OECD-Annex, and those according to the 2-degrees climate protection scenario 2030 for the variants *a* and *b*, **for each consumer sector.**

The four consumer sectors are (total specific $CO_2$ emissions in 2011: 9.9 t/ capita):

- industry (1.6 t/capita)
- transport (2.8 t/capita)
- residential, services, agriculture etc. (2.3 t/capita)
- energy-sector losses (3.2 t/capita).

The energy and emissions data for 2011 correspond to the OECD-34 Annex A5.

Final-energy carriers emit $CO_2$ directly or indirectly (via electricity and district heating). The assumed distribution of these energy carriers within the consumer sectors, for both variants in 2030, is ***one way among others*** to meet the 2-degree condition.

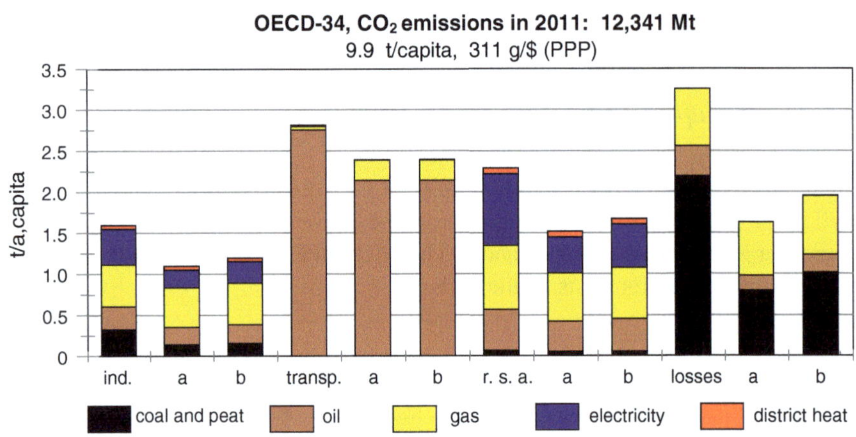


**Fig. 4.12** $CO_2$ emissions per capita of the 4 sectors: industry, transport, residential + services + agriculture, energy-sector losses (definitions see Appendix), each sector represented by three bars: *First bar* 2011 (data in the text and the diagram title, according to the OECD-Annex A5), *second bar* possible 2 °C scenario **variant *a*** for **2030**, 8,900 Mt, 6.6 t/capita, 171 g/$ (PPP), *third bar* possible 2 °C scenario **variant *b*** for **2030**, 9,600 Mt, 7.2 t/capita, 185 g/$ (PPP)

The target value 2030 is, for example, achieved when the per capita emissions:

- in the **electricity and district heat production sector, including energy-sector losses**, are reduced by 49 % (variant *a*) or 39 % (variant *b*) from 4.7 t/capita in 2011 to 2.4–2.9 t/capita (efficiency improvements, replacement of coal by gas, CCS, renewable energy and nuclear power)
- in the **heating sector** (industry, residential, services etc.) through efficiency improvements, reduction of coal and oil consumption and the use of renewable

energy sources (heat pump, waste management, solar energy, geothermal energy), are reduced by 25 % (*a*) or 20 % (*b*) (from 2.5 t/capita in 2011 to 1.8–2.0 t/capita)

- in the **transport sector** are reduced by 15 % (*a* and *b*), from 2.8 to 2.4 t/capita), through efficiency improvements, gas fuels, biofuels and electric mobility.

# Chapter 5
# European Union EU-27

## 5.1 Population and Gross Domestic Product 2011

The EU of 27 nations had, in 2011, over half a billion inhabitants (Fig. 5.1). Taking account of purchasing power parity (PPP), this generates a GDP of 14,500 billion (of year-2007 dollars), which is about 20 % of world GDP (PPP). The six countries with the greatest populations (Germany, France, United Kingdom, Italy, Spain and Poland), with around 70 % of the population of the EU, account for 73 % of the GDP. The 15 Western European countries (EU-15) that are members of the OECD, with 79 % of the population, account for around 88 % of the GDP of the EU-27.

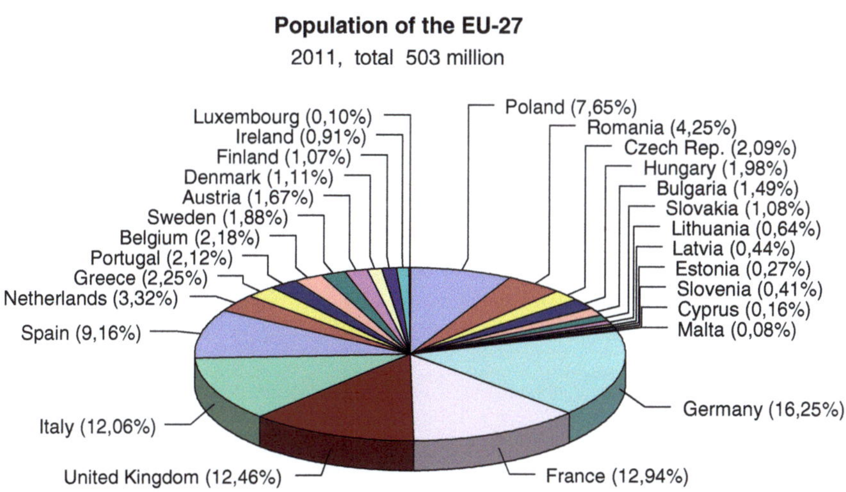


**Fig. 5.1** Per cent partition of the population of the EU-27



V. Crastan, *Global Energy Demand and 2-degree Target, Report 2014*,
DOI 10.1007/978-3-319-12310-3_5

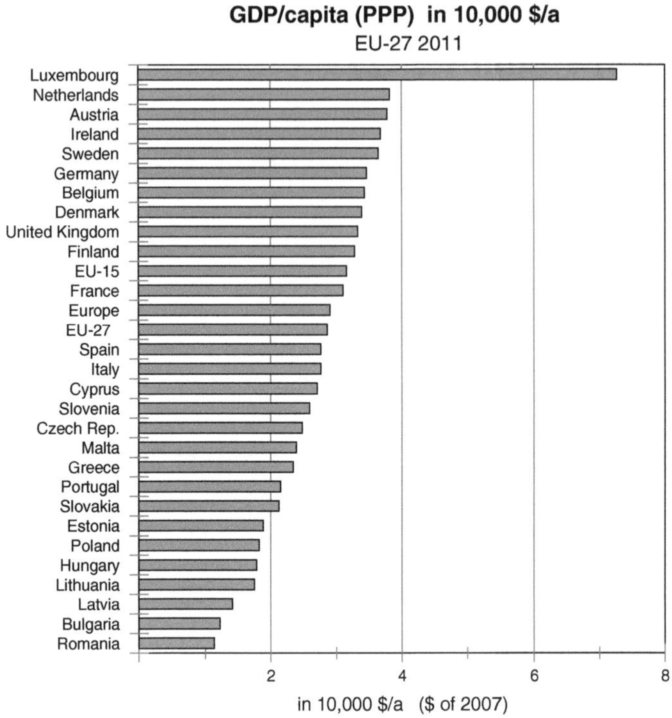


**Fig. 5.2** GDP (PPP) per capita of the EU-27 countries

The per capita GDP (PPP) (Fig. 5.2) has an average of $29,000/a and varies (excluding the special case of Luxembourg, with a large share of cross-border commuters) between $38,200/a in the Netherlands and $11,400/a in Romania. The six major countries are, with the exception of Poland ($18,000/a), all in the range 28,000–35,000 $/a. (We denote as "Europe" the EU-27 + Switzerland + Norway + Iceland)

## 5.2 Energy Intensity 2011

The energy intensity is a measure of the gross energy required for producing the GDP (PPP). The mean for the EU is 1.25 kWh/$. The group of EU-15 countries, at 1.19 kWh/$, is slightly more efficient (Fig. 5.3). Energy intensity is also influenced by climate. The more southerly countries (such as Italy, Greece, Spain and Portugal) have lower average energy intensity than the Nordic countries (such as Finland, Estonia and Sweden). Differences in energy efficiency are considerable. Ireland, for example, needs less than 1 kWh/$, whereas Estonia requires about 2.6 kWh/$. The target values required by 2018 for climate protection (based on the GDP forecast of the IMF) and for 2030 are also shown (see Sect. 5.5).

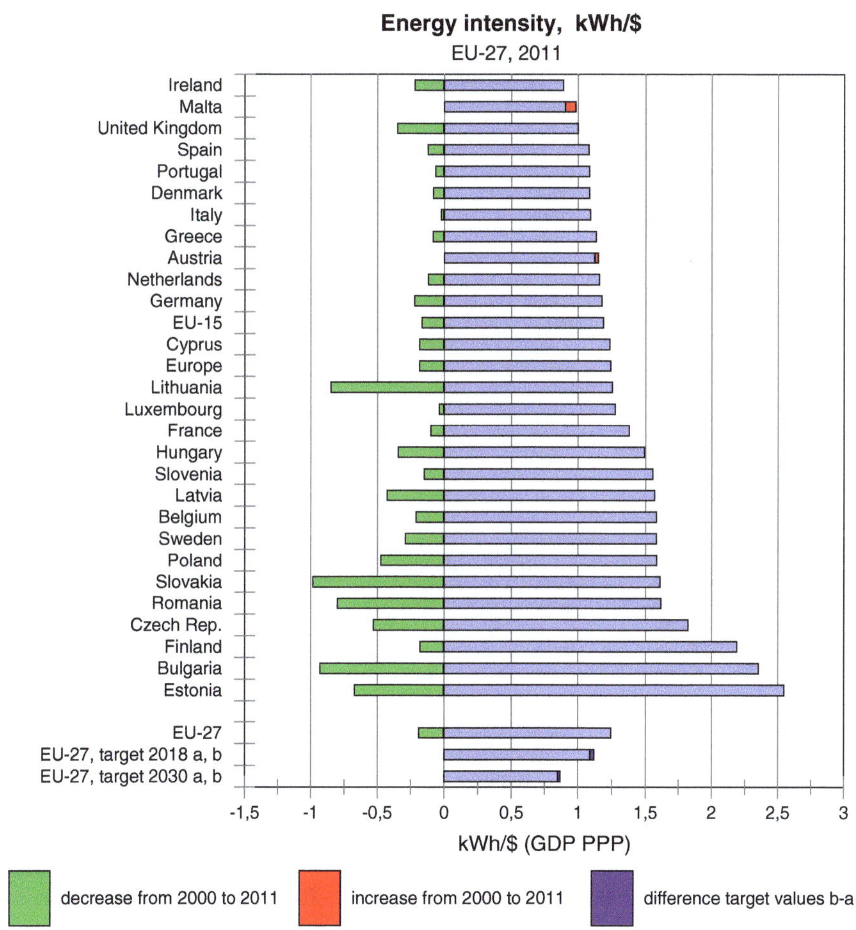


**Fig. 5.3** Energy intensity of EU-27 countries and changes from 2000 to 2011

## 5.3 $CO_2$ Intensity of Energy 2011

Energy is only harmful to the climate when the associated $CO_2$ intensity is large. The mean value for this in the EU is 197 g $CO_2$/kWh. In the EU-15 it is slightly lower, at 190 g $CO_2$/kWh. The differences are quite large, ranging from 86 g $CO_2$/kWh in Sweden to 298 g $CO_2$/kWh in Estonia. The way that electricity is generated is a very important factor. Countries where electricity is generated by water power and nuclear energy (Sweden, France) come off well in this respect, and those where the power stations are fuelled predominantly by coal, oil and gas (Poland, Italy, Germany, United Kingdom) are worse than average. The significant deterioration in Lithuania is a result of the shutdown of the nuclear power plant, partly compensated by improved energy intensity (Fig. 5.4).

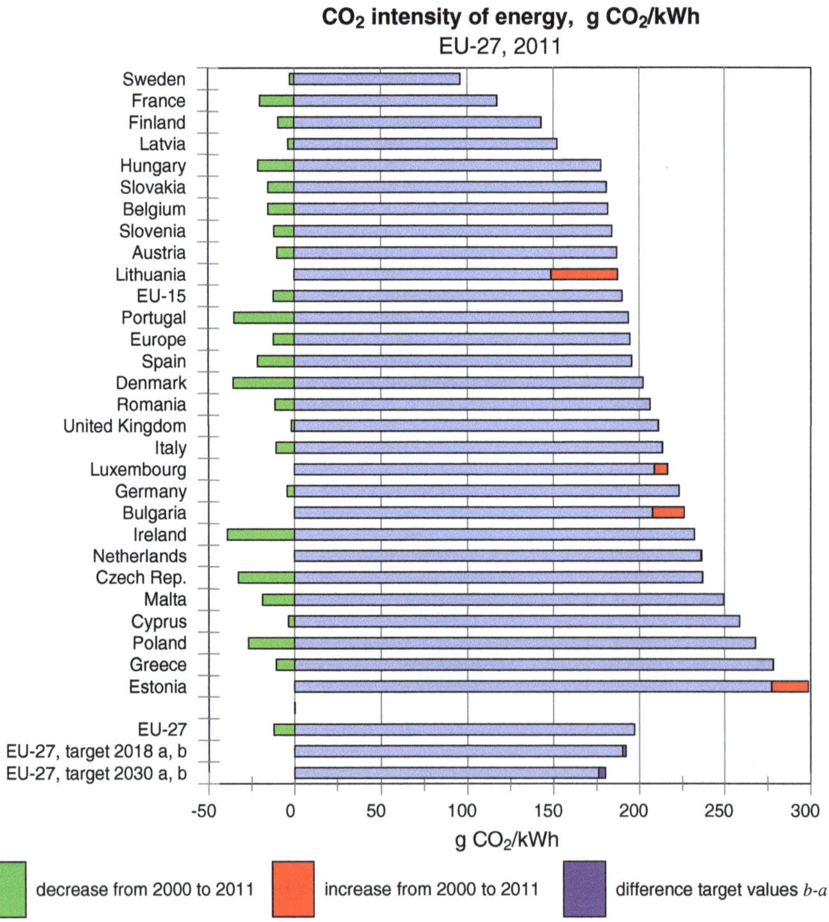


**Fig. 5.4** $CO_2$ intensity of the EU-27 countries and changes from 2000 to 2011

## 5.4 Indicator of $CO_2$ Sustainability 2011

To assess the sustainability of energy consumption in climate terms, both factors are significant: energy intensity and $CO_2$ intensity. The product of these two quantities gives the $CO_2$ indicator in g $CO_2$/\$. The value for EU-27 is 245 g $CO_2$/\$ and that for EU-15 is 225 g $CO_2$/\$. Sweden and France are significantly below 200 g $CO_2$. Germany's energy has a poorer sustainability rating with 263 g $CO_2$/\$ and the figures for some east European countries are very bad, such as for Poland and the Czech Republic (>400 g $CO_2$/\$), Bulgaria (>500 g $CO_2$/\$) and Estonia (over 750 g $CO_2$/\$) (Fig. 5.5).

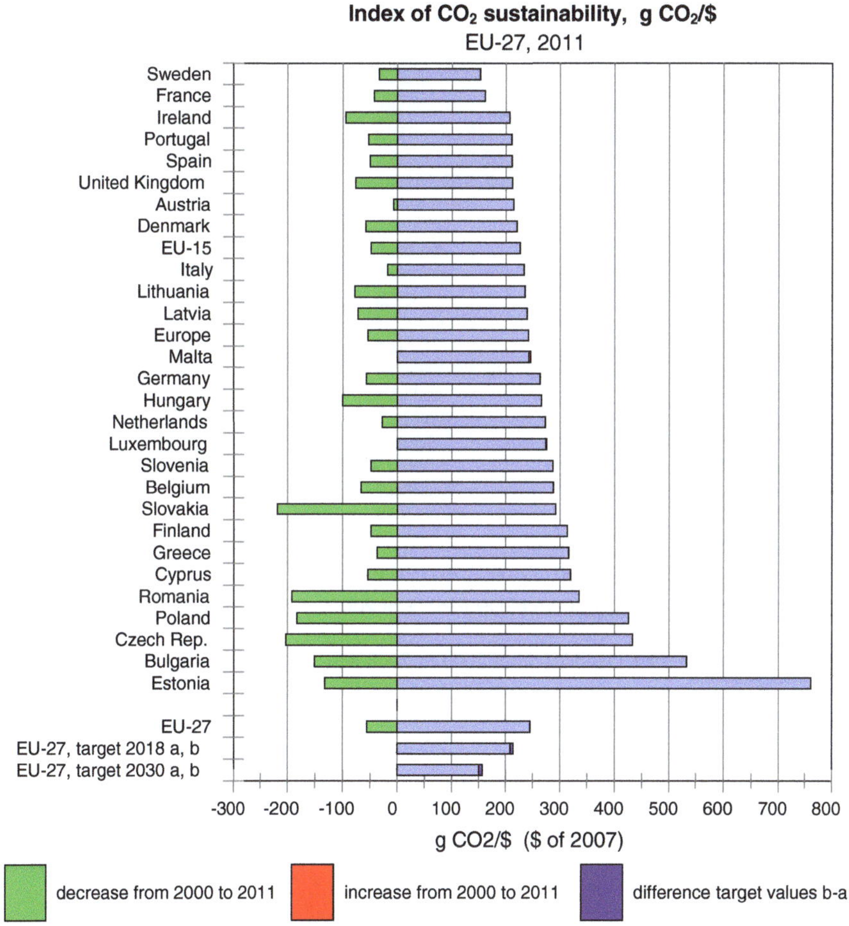


**Fig. 5.5** Index of $CO_2$ sustainability of the UE-27 countries and progress from 2000 to 2011

The change in $CO_2$ sustainability index is shown in more detail in Fig. 5.6 for the period 2000–2008 and 2008–2011. The general trend from 2000 to 2008 is good. Some countries have taken a step backwards in the last three years.

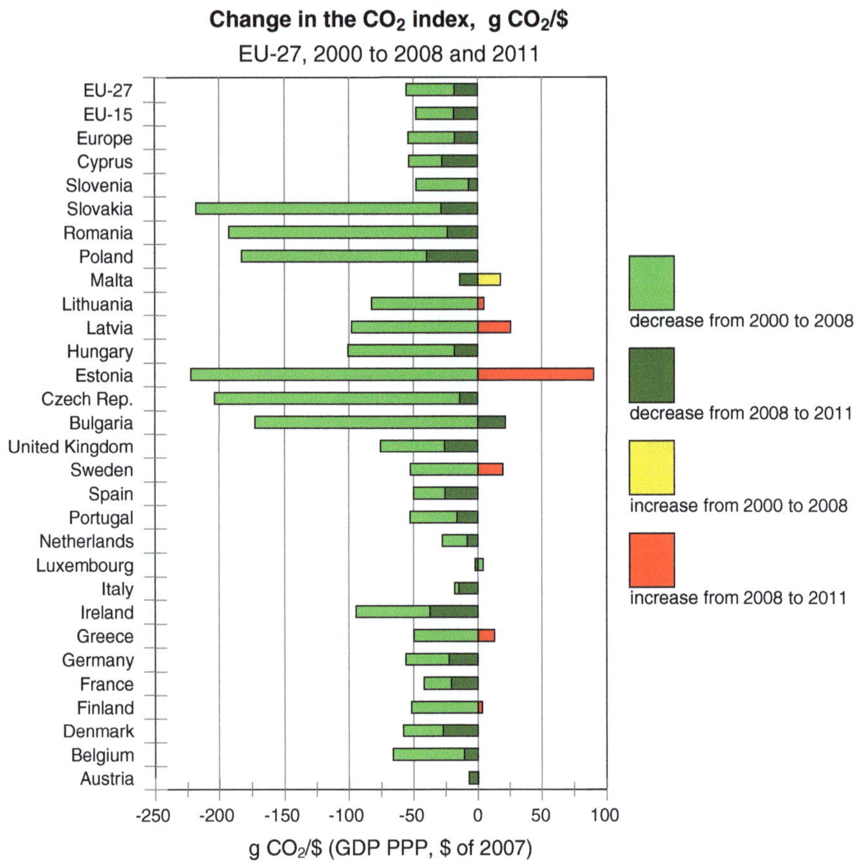


**Fig. 5.6** Change in the $CO_2$-index from 2000 to 2008 and 2008 to 2011 for EU-27 countries

## 5.5 Actual Indicators and $CO_2$ Emissions from 1990 to 2011 and Values Required from 2011 to meet the 2 °C Target

### 5.5.1 *Energy Intensity, $CO_2$ Intensity and $CO_2$ Sustainability up to 2030*

For 2011, the EU-27 has, at 245 g $CO_2$/\$, the best sustainability index of the world's regions and the fourth-best of the G-20 group. The big role of coal in electricity production is still a concern (see Appendix). The actual indicators from 1990 to 2011 and the graph up to 2030 necessary for the 2-degree target are shown in Fig. 5.7.

The climate target to be aimed at for 2030 is 150–157 g $CO_2$/\$ for a $CO_2$ emission share of the EU-27 of 2,700–2,830 Mt from the G-20 group. The values for 2018 take into account the GDP (PPP) forecast by the International Monetary Fund IMF of around \$15,650 billion (\$¨of 2007). The GDP for 2030 is estimated at \$18,000 billion.

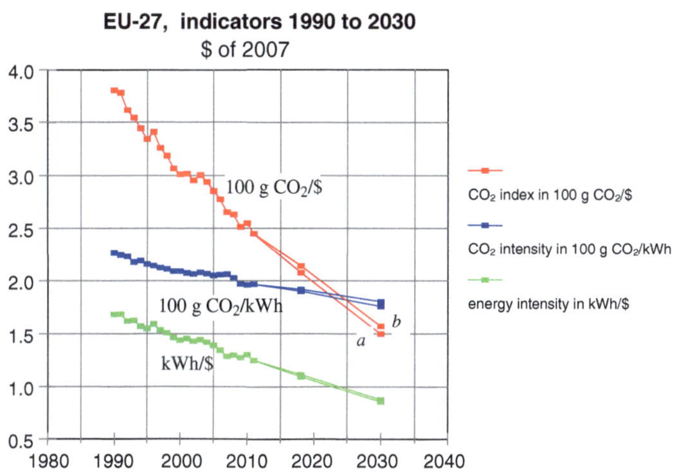


**Fig. 5.7** Indicators of the EU-27, 1990–2011 and climate protection scenario up to 2030

With this scenario the gross energy demand in 2030 is 1,320–1,350 Mtoe (only energy use). The per capita indicators for energy and $CO_2$ emissions would then be (variants *a* and *b*) $e = 3.4$ kW/capita (a decrease of about 17 % relative to 2011), and $\alpha = 5.2–5.5$ t $CO_2$/capita (−26 to −23 % relative to 2011).

The necessary change in the trend of the indicators relative to the period 2000–2011 is illustrated in Fig. 5.8 (variant *a*) and Fig. 5.9 (variant *b*). The European Union is well on track. The trend of $CO_2$ intensity of energy must be maintained that of the energy intensity strengthened. With variant *b* both trends are a little gentler before 2018.

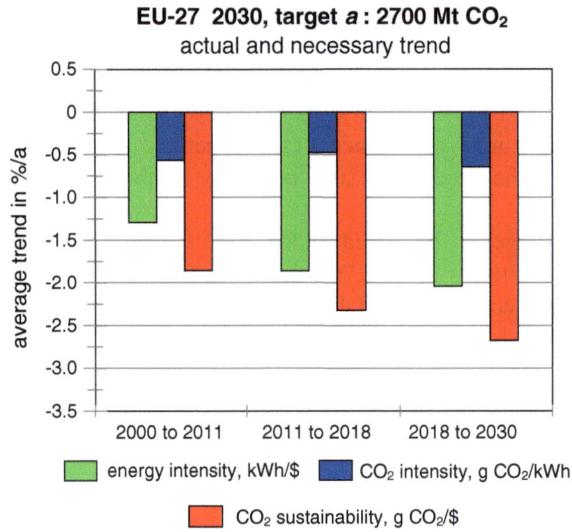


**Fig. 5.8** Trend from 2000 to 2011 and necessary change in the trend from 2011 to achieve the two-degree target, variant *a*

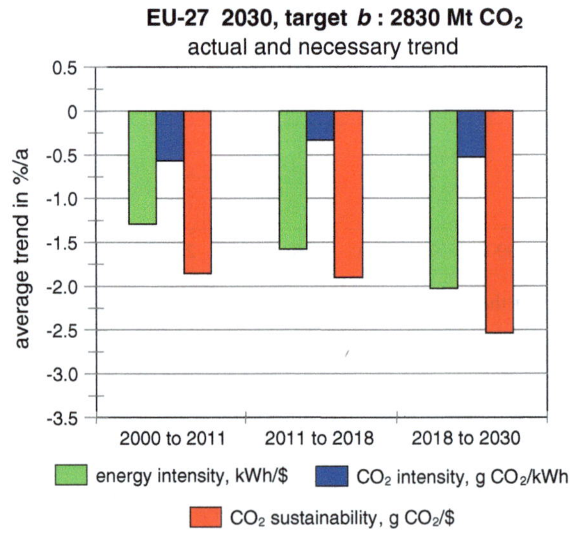


**Fig. 5.9** Trend from 2000 to 2011 and necessary change in the trend from 2011 to achieve the two-degree target, variant *b*

### 5.5.2 $CO_2$ Emissions until 2050

Figure 5.10 shows the actual $CO_2$ emissions from 1990 to 2011 and the permissible values for compliance with the 2-degree target for both variants.

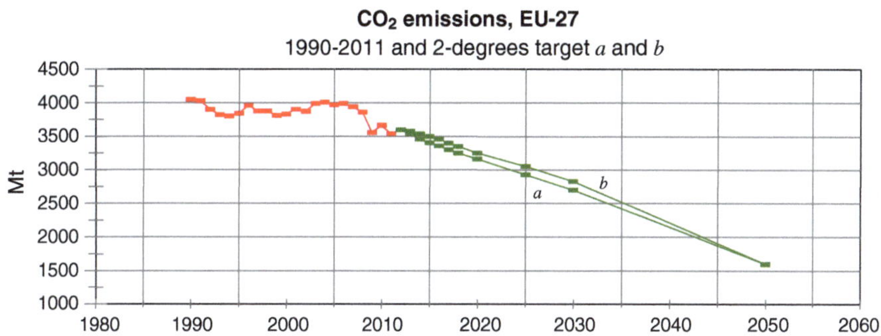


**Fig. 5.10** $CO_2$ emissions of EU-27 countries from 1990 to 2010 and climate protection scenario up to 2050

### 5.5.3 Per Capita Indicators up to 2030

Figure 5.11 shows the **per capita indicators** of the EU-27 from 1980 to 2011 as well as the profile up to 2030 resulting from the above considerations to meet the two-degree target, for both variants *a* and *b*. The GDP (PPP) data up to 2018 correspond to the statistics and forecasts of the IMF.

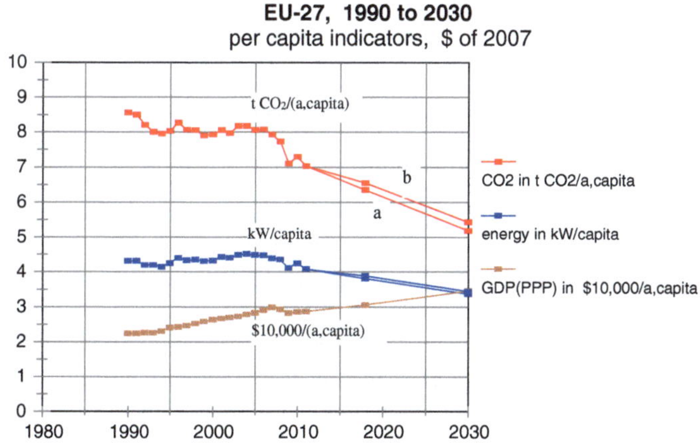


**Fig. 5.11** Per capita indicators of the EU-27 from 1990 to 2011 and two-degrees-scenario until 2030

The following Fig. 5.12 shows, in more detail **per consumer sector,** the per capita emissions of the resident population in 2011, and those corresponding to the two-degree scenario for 2030 (variant *a* and *b*).

The four consumer sectors are (total specific emissions of $CO_2$ in 2011: 7.0 t/capita):

- Industry (1.2 t/capita).
- Transport (1.8 t/capita).
- Residential, services, agriculture, etc. (1.9 t/capita).
- Energy-sector losses (2.1 t/capita).

The energy and emissions data for 2011 correspond to the EU-27-Annex A5.

Final-energy carriers emit $CO_2$ directly or indirectly (via electricity and district heating). The assumed distribution of these energy carriers within the consumer sectors, for both variants in 2030, is ***one way among others*** to meet the 2-degree condition.

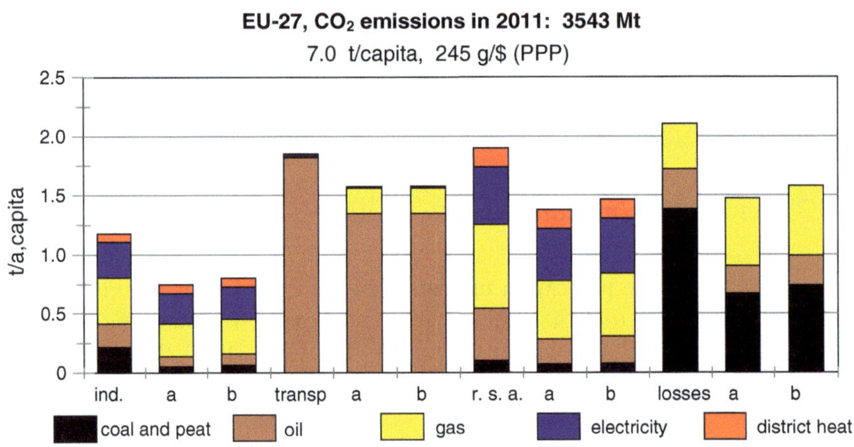


**Fig. 5.12** $CO_2$ emissions per capita of the 4 sectors: industry, transport, residential + services + agriculture, energy-sector losses (definitions see Appendix), each sector represented by three bars: *First bar* 2011 (data in the text and the diagram title, according to the EU-27-Annex A5), *Second bar* possible 2 °C scenario **variant *a*** for **2030**, 2,700 Mt, 5.2 t/capita, 150 g/$ (PPP), *Third bar* possible 2 °C scenario **variant *b*** for **2030**, 2,830 Mt, 5.4 t/capita, 157 g/$ (PPP))

The 2030 target value is, for example, achieved when the per capita emissions:

- in the **electricity and district heat production sector, including energy-sector losses**, are reduced by 23 % (variant *a*) or 18 % (variant *b*) from 3.1 t/capita in 2011 to 2.4–2.6 t/capita (efficiency improvements, replacement of coal by gas, CCS, renewable energy and nuclear power).
- in the **heating sector** (industry, residential, services etc.) through efficiency improvements (better insulation), reduction of coal and oil consumption and the use of renewable energy sources (heat pump, waste management, solar energy, geothermal energy), are reduced by 42 % (*a*) or 37 % (*b*) (from 2.1 to 1.2–1.3 t/capita).
- in the **transport sector** are reduced by 15 % (*a* and *b*), from 1.85 to 1.6 t/capita, through efficiency improvements, gas fuels, biofuels and electric mobility.

# Chapter 6
# Middle East

## 6.1 Population and Gross Domestic Product 2011

With a population of 209 million inhabitants, the Middle East generates a GDP (PPP) of around $2,720 billion ($ of 2007). About 50 % of this is produced by Saudi Arabia and the Gulf States (from Oman to Kuwait), which together have 21 % of the Middle East's population (Fig. 6.1). Iran, the state with the greatest population, generates a further 34 %. The remaining 16 % come from the other countries Iraq, Syria, Yemen, Jordan and Lebanon, with a population share of around 43 %. Israel is analyzed in Chap. 4, as a member of the OECD.

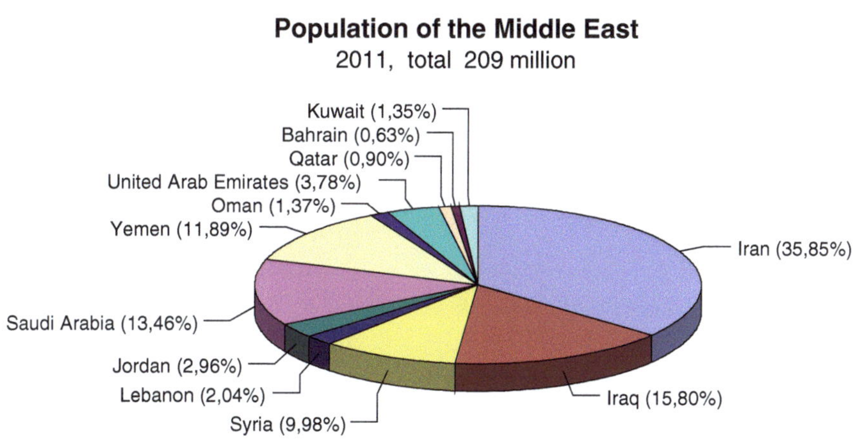


**Fig. 6.1** Percent partition of the population of the Middle East



V. Crastan, *Global Energy Demand and 2-degree Target, Report 2014*,
DOI 10.1007/978-3-319-12310-3_6

Fig. 6.2 GDP (PPP) per capita of the countries of the Middle East

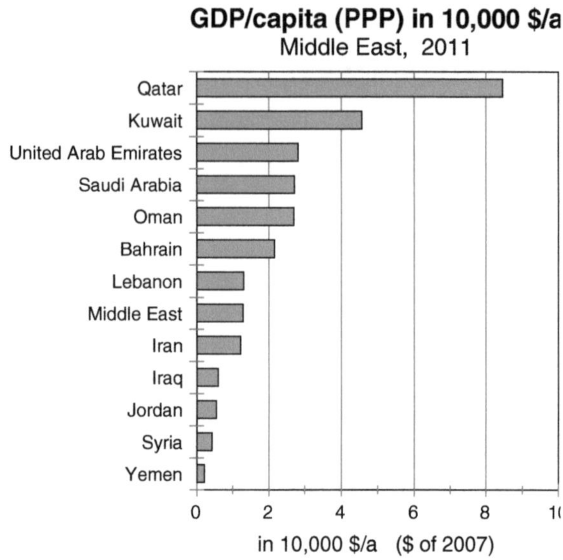


The wealth gap is therefore very large. The GDP (PPP) per capita of the region (Fig. 6.2) is $13,000/a, i.e. slightly above the world average. Saudi Arabia and the Gulf States have a GDP (PPP) of 20,000–50,000$/a, which is similar to the level in the EU or the OECD, with the exception of Qatar which is considerably higher. The purchasing power of other countries is however in the range $2,000–$13,000/a.

## 6.2 Energy Intensity 2011

The mean energy intensity of 2.5 kWh/$ is well above the world average, indicating an enormous waste of energy. This applies to almost all countries and particularly to some Gulf States. Only Lebanon has an energy intensity below 1.5 kWh/$ and thus an energy efficiency, comparable to that of Europe (Fig. 6.3).

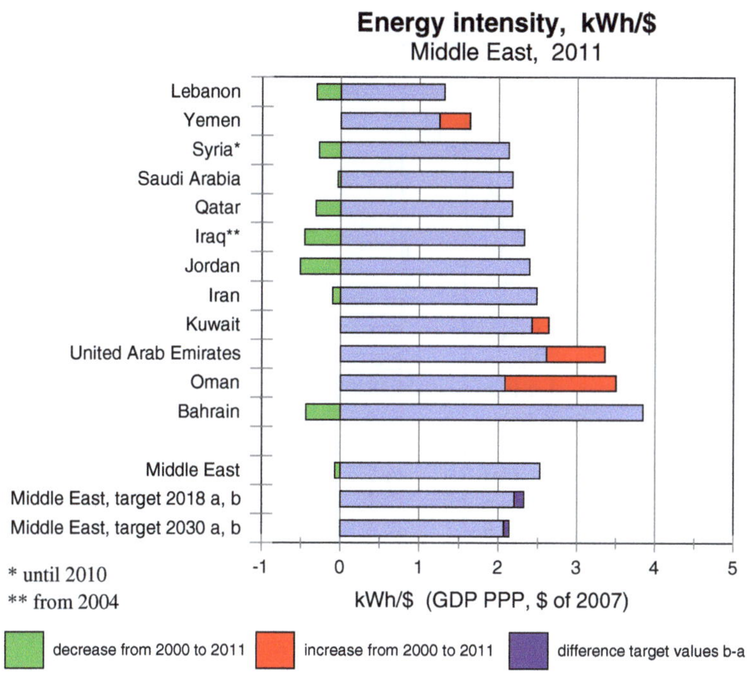


**Fig. 6.3** Energy intensity of the countries of the Middle East and changes from 2000 to 2011

## 6.3 $CO_2$ Intensity of Energy 2011

The average $CO_2$ intensity of 241 g $CO_2$/kWh is slightly above the world average of 220 g $CO_2$/kWh. The differences from one country to another are not very large. Saudi Arabia lies at the lower end of the scale, with 276 g $CO_2$/kWh, and is a large contributor to the overall negative development of the region. Between 2020 and 2030 a significant improvement must be made for the 2-degree target (Fig. 6.4).

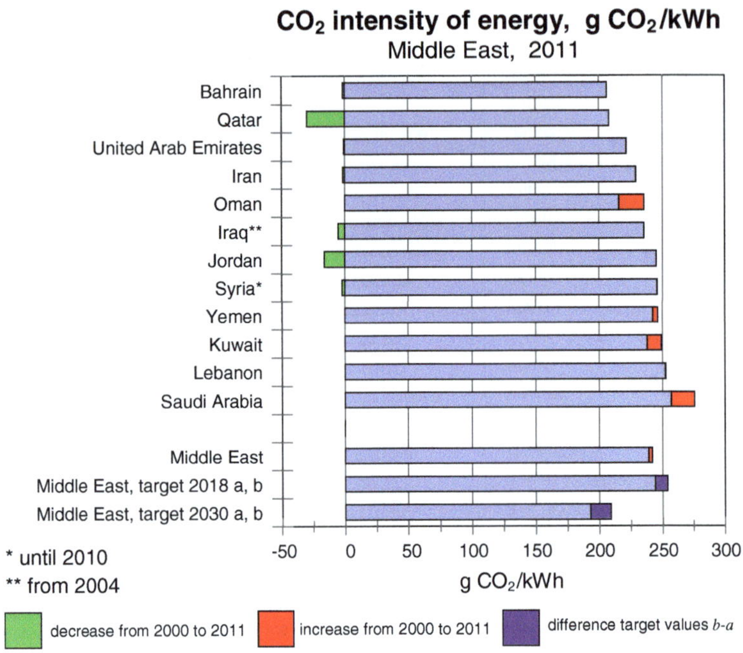


**Fig. 6.4** $CO_2$ intensity of the countries of the Middle East and changes from 2000 to 2011

## 6.4 Indicator of $CO_2$ Sustainability 2011

With a mean value of 610 g $CO_2$/$ (for comparison, OECD: 311 g $CO_2$/$), the energy system of the Middle East is anything but sustainable. Especially in the Gulf States, but also in Saudi Arabia and Iran, a more reasonable use of energy should be aimed at in the medium-term, despite their oil riches, with a big increase in energy efficiency. By using CCS and solar energy, abundant in this region, and perhaps nuclear energy, the $CO_2$ intensity could also be greatly improved (Fig. 6.5).

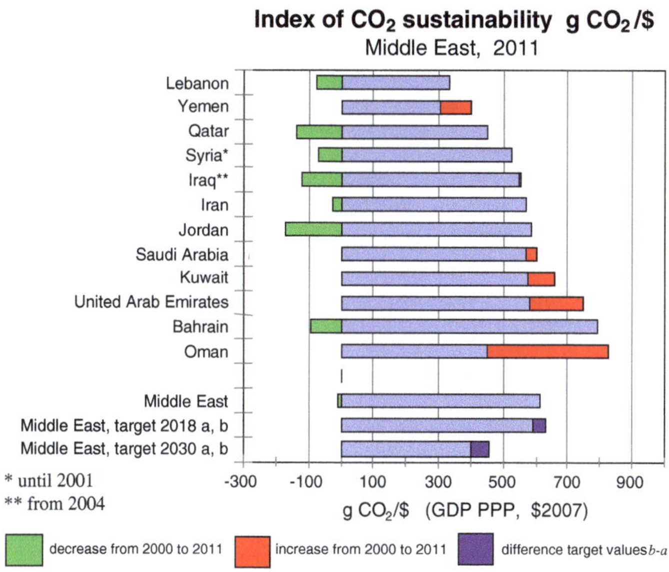


**Fig. 6.5** $CO_2$ sustainability of the countries of the Middle East and changes from 2000 to 2011

The change in $CO_2$ sustainability index is shown in more detail in Fig. 6.6 for the periods 2000–2008 and 2008–2011. Development is seen to be negative overall, even if some countries (Jordan, Qatar, Bahrain and Lebanon) are improving the picture. Saudi Arabia and most of the Gulf States stand out as negative.

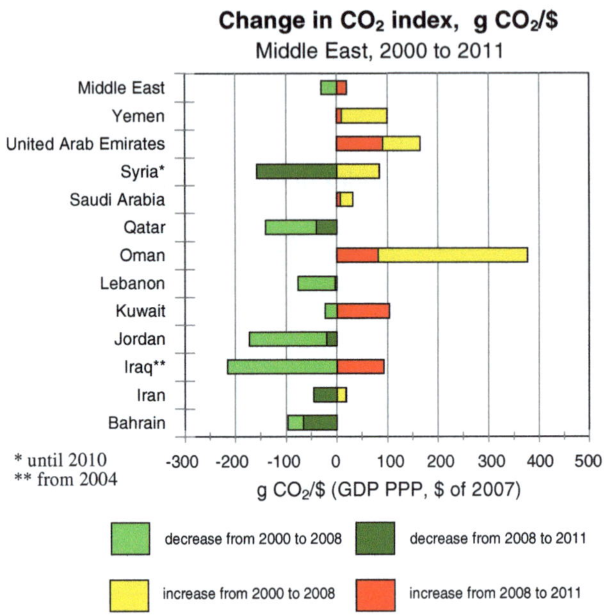


**Fig. 6.6** Change in the index of $CO_2$ sustainability of the Middle East countries from 2000 to 2008 and 2008 to 2011

# 6.5 Actual $CO_2$ Emissions and Indicators from 1980 to 2011 and Values Required from 2011 to meet the 2 °C Target

## *6.5.1 Energy Intensity, $CO_2$ Intensity and $CO_2$ Sustainability up to 2030*

The Middle East in 2011, with an indicator of $CO_2$ sustainability of 612 g $CO_2$/\$, has the third lowest ranking of the regions of the world (Fig. 3.5), well above the global average of 429 g $CO_2$/\$. This is primarily because of electricity production solely based on petroleum (see Appendix) and an insufficient energy efficiency. The graph of the indicators 1980–2011 and the necessary evolution to achieve the 2-degree target by 2030, for variants *a* and *b* is shown in Fig. 6.7.

The climate protection target to be aimed at for 2030 is 400–460 g $CO_2$/\$ with a share of 1,800–2,070 Mt of global $CO_2$ emissions. The values for 2018 take into account the forecast of the International Monetary Fund IMF with a GDP (PPP) of \$3,330 billion (\$ 2007). The GDP for 2030 is estimated at \$4,500 billion.

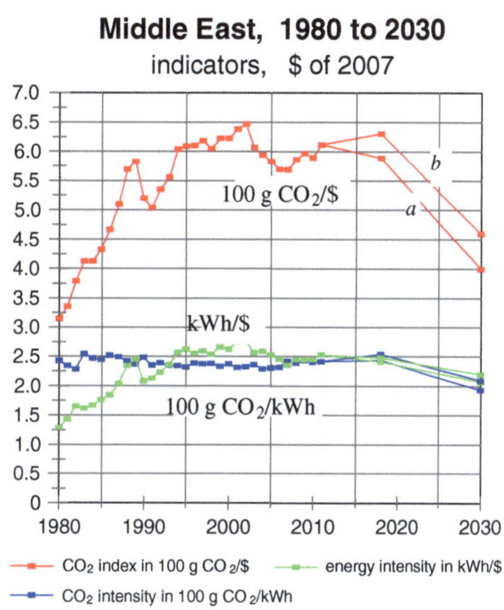


**Fig. 6.7** Indicators of the Middle East, 1980–2011 and climate protection scenario up to 2030

The gross energy demand in 2030 with this scenario is 800–850 Mtoe (only energy use). The per capita indicators for energy and $CO_2$ emissions need to be (variants *a* to *b*): e = 3.8–4.0 kW/capita (increase of 4–11 % relative to 2011) and $\alpha$ = 6.4–7.4 t $CO_2$ per capita (decrease of 16–4 % relative to 2011).

The necessary change in the trend of the indicators is illustrated in Figs. 6.8 and 6.9.

- The reduction of energy intensity to be aimed at, comes before 2018 in scenario *a*, and in scenario *b* only begins at that time.

- The turnaround in the trend of $CO_2$ intensity of energy is possible only from 2018, but must then be correspondingly vigorous. Scenario *b* even allows an increase of $CO_2$ intensity before this date.

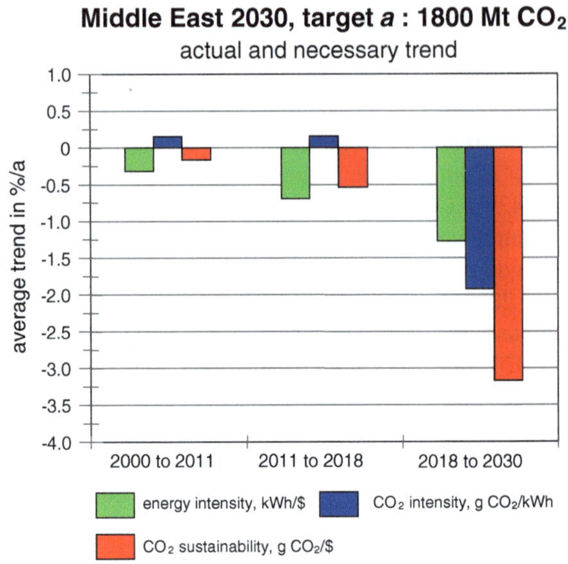


**Fig. 6.8** Trend from 2000 to 2011 and necessary change in the trend from 2011 to achieve the two-degree target, variant *a*

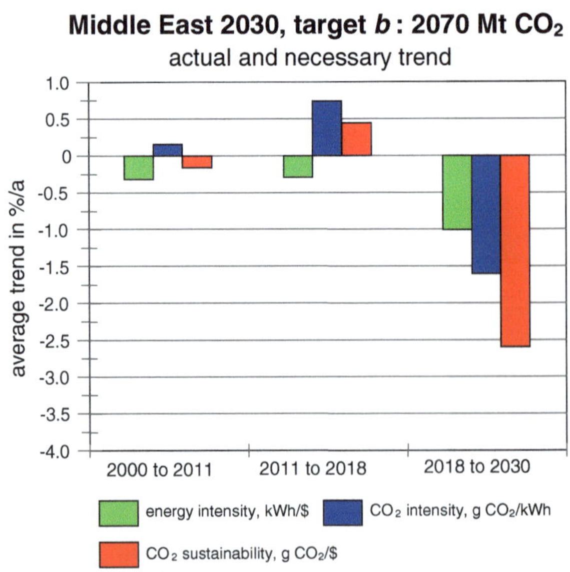


**Fig. 6.9** Trend from 2000 to 2011 and necessary change in the trend from 2011 to achieve the 2-degree target, variant *b*

### 6.5.2 $CO_2$ Emissions until 2050

Figure 6.10 shows the actual $CO_2$ emissions from 1970 to 2011, and the permissible levels up to 2030 for meeting the 2-degree limit, and the further extrapolation to 2050, for both variants *a* and *b*.

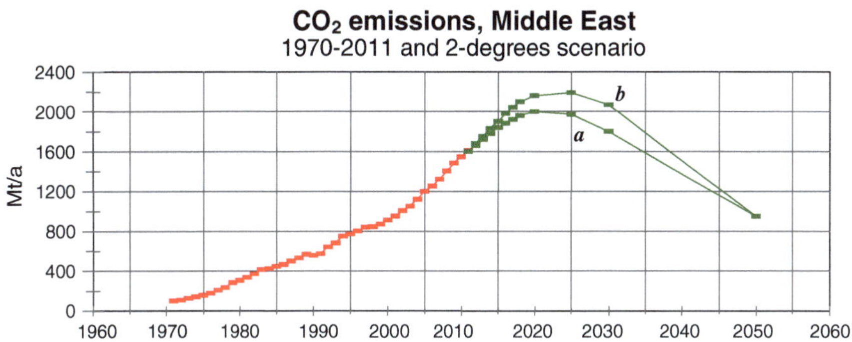


**Fig. 6.10** $CO_2$ emissions of the Middle East, 1970–2010 and climate protection scenario until 2050

### 6.5.3 Per Capita Indicators up to 2030

Figure 6.11 shows the **per capita indicators** of the Middle East from 1980 to 2011 as well as the profile up to 2030 resulting from the above considerations to meet the 2-degree target, for both variants *a* and *b*. The GDP (PPP) data to 2018 correspond to the statistics and forecasts of the IMF.

**Fig. 6.11** Per capita indicators of the Middle East from 1980 to 2011 and two-degrees-scenario until 2030

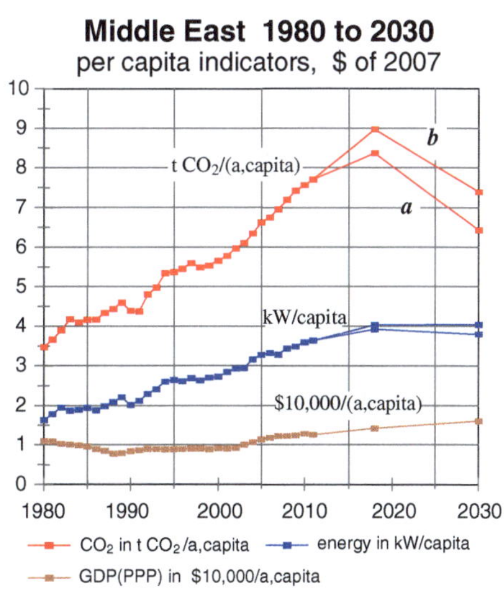

Figure 6.12 shows the per capita emissions of the resident population in 2011, and those corresponding to the 2-degree scenario for 2030 (variant *a* and *b*), in more detail **per consumer sector.**

The four consumer sectors are (total specific emissions of $CO_2$ in 2011: 7.7 t/capita):

- industry (1.6 t/capita)
- transport (1.8 t/capita)
- residential, services, agriculture, etc. (1.5 t/capita)
- energy-sector losses (2.8 t/capita).

The energy and emissions data for 2011 correspond to the Middle East-Annex A5.

Final-energy carriers emit $CO_2$ directly or indirectly (via electricity and district heating). The assumed distribution of these energy carriers within the consumer sectors, for both variants in 2030, is ***one way among others*** to meet the 2-degree condition.

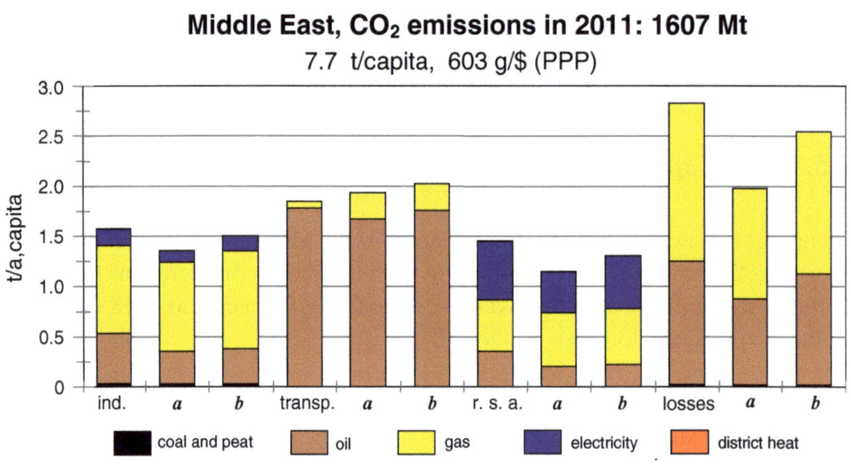


**Fig. 6.12** $CO_2$ emissions per capita of the 4 sectors: industry, transport, residential + services + agriculture, energy-sector losses (definitions see Appendix), each sector represented by three bars: *First bar* 2011 (data in the text and the diagram title, according to the Middle East-Annex A5), *Second bar* possible 2 °C scenario **variant *a*** for **2030**, 1,800 Mt, 6.4 t/capita, 400 g/$ (PPP), *Third bar* possible 2 °C scenario **variant *b*** for **2030**, 2,070 Mt, 7.4 t/capita, 460 g/$ (PPP)

The 2030 target value is, for example, achieved when the per capita emissions:

- in the **electricity and district heat production sector**, **including energy-sector losses**, decrease by 30 % (variant *a*) or 10 % (variant *b*) from 3.6 t/capita in 2011 to 2.5–3.2 t/capita (efficiency improvements, replacement of oil by gas, CCS, renewable energy and nuclear power)
- in the **heating sector** (industry, residential, services etc.) through efficiency improvements, reduction of oil consumption and the use of renewable energy

sources (heat pump, waste management, solar energy, geothermal energy), are reduced by 13 % (*a*) or 6 % (*b*) from 2.3 t/capita in 2011 to 2.0–2.1 t/capita

- in the **transport sector** increase by a maximum of 5 % (*a*) to 10 % (*b*), from 1.85 in 2011 to 1.94–2.03 t/capita), through efficiency improvements, gas fuels, biofuels and electric mobility.

# Chapter 7
# Eurasia+

## 7.1 Population and Gross Domestic Product 2011

The term Eurasia+ is used to indicate the 14 countries of the former Soviet Union (excluding Estonia, member of the OECD), and a further 12 European countries that are not members of the OECD, with a total population of 340 million (Fig. 7.1).

By far the most significant country is Russia, with 42 % of the population and about 58 % of GDP (PPP). Together with Ukraine and Belarus, the population share is around 58 % and the share of the GDP (PPP) is 69 %. The Baltic countries Lithuania and Latvia are now also members of the EU-27.

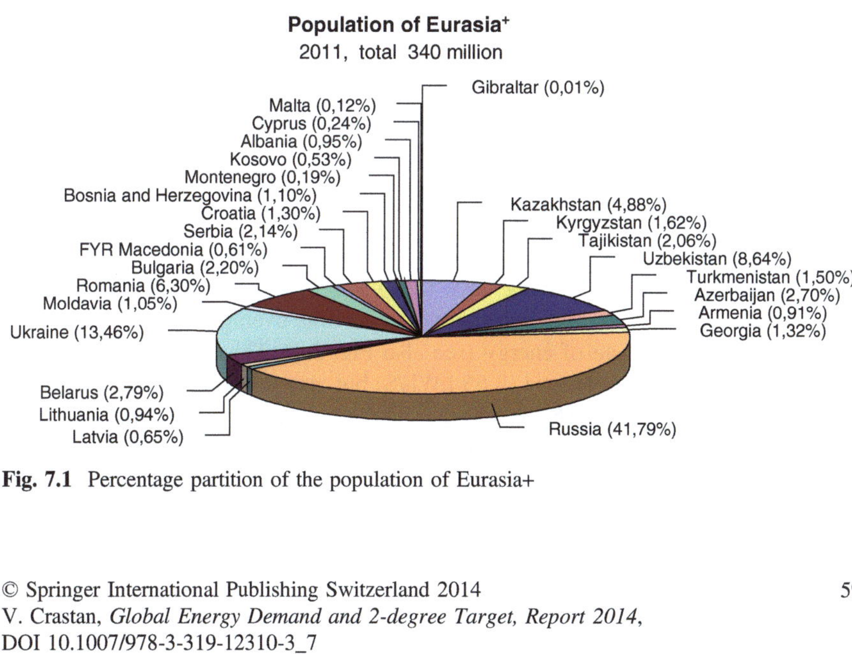


**Fig. 7.1** Percentage partition of the population of Eurasia+



V. Crastan, *Global Energy Demand and 2-degree Target, Report 2014*,
DOI 10.1007/978-3-319-12310-3_7

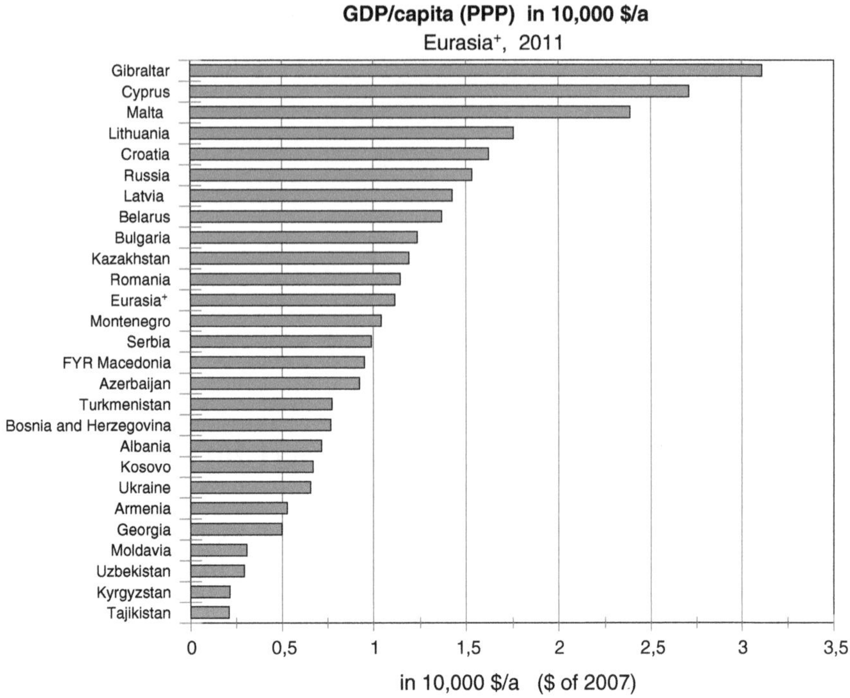


**Fig. 7.2** GDP (PPP) per capita of the countries of Eurasia+

Among the 12 non-OECD countries of Europe, Romania and Bulgaria (which have become members of the EU-27) are also included.

The total GDP (PPP) in 2011 reached $3,780 billion in Eurasia+ ($ of 2007). The GDP (PPP) per capita (Fig. 7.2) has in Eurasia+ an average value of $11,100/a, in Russia $15,300/a; only Gibraltar, Cyprus and Malta are over $20,000/a.

## 7.2 Energy Intensity 2011

Overall, the countries of Eurasia have (despite progress) an energy intensity of 3.4 kWh/$, which is significantly above the world average and represents the worldwide greatest waste of energy (see also Fig. 3.3). This high value is mainly due to the high values in Russia (3.55 kWh/$) but also in Ukraine and Kazakhstan (>4.5 kWh/$). Uzbekistan (6 kWh/$) and Turkmenistan (7 kWh/$) have extreme values. However, in both these countries there was a clear improvement from 2000 to 2011, mainly thanks to a big increase in the GDP (Fig. 7.3). In the European non-OECD countries, the average is, at around 1.7 kWh/$, much lower, but higher than the EU mean value.

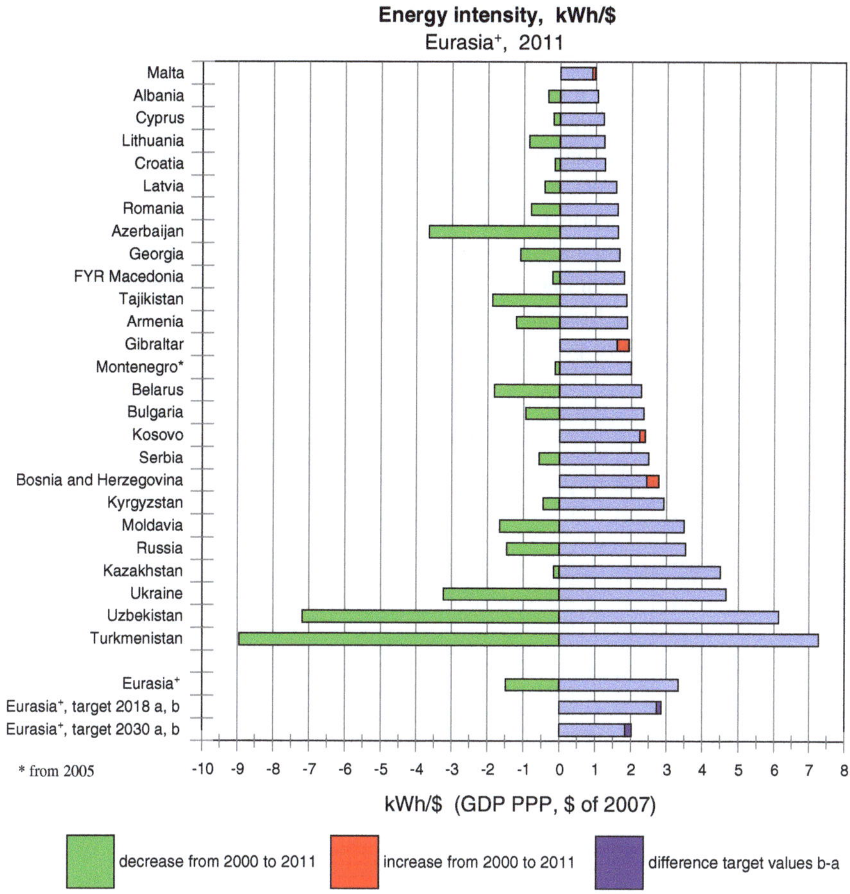


**Fig. 7.3** Energy intensity of the countries of Eurasia+ and progress from 2000 to 2011

## 7.3 $CO_2$ Intensity of Energy 2011

The $CO_2$ intensity of 215 g $CO_2$/kWh is not far from the world average and in the European non-OECD countries mostly higher than in the countries of the former Soviet Union (more coal and less gas consumption). The $CO_2$ intensity is particularly high (250 $CO_2$/kWh and more) in FYR Macedonia, Serbia, Bosnia-Herzegovina, Kosovo and Cyprus but also in Kazakhstan. The low values of Tajikistan, Latvia, Albania, Armenia and Georgia are due to the generation of electricity using hydropower and/or nuclear energy (the latter in Armenia). The switching off of the nuclear power station in Lithuania at the end of 2009 has brought this country some steps down from its third position (Fig. 7.4).

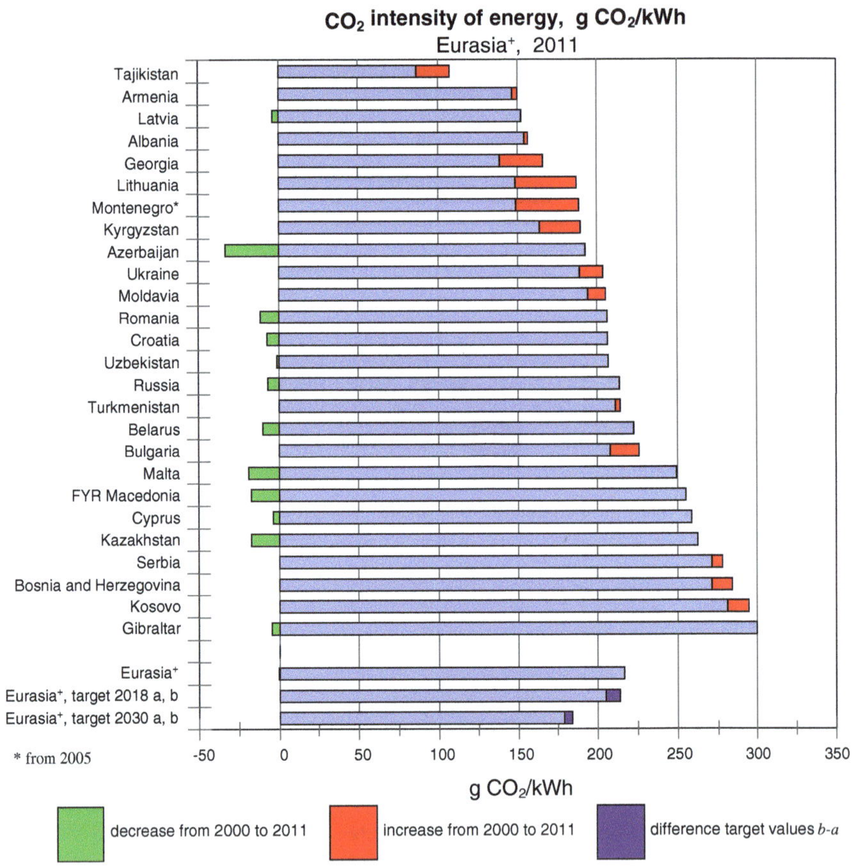


**Fig. 7.4** $CO_2$ intensity of the countries of Eurasia+ and changes in the last 11 years

## 7.4 Indicator of $CO_2$ Sustainability 2011

In the European non-OECD countries, the indicator of $CO_2$ sustainability has an average of 425 g $CO_2$/\$, in the former Soviet Union 780 g $CO_2$/\$, i.e. nearly twice as large due to poorer energy efficiency. The lack of efficiency of energy use in important countries of Eurasia (Fig. 7.3), has a negative impact on the overall sustainability of energy consumption (mean of Eurasia+ is 725 g $CO_2$/\$). Extreme values apply to the Ukraine (950 g $CO_2$/\$), Uzbekistan and Kazakhstan (about 1,200 g $CO_2$/\$), and especially (despite great progress) Turkmenistan (1,500 g $CO_2$/\$) (Fig. 7.5).

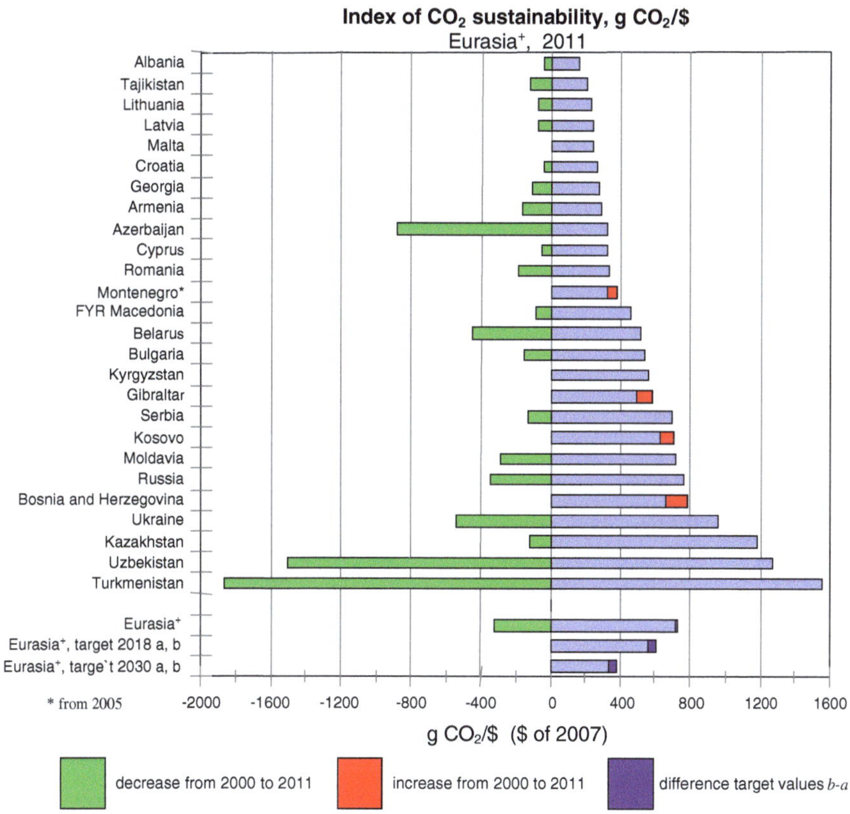


**Fig. 7.5** Index of $CO_2$ sustainability of the countries of Eurasia+ and progress from 2000 to 2011

The change in the $CO_2$ sustainability index is shown in more detail in Fig. 7.6 for the periods 2000–2008 and 2008–2011. In total, a positive development can be observed with the exception of the past 3 years, where in some small states there have been backward steps. In Russia and Ukraine too, the positive trend has unfortunately stopped.

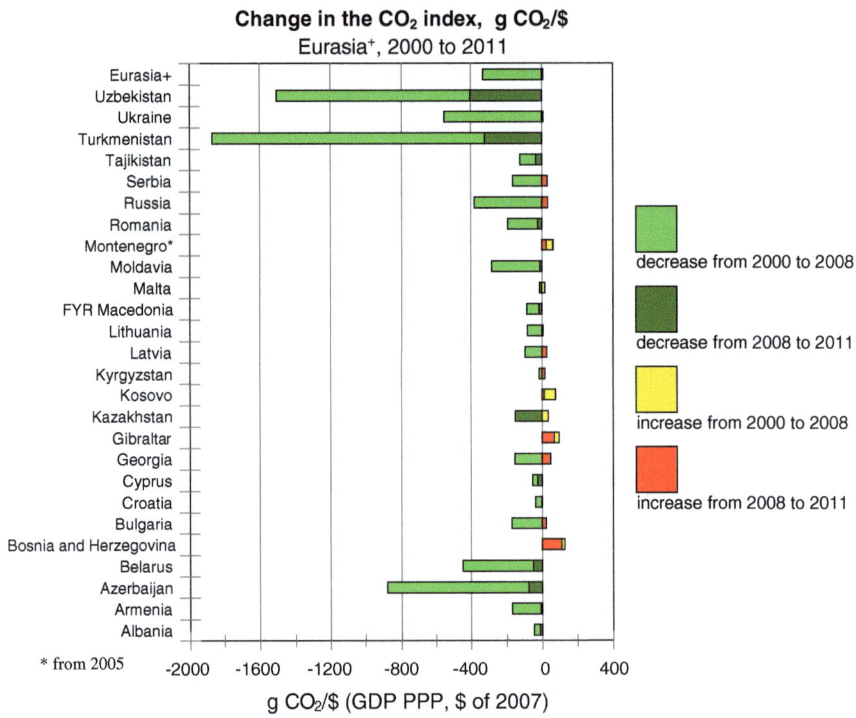


**Fig. 7.6** Change in the $CO_2$ sustainability indicator of the countries of Eurasia+ from 2000 to 2008 and 2008 to 2011

## 7.5 Actual Indicators and $CO_2$ Emissions from 1990 to 2011 and Values Required from 2011 to meet the 2 °C Target

### *7.5.1 Energy Intensity, $CO_2$ Intensity and $CO_2$ Sustainability up to 2030*

Eurasia+ in 2011, with a $CO_2$ sustainability index of 720 g $CO_2$/$, has the world's second-highest value, second only to China. The main reason is the world's worst energy efficiency (see Appendix), together with electricity production strongly dependent on coal. The graph of the past indicators from 1990 to 2011 and those necessary by 2030 to achieve the 2-degree target, for variants *a* and *b* are shown in Fig. 7.7.

The climate target to be aimed at for 2030 is 330–375 g $CO_2$/$ with a share of 2,080–2,360 Mt of global $CO_2$ emissions. The values for 2018 take into account the GDP (PPP) forecast by the International Monetary Fund of around $4,700 billion ($ of 2007). The GDP for 2030 is estimated at $6,300 billion.

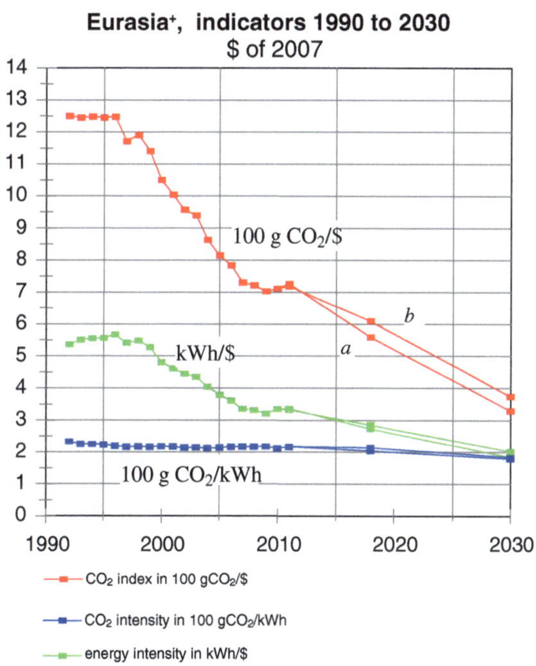


**Fig. 7.7** Indicators of Eurasia+, 1990–2010 and climate protection scenario up to 2030

The gross energy demand in 2030 with this scenario is 1,000–1,100 Mtoe (only energy use). The per capita indicators for energy and $CO_2$ emissions need to be at this time, for the two variants: e ≈ 4.0–4.4 kW/capita (decrease of 3 % to increase of 4 % relative to 2011) and $\alpha$ = 6.3–7.2 t $CO_2$ per capita (decrease of 18–12 % relative to 2011).

The necessary change in the trend of the indicators is illustrated in Fig. 7.8 for variant *a*.

- The good average trend 2000–2011 of energy intensity has to be maintained.
- The stagnation in the $CO_2$ intensity of energy need to be overcome.

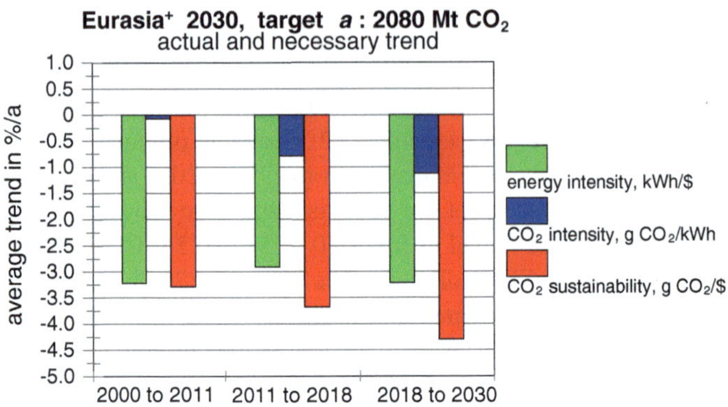


**Fig. 7.8** Trend from 2000 to 2011 and necessary change in the trend from 2011 to achieve the 2-degree target, variant *a*

If the pause of the last 3 years is overcome, this would be overall on the right course. If it continues, variant b is more likely (Fig. 7.9) with a significantly weaker improvement in the indicators by 2018.

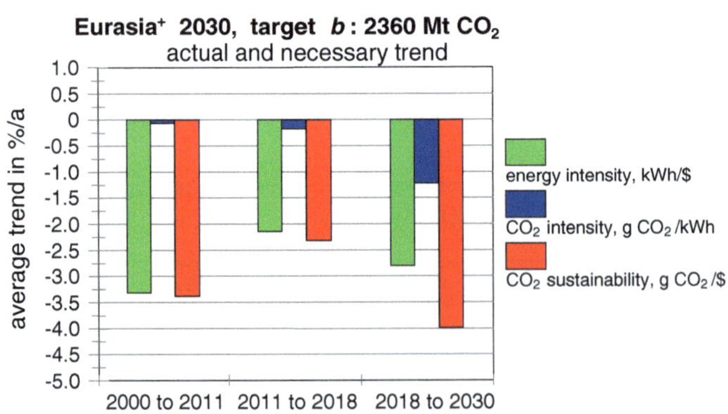


**Fig. 7.9** Trend from 2000 to 2011 and necessary change in the trend from 2011 to achieve the 2-degree target, variant *b*

### 7.5.2 $CO_2$ Emissions until 2050

Figure 7.10 shows the actual $CO_2$ emissions from 1970 to 2011, and the permissible levels up to 2030 for meeting the 2-degree limit, and the further extrapolation to 2050, for both variants *a* and *b*. The part of this graph from 1990 to 2000 clearly shows the large reduction in emissions as a result of the collapse of the Soviet Union.

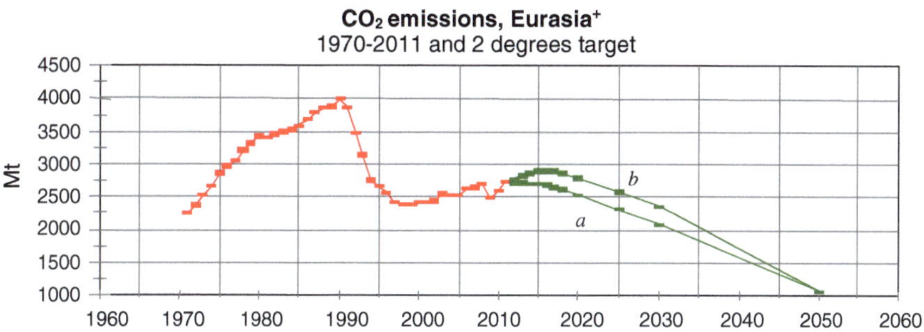


**Fig. 7.10** $CO_2$ emissions of Eurasia+, from 1970 to 2011 and climate protection scenario until 2050

### 7.5.3 Per Capita Indicators up to 2030

Figure 7.11 shows the **per capita indicators** of Eurasia+ from 1990 to 2011 as well as the profile up to 2030 resulting from the above considerations to meet the 2-degree target, for both variants *a* and *b*. The GDP (PPP) data up to 2018 correspond to the statistics and forecasts of the IMF.

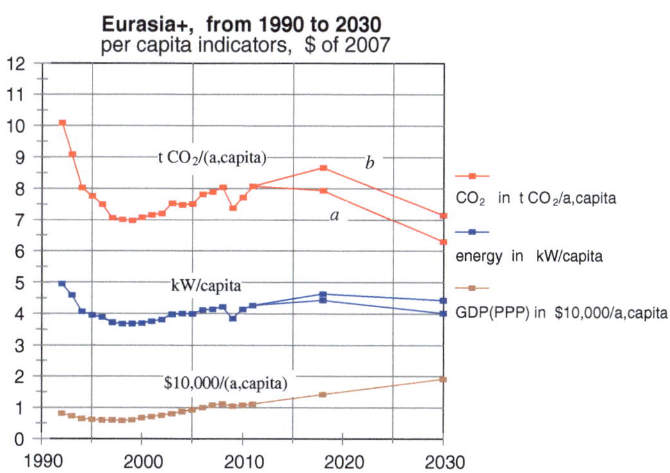


**Fig. 7.11** Per capita indicators of Eurasia+ from 1990 to 2011 and 2-degrees-scenario until 2030

The following Fig. 7.12 shows, in more detail **per consumer sector,** the per capita emissions of the resident population in 2011, and those for the 2-degree scenario 2030 (variant *a* and *b*).

The four consumer sectors are (total specific emissions of $CO_2$ in 2011: 8.1 t/capita):

- industry (1.7 t/capita)
- transport (1.2 t/capita)
- residential, services, agriculture, etc. (2.0 t/capita)
- energy-sector losses (3.2 t/capita).

The energy and emissions data for 2011 correspond to the Eurasia+, Annex A5

Final-energy carriers emit $CO_2$ directly or indirectly (via electricity and district heating). The assumed distribution of these energy carriers within the consumer sectors, for both variants in 2030, is ***one way among others*** to meet the 2-degree condition.

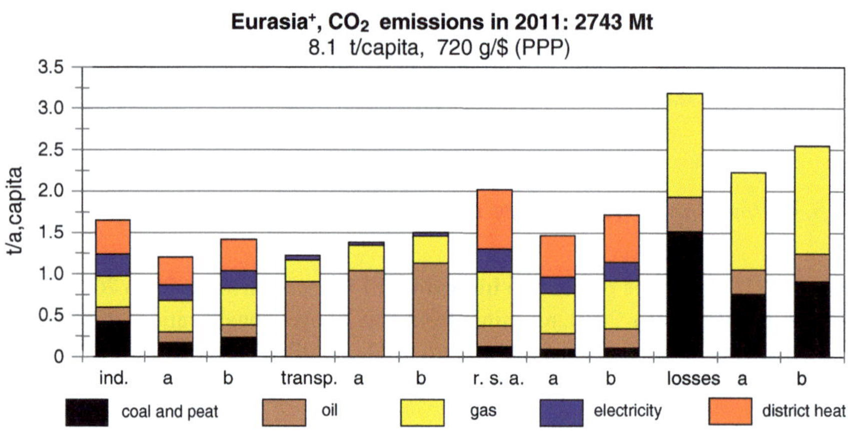


**Fig. 7.12** $CO_2$ emissions per capita of the 4 sectors: industry, transport, residential + services + agriculture, energy- sector losses (definitions see Appendix), each sector represented by three bars: *First bar* 2011 (data in the text and the diagram title, according to the Eurasia+, Annex A5), *Second bar* possible 2 °C scenario **variant *a*** for **2030**, 2,080 Mt, 6.3 t/capita, 330 g/$ (PPP) *Third bar* possible 2 °C scenario **variant *b*** for **2030**, 2,360 Mt, 7.2 t/capita, 375 g/$ (PPP)

The 2030 target value is, for example, achieved when the per capita emissions:

- in the **electricity and district heat production sector, including energy-sector losses**, are reduced by 29 % (variant *a*) or 19 % (variant *b*) from 4.9 t/capita in 2011 to 3.5 or 4.0 t/capita (efficiency improvements, replacement of coal by gas, CCS, renewable energy and nuclear power)
- in the **heating sector** (all end user: industry, residential, services etc.) through efficiency improvements (better insulation), reduction of coal and oil consumption and the use of renewable energy sources (heat pump, waste

management, solar energy, geothermal energy), are reduced between 27 % (*a*) and 13 % (*b*) (from 2.0 to 1.45–1.75 t/capita)

- in the **transport sector** increase by 13 % (a) to not more than 23 % (b), from 1.2 to 1.4–1.5 t/capita), through efficiency improvements, gas fuels, biofuels and electric mobility.

# Chapter 8
# Rest of Asia/Oceania

## 8.1 Population and Gross Domestic Product 2011

The rest of Asia/Oceania is understood as the continents of Asia and Oceania without China, India, Middle East, Eurasia (Asian part), Japan and all further member states of the OECD such as South Korea, Australia and New Zealand. The total population is 1,071 million (Fig. 8.1) and the GDP (PPP) $4,940 billion ($ of 2007) or about 6.8 % of world GDP. The six most populous countries are Indonesia, Pakistan, Bangladesh, Philippines, Vietnam and Thailand. With 77 % of the population, they produce 60 % of the GDP (PPP).

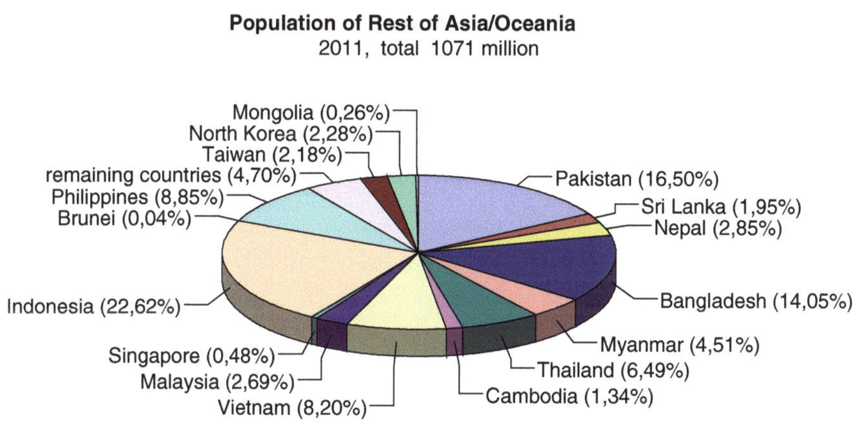


**Fig. 8.1** Percent partition of the population of the rest of Asia/Oceania



V. Crastan, *Global Energy Demand and 2-degree Target, Report 2014*,
DOI 10.1007/978-3-319-12310-3_8

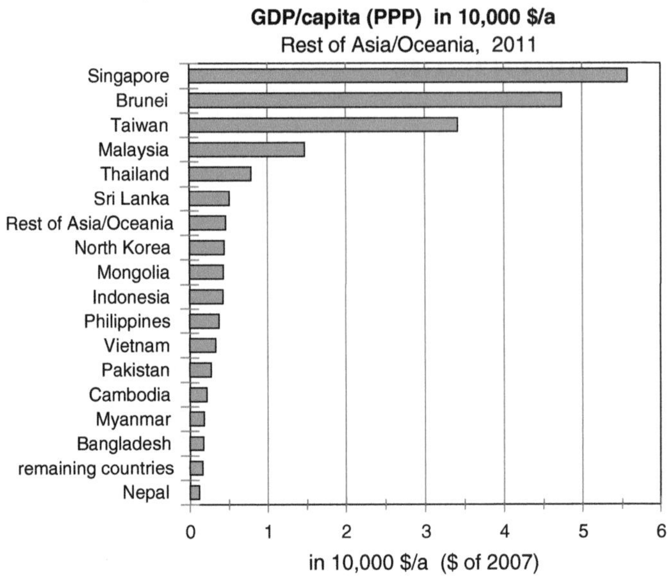


**Fig. 8.2** GDP (PPP) per capita of the rest of Asia/Oceania

The GDP (PPP) per capita (Fig. 8.2) has a mean value of $4,600/a, that of the six most populous countries varying between 1,700 in Bangladesh and 7,900 in Thailand. OECD level are attained only in Singapore, Brunei and Taiwan with a very high GDP (PPP) between $30,000 and $60,000/a.

## 8.2 Energy Intensity 2011

Thanks to the underdevelopment of many countries, the gross energy intensity, at 1.80 kWh/\$, is slightly below the world average. The energy consumptions of Indonesia, Pakistan, Malaysia, Thailand and Vietnam have low efficiency, at around 2–3 kWh/\$. Only Singapore, Taiwan and the Philippines have OECD level. Energy use in Mongolia (about 3.5 kWh/\$), based on coal, is extremely inefficient. In Nepal (also 3.5 kWh/\$) it is based on biomass and hydropower (Fig. 8.3).

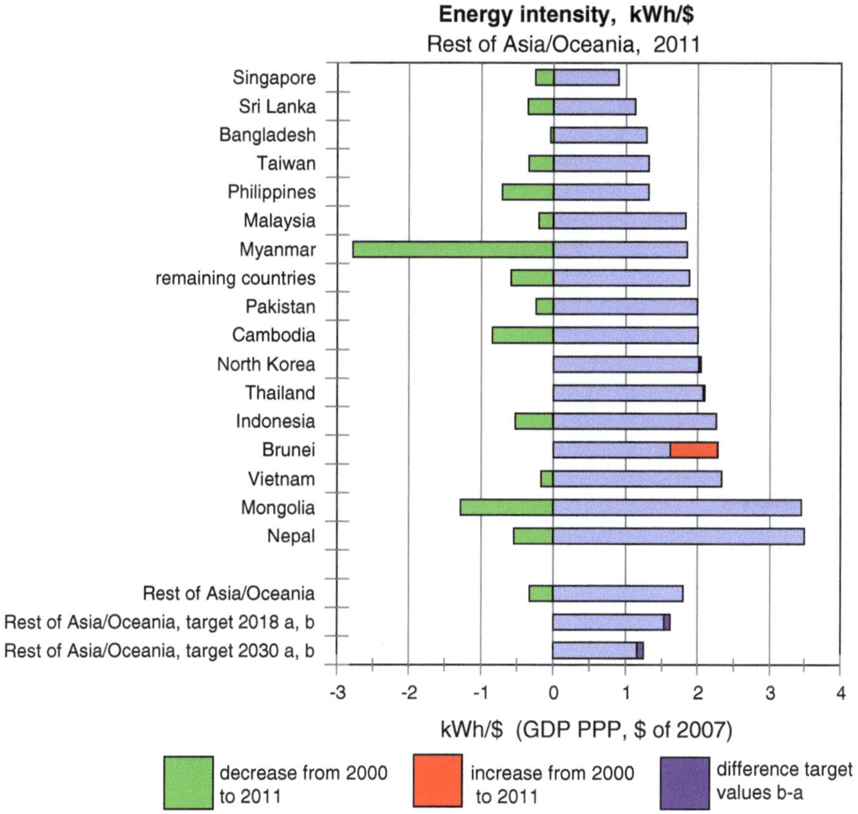


**Fig. 8.3** Energy intensity of the rest of Asia/Oceania and changes from 2000 to 2011

## 8.3 $CO_2$ Intensity of Energy 2011

The average $CO_2$ intensity, at 195 g $CO_2$/kWh, is below the world average. Most underdeveloped countries are lower (high biomass use), with the exception of North Korea and Mongolia, whose energy sectors, evidently dominated by coal, have a high $CO_2$ intensity at about 300 g $CO_2$/kWh. The low values in Nepal and Myanmar are due to the exclusive or high share of electricity produced by hydropower. Among developed countries, the specific $CO_2$ emissions of Singapore and Taiwan are significantly higher than the OECD average. In the last 11 years, the $CO_2$ intensity has increased significantly in all the most populous countries of the rest of Asia/Oceania (Fig. 8.4).

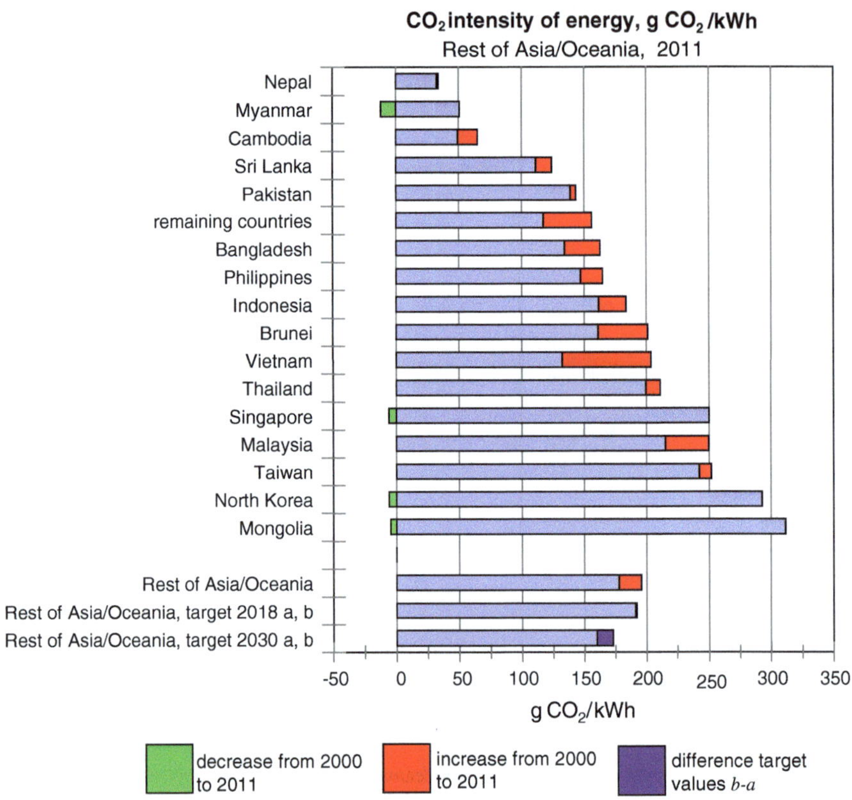


**Fig. 8.4** $CO_2$ intensity of the rest of Asia/Oceania and changes in the last 11 years

## 8.4 Indicator of $CO_2$ Sustainability 2011

The indicator of $CO_2$ sustainability for the rest of Asia/Oceania, with an average of 352 g $CO_2$/$ is significantly lower than the world average (428 g $CO_2$/$) but higher than the OECD average (318 g $CO_2$/$). The differences between countries are enormous. Among the countries with greater populations and therefore greater impact, Sri Lanka and Bangladesh, because of underdevelopment but also thanks to hydroelectricity and natural gas, and the Philippines, thanks to hydropower and geothermal energy, have values significantly below 200 g $CO_2$/$ or slightly above. Indonesia is near to the world average, Vietnam and Thailand significantly above. Singapore, which is economically developed, at 224 g $CO_2$/$, is in the same range as the EU-15. Taiwan, at 345 g $CO_2$/$, is significantly less sustainable. Mongolia with more than 1,100 g $CO_2$/$ is, despite progress, bottom of the league (Fig. 8.5).

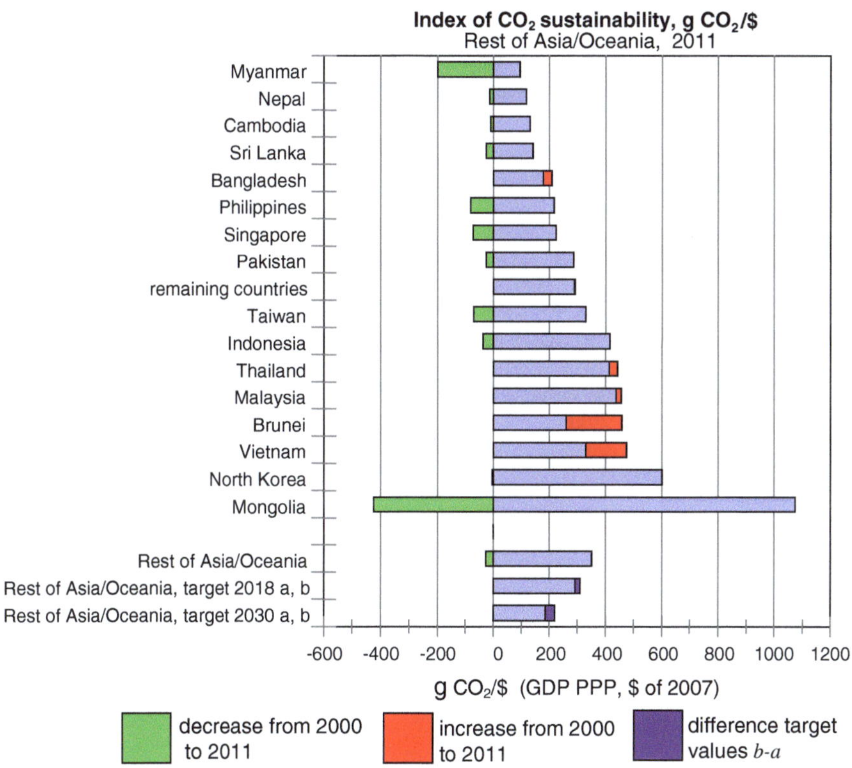


**Fig. 8.5** $CO_2$ sustainability indicator of the rest of Asia/Oceania and changes from 2000 to 2011

The change in $CO_2$ sustainability index is shown in more detail in Fig. 8.6 for the period 2000–2008 and 2008–2011. In addition to some overall small progress, one also notes regress, especially in Brunei and Vietnam.

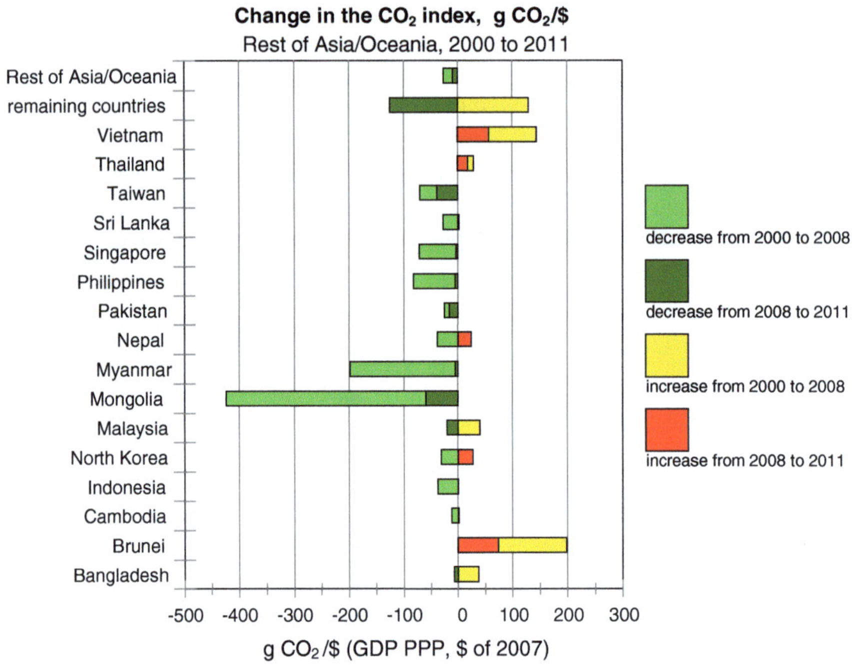


**Fig. 8.6** Changes in the $CO_2$ sustainability indicator of the countries of Rest Asia/Oceania from 2000 to 2008 and 2008 to 2011

## 8.5 Actual Indicators and $CO_2$ Emissions from 1980 to 2011 and Values Required from 2011 to meet the 2 °C Target

### *8.5.1 Energy Intensity, $CO_2$ Intensity and $CO_2$ Sustainability up to 2030*

The $CO_2$ sustainability indicator of Rest Asia/Oceania in 2011 is despite under-development relatively high at 352 g $CO_2$/\$, mainly because of the use of fossil fuel, particularly coal, for electricity production (see Appendix). The actual value of the indicators from 1980 to 2011 and the values required for the 2-degree target by 2030, for variants *a* and *b* are shown in Fig. 8.7.

The climate target to be aimed at for 2030 is 186–220 g $CO_2$/\$ with a share of 1,950–2,310 Mt of global $CO_2$ emissions. The values for 2018 take into account the value of GDP (PPP) of around \$6,800 billion (2007) forecast by the International Monetary Fund IMF. The GDP for 2030 is estimated at \$10,500 billion.

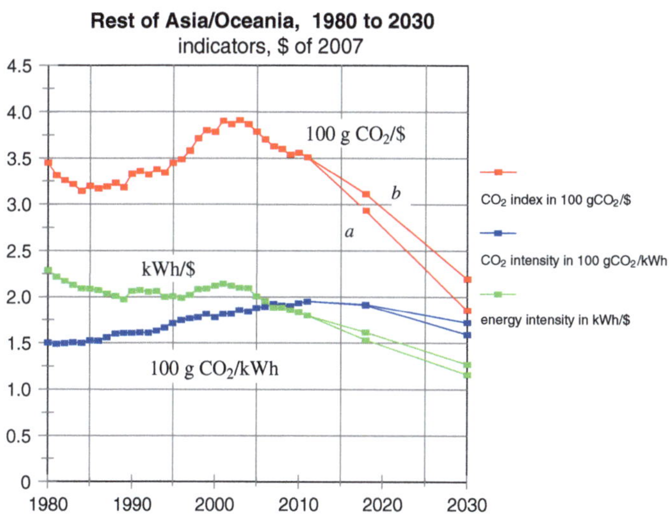


**Fig. 8.7** Indicators of Rest Asia/Oceania, from 1980 to 2011 and climate protection scenario up to 2030

The gross energy demand in 2030 with this scenario is 1,050–1,150 Mtoe. The per capita indicators for energy and $CO_2$ emissions need to be (variants *a* to *b*): e ≈ 1.06–1.16 kW/capita (increase of 12–22 % relative to 2030) and $\alpha$ = 1.5–1.75 t $CO_2$ per capita (decrease of 9 % to increase of 8 % relative to 2030).

The necessary change in the trend of the indicators to meet the 2-degree target is illustrated in Fig. 8.8 for variant *a*.

- The good trend in energy intensity must be maintained and strengthened.
- A turnaround has to take place in the $CO_2$ intensity and must be significantly enhanced after 2018.

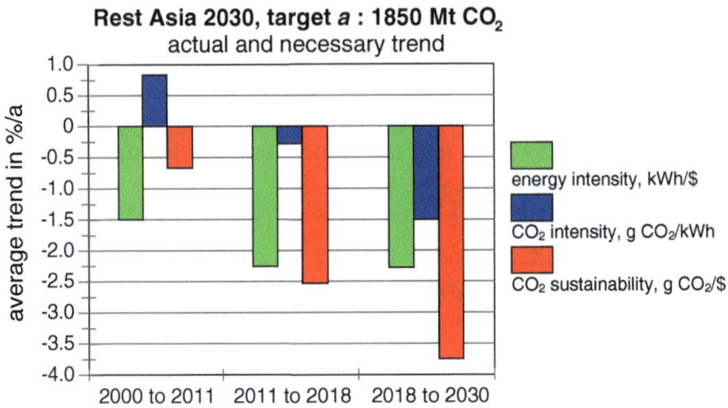


**Fig. 8.8** Trend from 2000 to 2011 of the rest of Asia/Oceania and necessary change in the trend from 2011 to achieve the 2-degree target, variant *a*

With variant *b* (Fig. 8.9):

- The improvement in the energy intensity trend only starts after 2018.
- The trend reversal of $CO_2$ intensity is milder after 2018.

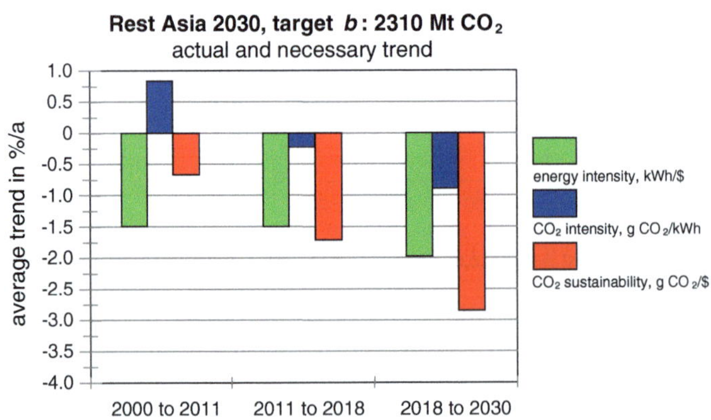


**Fig. 8.9** Trend from 2000 to 2011 of the rest of Asia/Oceania and necessary change in the trend from 2011 to achieve the 2-degree target, variant *b*

### 8.5.2 $CO_2$ Emissions until 2050

Figure 8.10 shows the actual $CO_2$ emissions of Rest Asia/Oceania from 1970 to 2011, and the permissible levels up to 2030 for meeting the 2-degree limit, and the further extrapolation to 2050, for both variants *a* and *b*.

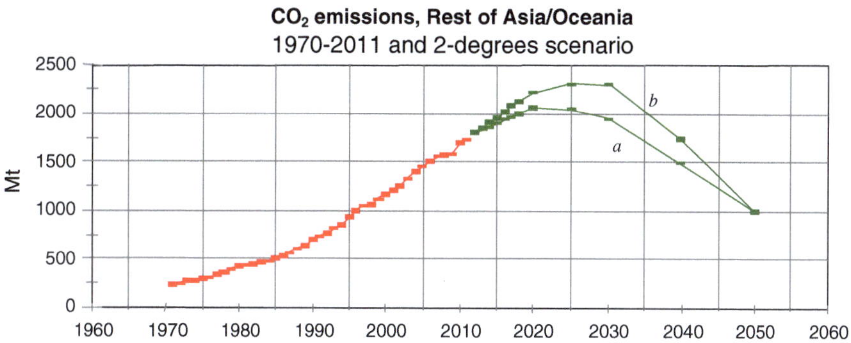


**Fig. 8.10** $CO_2$ emissions of the rest of Asia/Oceania from 1970 to 2011 and 2-degrees scenario until 2050

### 8.5.3 Per Capita Indicators up to 2030

Figure 8.11 shows the **per capita indicators** of Rest Asia/Oceania from 1980 to 2011 as well as the profile up to 2030 resulting from the above considerations to meet the 2-degree target, for both variants *a* and *b*. The GDP (PPP) data until 2018 correspond to the statistics and forecasts of the IMF.

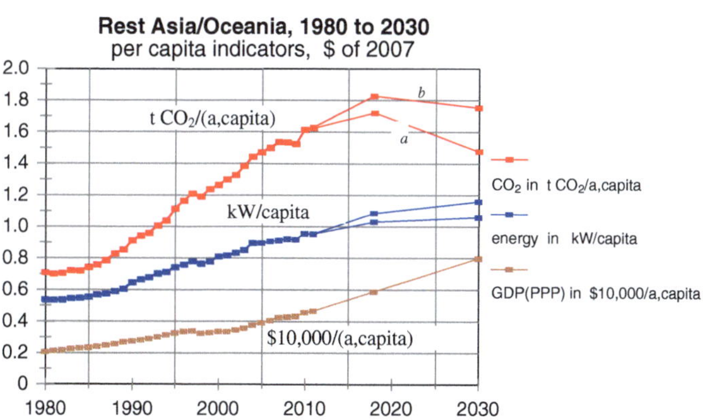


**Fig. 8.11** Per capita indicators of Rest Asia/Oceania from 1980 to 2011 and 2-degrees-scenario until 2030

The following Fig. 8.12 shows, in more detail **per consumer sector,** the per capita emissions of the resident population in 2011, and those of the 2-degree scenario for 2030 (variant *a* and *b*).

The four consumer sectors are (total specific emissions of $CO_2$ in 2011: 1.62 t/capita):

- Industry (0.48 t/capita)
- Transport (0.38 t/capita)
- Residential, services, agriculture, etc. (0.25 t/capita)
- Energy-sector losses (0.51 t/capita).

The energy and emissions data for 2011 correspond to the Rest Asia/Oceania-Annex A5.

Final-energy carriers emit $CO_2$ directly or indirectly (via electricity and district heating). The assumed distribution of these energy carriers within the consumer sectors, for both variants in 2030, is ***one way among others*** to meet the 2-degree condition.

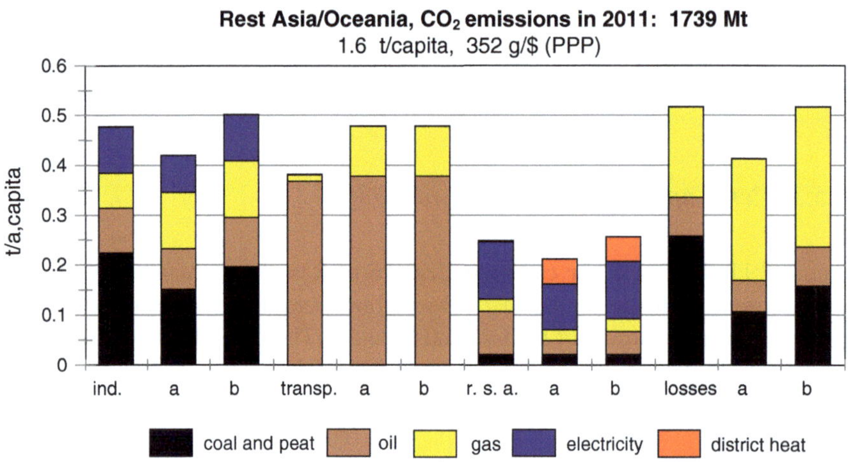


**Fig. 8.12** $CO_2$ emissions per capita of the four sectors: industry, transport, residential + services + agriculture, losses of the energy sector (Definitions see Appendix), each sector represented by three bars: *first bar* 2011 (data in the text and the diagram title, according to the rest of Asia/Oceania-Annex A5), *second bar* possible 2 °C scenario **variant *a*** for **2030**, 1,950 Mt, 1.5 t/capita, 186 g/$ (PPP), *third bar* possible 2 °C scenario **variant *b*** for **2030**, 2,310 Mt, 1.75 t/capita, 200 g/$ (PPP)

The target value 2030 is, for example, achieved when the per capita emissions:

- In the **electricity and district heat production sector, including energy-sector losses**, are reduced by 44 % (variant *a*) or 29 % (variant *b*) from 0.73 t/capita in 2011 to 0.41–0.52 t/capita (efficiency improvements, replacement of coal by gas, geothermal energy, CCS, other renewable energies and nuclear power)
- In the **heating sector** (industry, residential, services etc.) through efficiency improvements, reduction of coal and oil consumption and the use of renewable

energy sources (heat pump, waste management, solar energy, geothermal energy), are reduced by 19 % (*a*) or 4 % (*b*) (from 0.52 t/capita in 2011 to 0.42–0.50 t/capita)

- In the **transport sector** increase by not more than 26 % (a and b), from 0.4 to 0.5 t/capita, through efficiency improvements, gas fuels, bio-fuels and electric mobility.

# Chapter 9
# Non-OECD America

## 9.1 Population and Gross Domestic Product 2011

Non-OECD America (Central and South America, excluding Chile), with 460 million inhabitants, have 91 % of the population of the EU-27, but only 32 % of its GDP (PPP). Brazil is with 43 % of the population and 46 % of the GDP (PPP), the most significant on the continent (Fig. 9.1). The 5 most populous countries (Brazil, Colombia, Argentina, Peru and Venezuela) account for 75 % of the population and produce 83 % of the GDP (PPP).

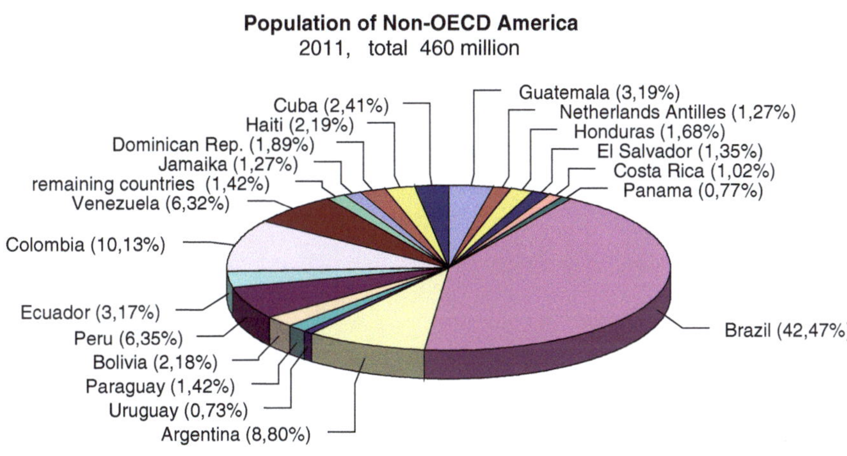


**Fig. 9.1** Percent partition of the population of non-OECD America



V. Crastan, *Global Energy Demand and 2-degree Target, Report 2014*,
DOI 10.1007/978-3-319-12310-3_9

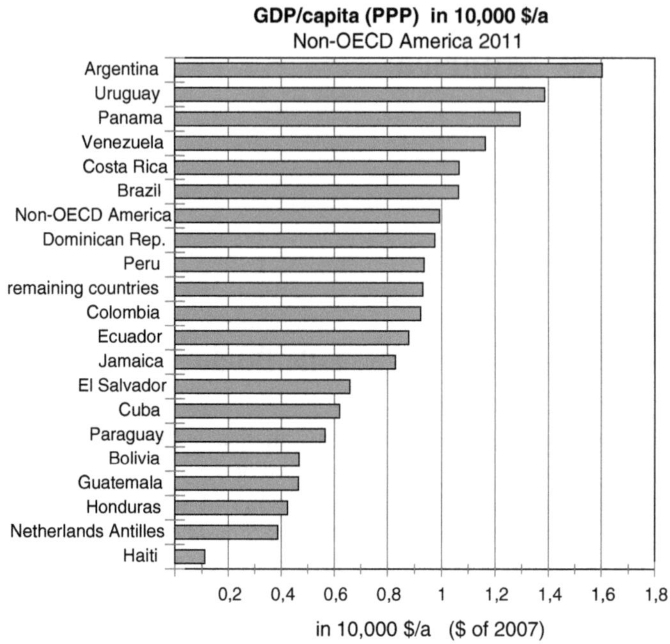


**Fig. 9.2** GDP (PPP) per capita of the countries of non-OECD America

The GDP (PPP) per capita (Fig. 9.2) has an average value of $9,900/a. Slightly above average is Brazil with $10,600/a. Among the larger countries are Argentina ($16,000/a) and Venezuela ($11,700/a) above average, Columbia and Peru below average ($9,300/a).

## 9.2 Energy Intensity 2011

The overall energy intensity is 1.43 kWh/$, which is about 72 % of world average and 13 % above that of the EU-15. With a development stage of about 95 % of the world average, the energy efficiency can therefore be assessed as relatively good. Of the 5 most populous countries, two (Peru and Colombia) are below average in their energy intensity. Less efficient, however, is the energy economy of Venezuela, with 2.36 kWh/$ significantly above the world average (Fig. 9.3).

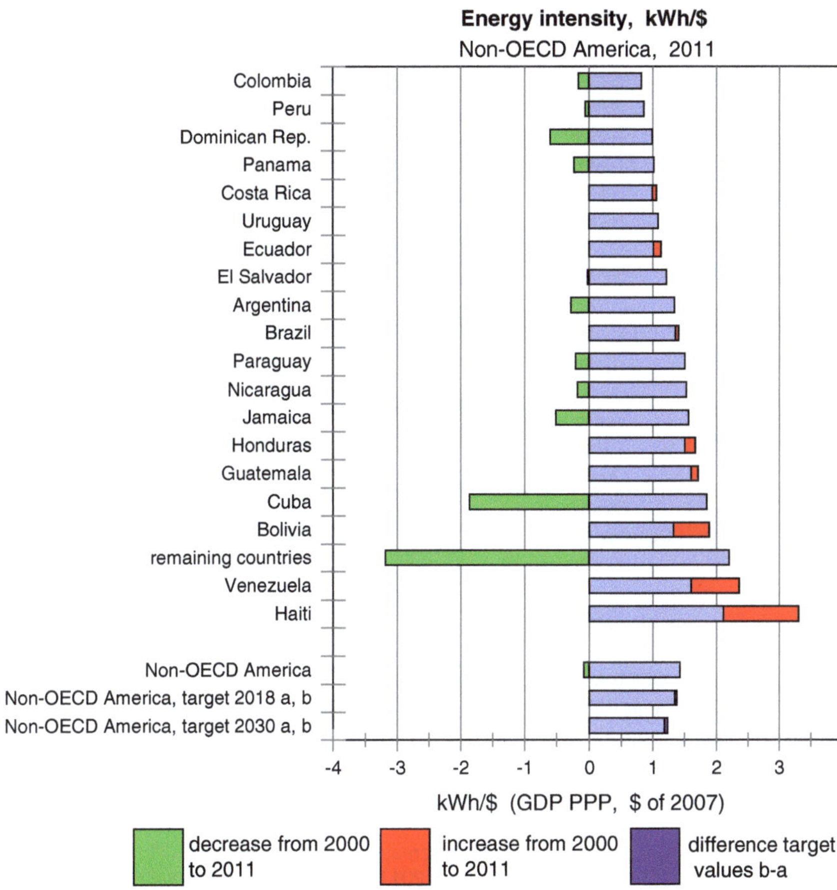


**Fig. 9.3** Energy intensity of non-OECD America and changes from 2000 to 2011

## 9.3 $CO_2$ Intensity of Energy 2011

The $CO_2$ intensity is 168 $CO_2$/kWh, which is significantly below both the world average and the EU-15 average. This is partly due to the big role of water power in electricity production. The $CO_2$ intensity of Brazil (134 $CO_2$/kWh) is also below average, partly thanks to bio-fuels. The highest $CO_2$ intensities (in the 5 most populous countries) are, at about 200 g $CO_2$/kWh, those of Argentina and Venezuela, with Venezuela making notable progress in the past 11 years (Fig. 9.4).

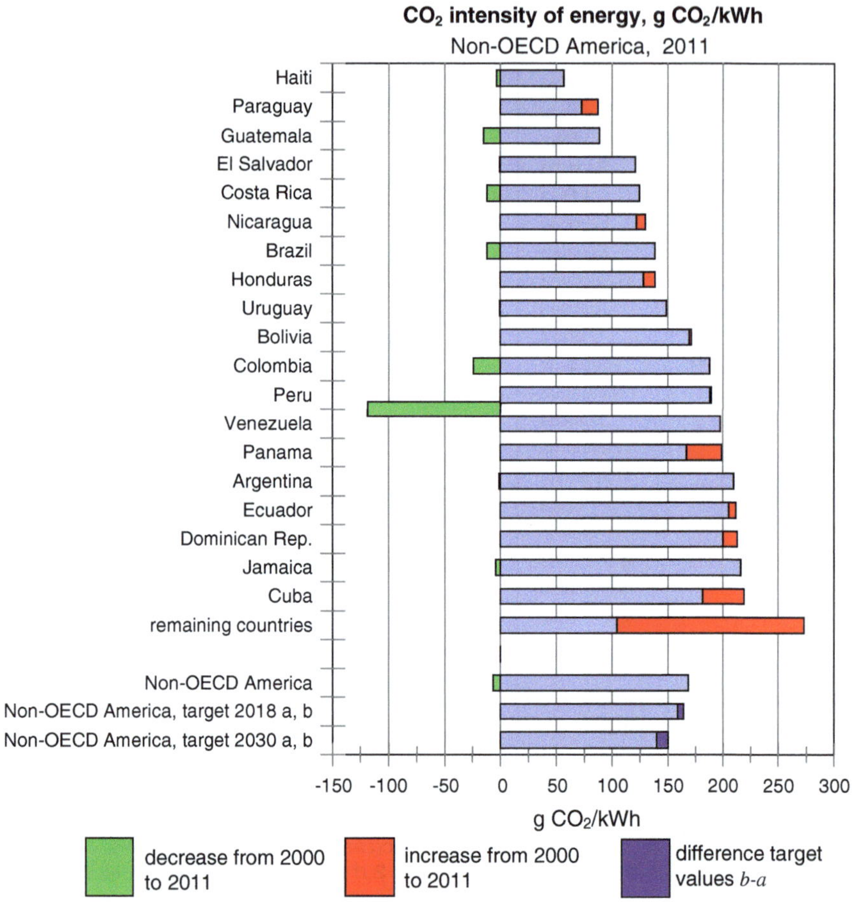


**Fig. 9.4** $CO_2$ intensity of non-OECD America and changes from 2000 to 2011

## 9.4 Indicator of $CO_2$ Sustainability 2011

With its moderate energy intensity and also $CO_2$ intensity, the sustainability of the energy consumption of non-OECD America is, at 240 g $CO_2$/\$, quite reasonable. This value even puts Central and South America together with the EU-15 in the lead (see Fig. 3.5). Three of the 5 major countries (Peru, Colombia and Brazil) are below this value. Venezuela is, despite progress, clearly out of line (disregarding some small Antillean countries), at about 470 g $CO_2$/\$ (Fig. 9.5).

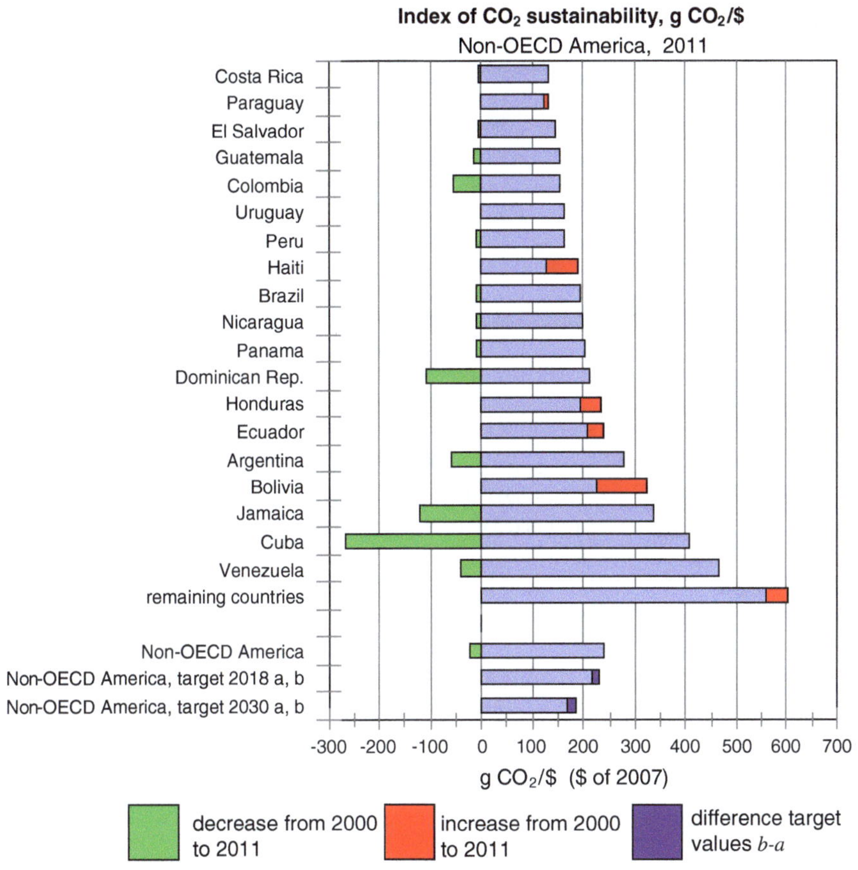


**Fig. 9.5** $CO_2$ sustainability indicator of non-OECD America and changes from 2000 to 2011

The change in $CO_2$ sustainability index is shown in more detail in Fig. 9.6 for the periods 2000–2008 and 2008–2011. The picture is very inconsistent but shows a small overall trend of improvement.

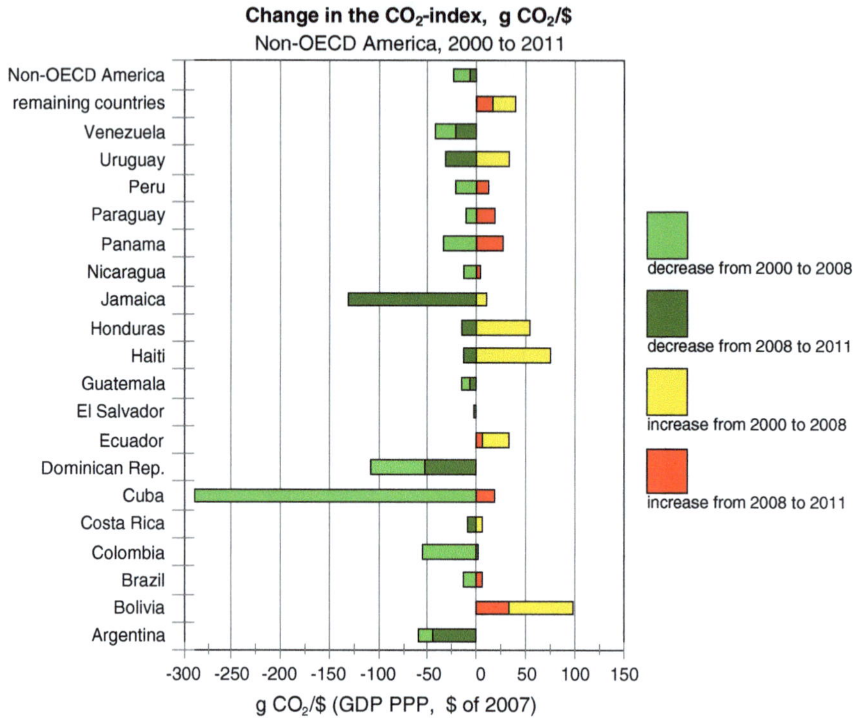


**Fig. 9.6** Change in the index of $CO_2$ sustainability for the countries of non-OECD America from 2000 to 2008 and 2008 to 2011

## 9.5 Actual Indicators and $CO_2$ Emissions from 1980 to 2011 and Values Required from 2011 to meet the 2 °C Target

### *9.5.1 Energy Intensity, $CO_2$ Intensity and $CO_2$ Sustainability up to 2030*

Non-OECD America in 2011, at 242 g $CO_2$/\$, has the best $CO_2$ sustainability indicator of all regions of the world. This is due to the highly water- power based electricity production (see Appendix). The actual indicator values from 1980 to 2011 and the necessary one for the 2-degree target by 2030, for variants a and b, are shown in Fig. 9.7.

The climate target to be aimed at for 2030 is 167–187 g $CO_2$/\$ with a share of 1,250–1,400 Mt of global $CO_2$ emissions. The values for 2018 take into account the GDP (PPP) forecast of the International Monetary Fund IMF of around \$5,600 billion (2007). The GDP for 2030 is estimated at \$7,500 billion.

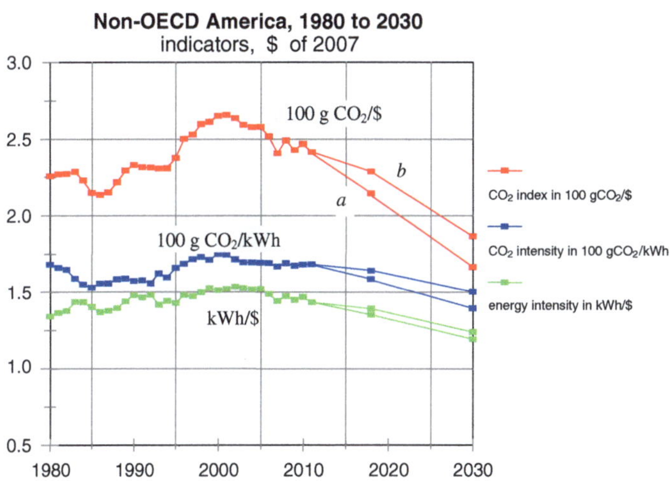


**Fig. 9.7** Indicators of non-OECD America from 1980 to 2011 and climate protection scenario until 2030

The gross energy demand in 2030 with this scenario is 770–800 Mtoe (variants *a* and *b*, only energy use). The per capita indicators for energy and $CO_2$ emissions need to be: e = 1.9–2.0 kW/capita (increase of 18–24 % relative to 2010) and $\alpha$ = 2.3–2.6 t $CO_2$ per capita (decrease of 2 % to increase of 10 % relative to 2011).

The necessary change in the trend of the indicators to meet the 2-degree target is illustrated in Fig. 9.8 for variant *a* and in Fig. 9.9 for variant *b*. For variant *a*, the increased trend must begin immediately, for variant *b*, from 2018.

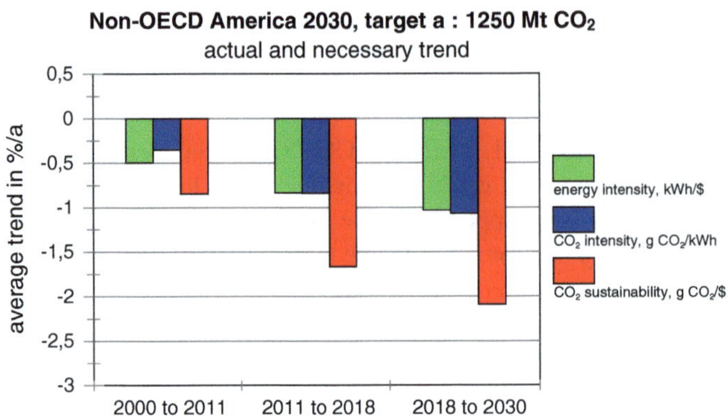


**Fig. 9.8** Trend from 2000 to 2011 of non-OECD America and necessary change in the trend from 2011 to achieve the 2-degree target, variant *a*

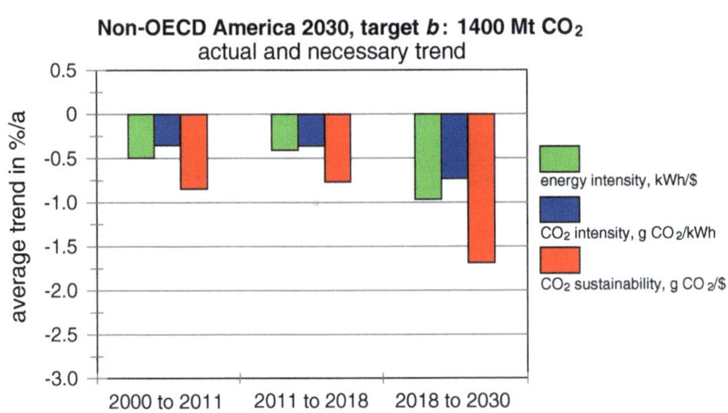


**Fig. 9.9** Trend from 2000 to 2011 of non-OECD America and necessary change in the trend from 2011 to achieve the 2-degree target, variant *b*

### 9.5.2 $CO_2$ Emissions until 2050

Figure 9.10 shows the actual $CO_2$ emissions of non-OECD America from 1970 to 2011, and the permissible levels up to 2030 for meeting the 2-degree limit, and the further extrapolation to 2050, for both variants *a* and *b*.

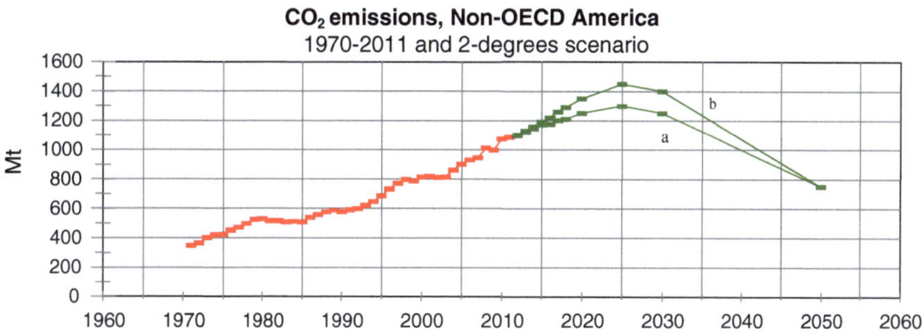


**Fig. 9.10** $CO_2$ emissions of non-OECD America, 1980–2011 and climate protection scenario until 2050

### 9.5.3 Per Capita Indicators up to 2030

Figure 9.11 shows the **per capita indicators** of non-OECD America from 1980 to 2011 as well as the profile up to 2030 resulting from the above considerations to meet the 2-degree target, for both variants *a* and *b*. The GDP (PPP) data until 2018 correspond to the statistics and forecasts of the IMF.

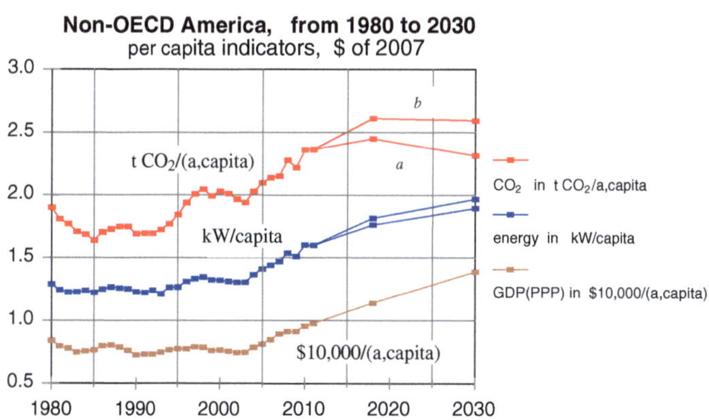


**Fig. 9.11** Per capita indicators of non-OECD America from 1980 to 2011 and 2-degrees-scenario until 2030

Figure 9.12 shows, in more detail **per consumer sector,** the per capita emissions of the resident population in 2011, and those corresponding to the 2-degree scenario for 2030 (variant *a* and *b*).

The four consumer sectors are (total specific emissions of $CO_2$ in 2011: 2.4 t/capita):

- industry (0.6 t/capita)
- transport (0.9 t/capita)
- residential, services, agriculture, etc. (0.3 t/capita)
- energy-sector losses (0.6 t/capita).

The energy and emissions data for 2011 correspond to the non-OECD America-Annex A5.

Final-energy carriers emit $CO_2$ directly or indirectly (via electricity and district heating). The assumed distribution of these energy carriers within the consumer sectors, for both variants in 2030, is ***one way among others*** to meet the 2-degree condition.

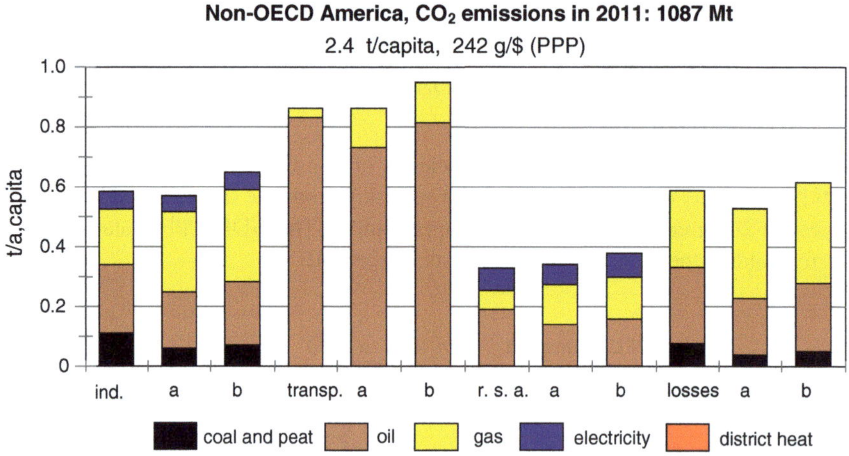


**Fig. 9.12** $CO_2$ emissions per capita of the 4 sectors: industry, transport, residential + services + agriculture, energy-sector losses (definitions see Appendix), each sector represented by three bars: *First bar* 2011 (data in the text and the diagram title, according to the non-OECD America-Annex A5), *Second bar* possible 2 °C scenario **variant *a*** for **2030**, 1,250 Mt, 2.3 t/capita, 167 g/$ (PPP), *Third bar* possible 2 °C scenario **variant *b*** for **2030**, 1,400 Mt, 2.6 t/capita, 187 g/$ (PPP)

The target value 2030 is, for example, achieved when the **per capita emissions**:

- in the **electricity and district heat production sector**, **including energy-sector losses**, are reduced by 26 % (variant *a*) or increase not more than 6 % (variant *b*) from 0.72 t/capita in 2011 to 0.53–0.76 t/capita (efficiency improvements, CCS and/or replacement of coal and oil by gas, renewable energy and nuclear power).

- in the **heating sector** (all end user: industry, residential, services etc.) through efficiency improvements, reduction of coal and the use of renewable energy sources (heat pump, waste management, solar energy, geothermal energy), remain unchanged (*a*) or increase not more than 14 % (*b*) (from 0.8 t/capita to a maximum of 0.9 t/capita).
- in the **transport sector** remain unchanged (*a*) or increase not more than 10 % (*b*), from 0.86 to not more than 0.95 t/capita), through efficiency improvements, gas fuels, bio-fuels and electric mobility.

# Chapter 10
# Africa

## 10.1 Population and Gross Domestic Product 2011

With a population of 1,045 million (Fig. 10.1) Africa has a GDP (PPP) of $ 2,900 billion ($ of 2007). The five countries with the largest GDP, namely South Africa, Algeria, Egypt, Nigeria and Morocco, together have 35 % of the population and produce 60 % of the GDP.

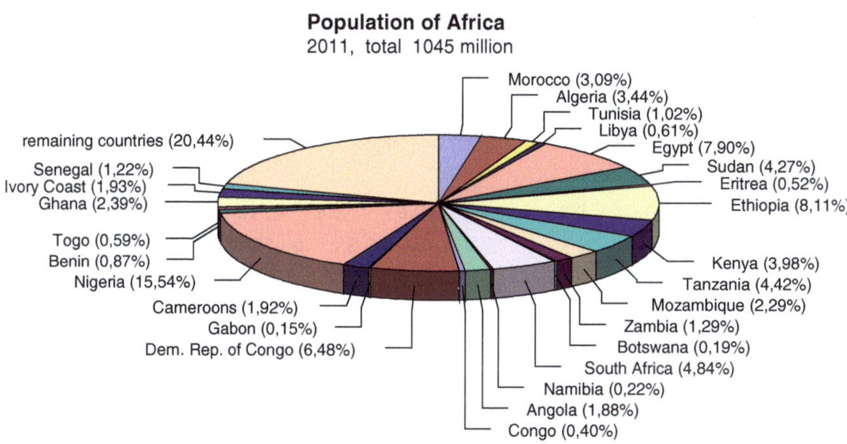


**Fig. 10.1** Percent partition of the population of Africa



V. Crastan, *Global Energy Demand and 2-degree Target, Report 2014*,
DOI 10.1007/978-3-319-12310-3_10

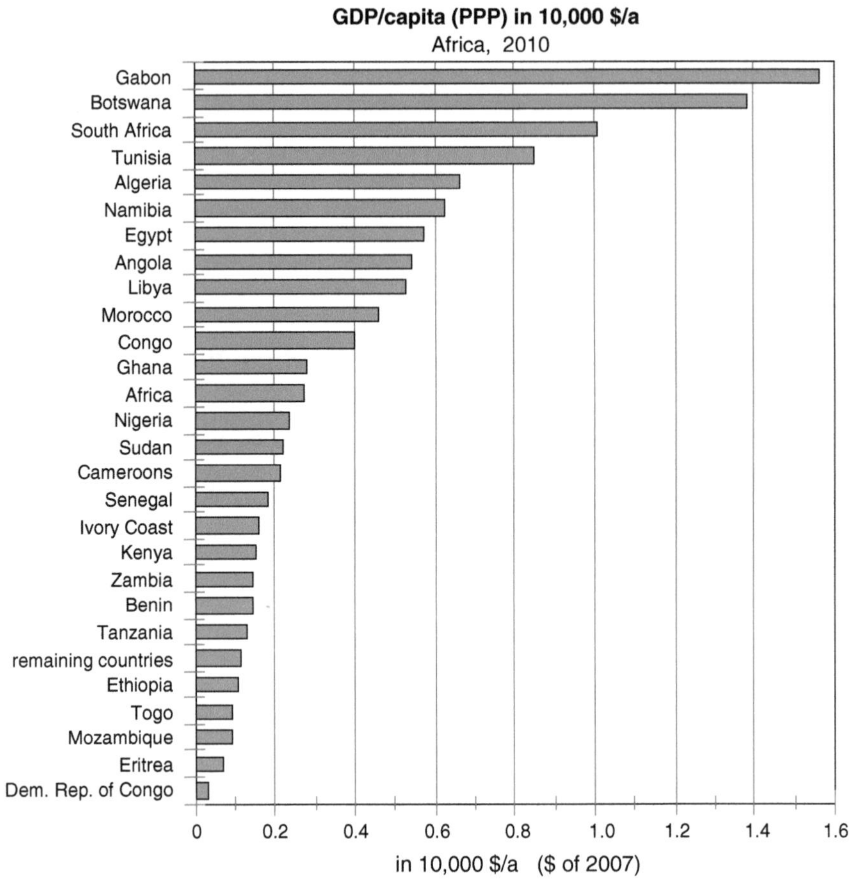


**Fig. 10.2** GDP (PPP) per capita of the African countries

The average per-capita GDP (PPP) (Fig. 10.2) is very low in the world comparison, amounting to $2,800/a. Only three countries (Gabon, Botswana and South Africa) reach or exceed $10,000/a in 2011. Seven further countries (Tunisia, Namibia, Algeria, Angola, Morocco, Egypt and Libya) are in the range $4,000–$9,000/a. All the others are below the average for the continent, some by a large amount.

## 10.2 Energy Intensity 2011

The very high mean energy intensity of 2.75 kWh/$, which is well above the worldwide mean of just 2 kWh/$ and indicates an inefficient use of energy, is the consequence of underdevelopment but also of the relative wealth in fossil energy carriers that are made available at low cost (coal in South Africa, oil in Nigeria and Libya, natural gas in Algeria and Egypt, etc.). In the past 11 years there has been some general progress (exceptions: particularly Libya and Ivory Coast) (Fig. 10.3).

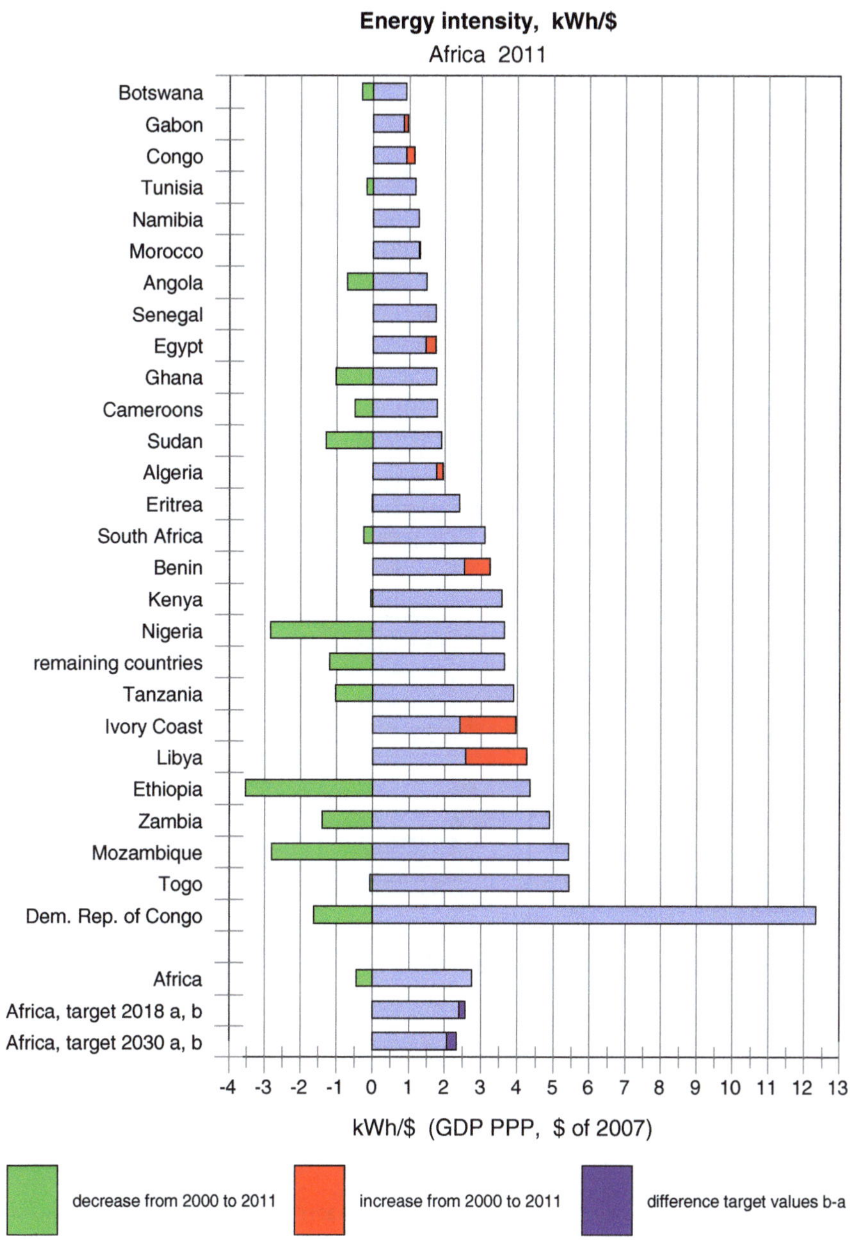


**Fig. 10.3** Energy intensity 2011 of the African countries and changes from 2000 to 2011

## 10.3 $CO_2$ Intensity of Energy 2011

Africa has the world's lowest $CO_2$ intensity, at 122 g $CO_2$/kWh. The reason is the large use of biomass, which is due to underdevelopment. The very low value in Mozambique, Ethiopia, Zambia and the Democratic Republic of Congo has its cause also in the production of electricity with only water power. The more developed countries have however a $CO_2$ intensity that is close to the world average of 200 g $CO_2$/kWh or exceeds it (Fig. 10.4).

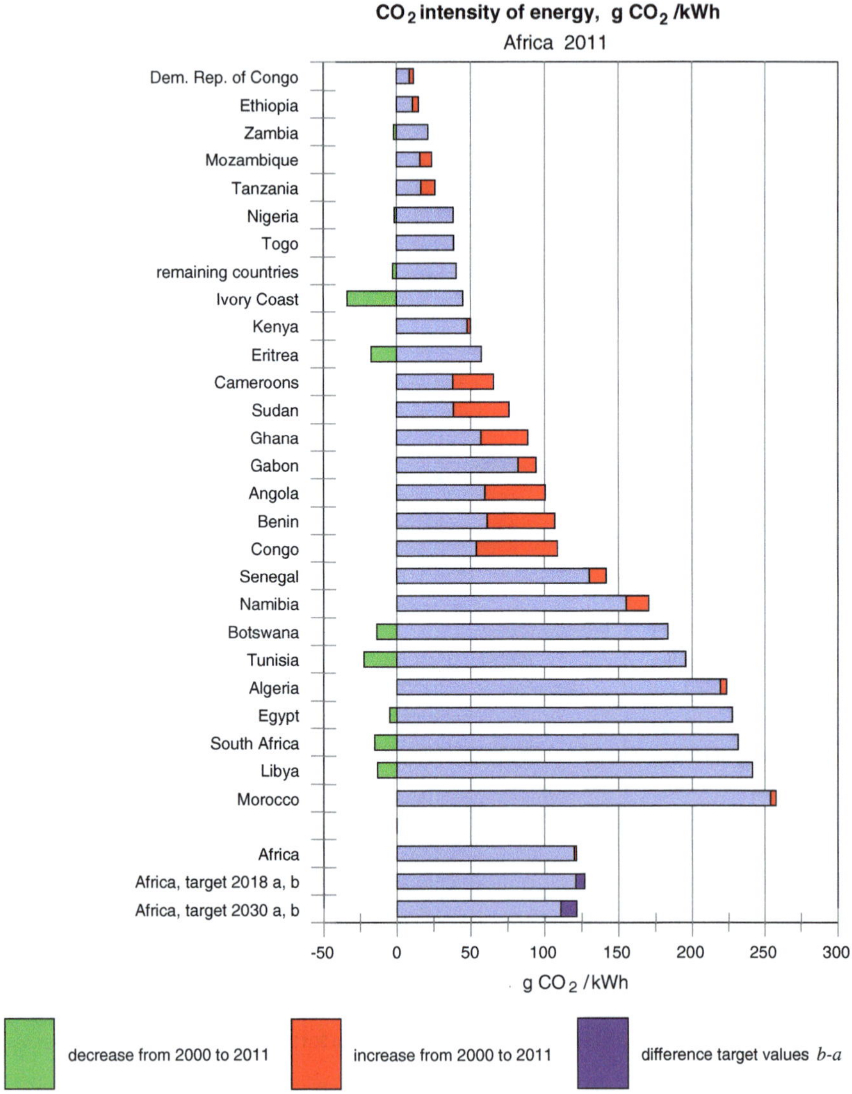


**Fig. 10.4** $CO_2$ intensity of energy 2011 of the African countries and changes from 2000 to 2011

## 10.4 Indicator of $CO_2$ Sustainability 2011

The very low $CO_2$ intensity of energy compensates the high energy intensity, so that the resulting $CO_2$ indicator of around 320 g $CO_2$/$ is close to the OECD average. However there are large differences between the countries: from less than 150 g $CO_2$/$ in highly underdeveloped countries to 400 g $CO_2$/$ and more in the countries with a strong oil and coal sector (Fig. 10.5). Among the more developed countries, Gabon has a very good $CO_2$ indicator and Botswana and Tunisia have a satisfactory one of less than 200 or 250 g $CO_2$/$. Far from sustainability are Libya (with >1,000 g $CO_2$/$), which is due to the political instability, and South Africa (>700 g $CO_2$/$ despite good progress).

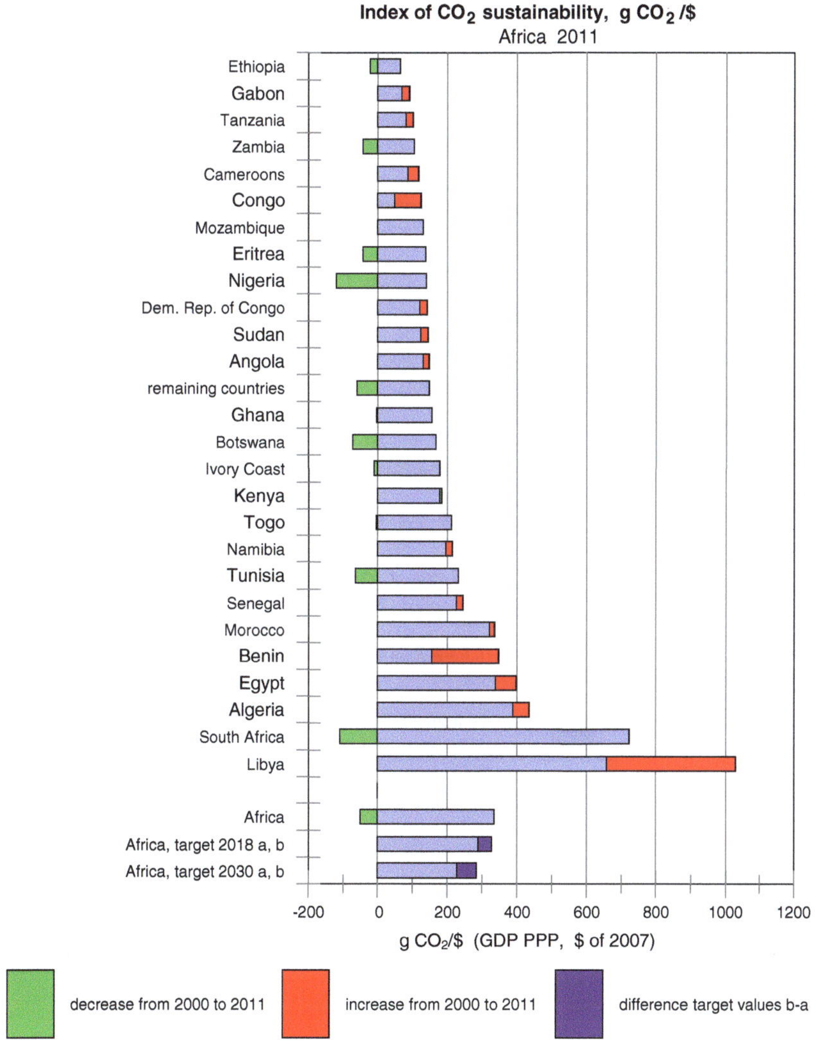


**Fig. 10.5** $CO_2$ sustainability indicator 2011 of the African countries and changes from 2000 to 2011

The change in $CO_2$ sustainability index is shown in more detail in Fig. 10.6 for the periods 2000–2008 and 2008–2011. The picture is rather inconsistent but shows a clear trend of improvement, particularly in the more significant countries such as South Africa and Nigeria. Libya is a special negative case as a consequence of the war.

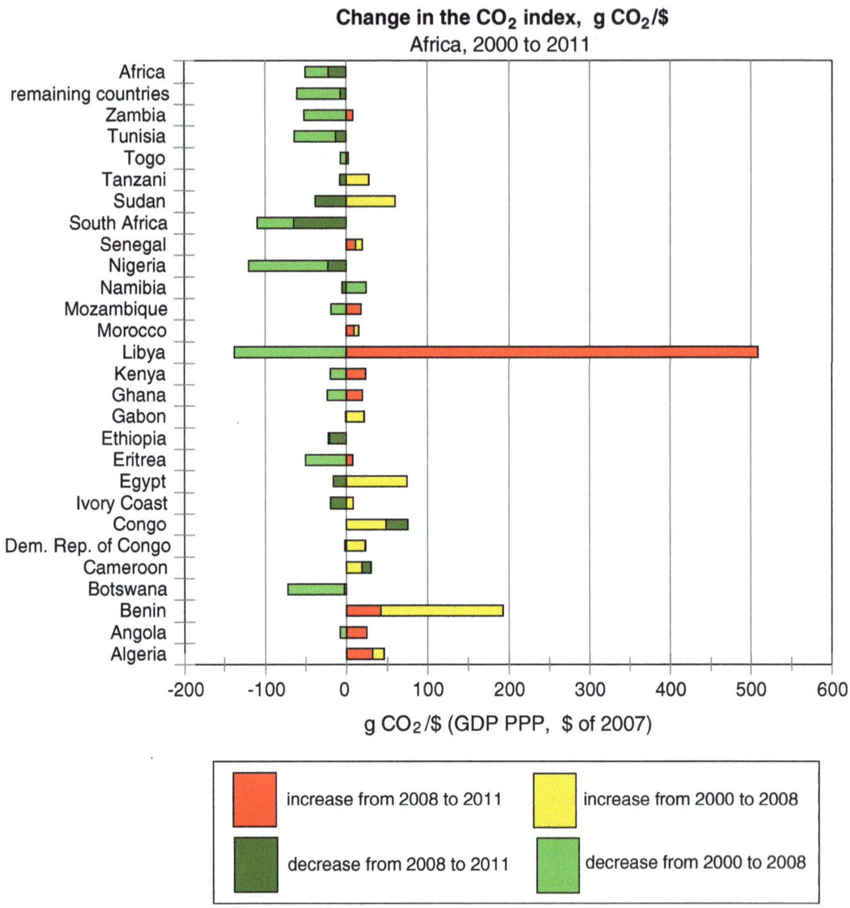


**Fig. 10.6** Change in the index of $CO_2$ sustainability from 2000 to 2008 and from 2008 to 2011 for the African countries

## 10.5 Actual Indicators and $CO_2$ Emissions from 1980 to 2011 and Values Required from 2011 to meet the 2 °C Target

### *10.5.1 Energy Intensity, $CO_2$ Intensity and $CO_2$ Sustainability up to 2030*

Africa had in 2011, given the low level of development, a relatively poor indicator of $CO_2$ sustainability at 335 g $CO_2$/$, which is primarily due to the low energy efficiency (see Appendix).

The actual value of the indicators from 1980 to 2011 and the necessary one to meet the 2-degree target by 2030, for the variants *a* and *b* are shown in Fig. 10.7.

The climate target to be aimed at for 2030 is 228–285 g $CO_2$/$ with a share of 1,480–1,850 Mt of global $CO_2$ emissions. The values for 2018 take into account the value of GDP (PPP) of around $ 4,100 billion (2007) forecast by the International Monetary Fund IMF. The GDP for 2030 is estimated at $6,500 billion.

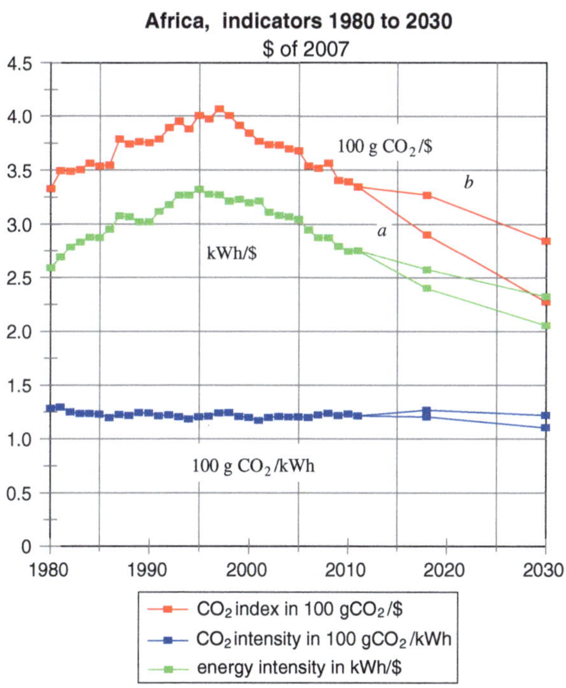


**Fig. 10.7** Indicators of Africa, 1980–2011 and climate protection scenario until 2030

The gross energy demand in 2030 with this scenario is 1,150–1,300 Mtoe (variants *a* to *b*). The per capita indicators for energy and $CO_2$ emissions need to be: $e$ = 0.96–1.08 kW/capita (increase of 9–24 % relative to 2011) and $\alpha$ = 0.93–1.2 t $CO_2$/capita (unchanged to increase of 25 % relative to 2011).

The necessary change in the trend of the indicators to meet the 2-degree target is illustrated in Fig. 10.8 for variant *a*.

- The good trend in energy intensity must be maintained and rapidly slightly reinforced.
- The stagnation of the $CO_2$ intensity of energy has to be overcome and especially after 2018 has to initiate a significant decrease.

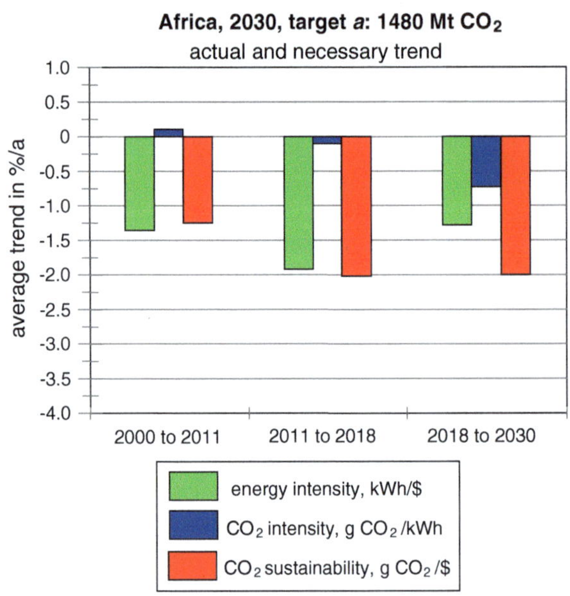


**Fig. 10.8** Trend from 2000 to 2011 of Africa and necessary change in the trend from 2011 to achieve the 2-degree target, variant *a*

Variant *b*, Fig. 10.9:

- permits a slight weakening of the trend reduction of energy intensity
- initially even permits a greater increase of $CO_2$ intensity, with a trend reversal only after 2018.

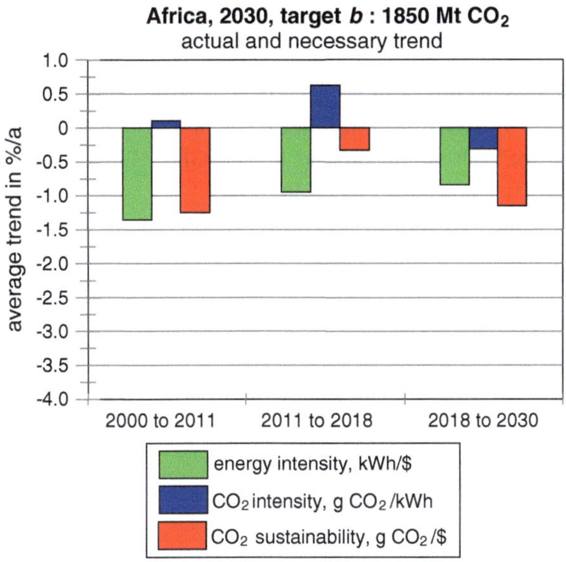


**Fig. 10.9** Trend from 2000 to 2011 of Africa and necessary change in the trend from 2011 to achieve the 2-degree target, variant *b*

## 10.5.2 $CO_2$ Emissions until 2050

Figure 10.10 shows the actual $CO_2$ emissions in Africa from 1970 to 2011, and the permissible levels up to 2030 for meeting the 2-degree limit, and the further extrapolation to 2050, for both variants *a* and *b*.

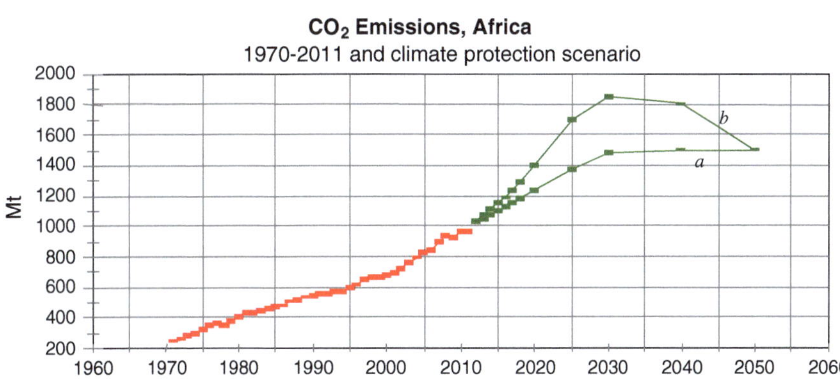


**Fig. 10.10** $CO_2$ emissions of Africa, 1970–2011 and climate protection scenario until 2050

### 10.5.3 Per Capita Indicators up to 2030

Figure 10.11 shows the **per capita indicators** of Africa from 1980 to 2011 as well as the profile up to 2030 resulting from the above considerations to meet the 2-degree target, for both variants *a* and *b*. The GDP (PPP) data until 2018 correspond to the statistics and forecasts of the IMF.

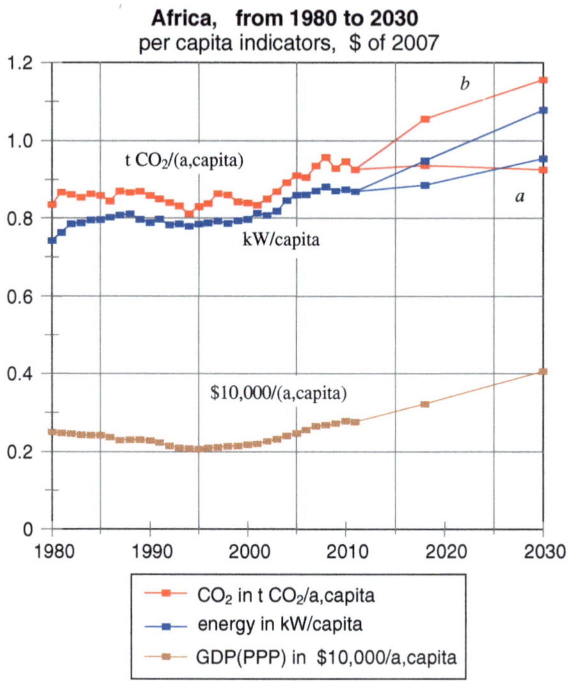


**Fig. 10.11** Per capita indicators of Africa from 1980 to 2011 and 2-degrees-scenario until 2030

The following Fig. 10.12 shows, in more detail **per consumer sector,** the per capita emissions of the resident population in 2011, and those of the 2-degree scenario for 2030 (variant *a* and *b*).

The four consumer sectors are (total specific emissions of $CO_2$ in 2011: 0.93 t/capita):

- industry (0.16 t/capita)
- transport (0.22 t/capita)
- residential, services, agriculture, etc. (0.16 t/capita)
- energy-sector losses (0.38 t/capita).

The energy and emissions data for 2011 correspond to the Africa-Annex A5.

Final-energy carriers emit $CO_2$ directly or indirectly (via electricity and district heating). The assumed distribution of these energy carriers within the consumer sectors, for both variants in 2030, is ***one way among others*** to meet the 2-degree condition.

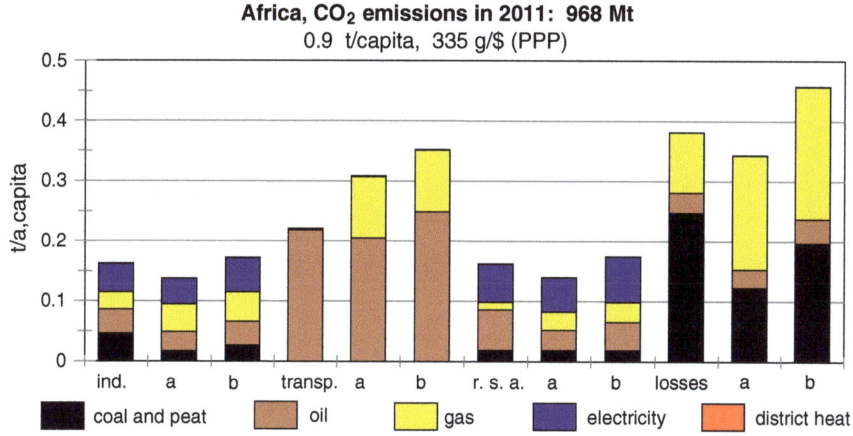


**Fig. 10.12** $CO_2$ emissions per capita of the 4 sectors: industry, transport, residential + services + agriculture, energy-sector losses (definitions see Appendix), each sector represented by three bars: *first bar* 2011 (data in the text and the diagram title, according to the Africa-Annex A5), *second bar* possible 2 °C scenario **variant *a*** for **2030**, 1,480 Mt, 0.93 t/capita, 228 g/$ (PPP), *third bar* possible 2 °C scenario **variant *b*** for **2030**, 1,850 Mt, 1.2 t/ capita, 285 g/$ (PPP)

The target value 2030 is, for example, achieved when the per capita emissions:

- in the **electricity and district heat production sector, including energy-sector losses**, are reduced by 10 % (variant *a*) or increase not more than 20 % (variant *b*) from 0.49 t/capita in 2011 to 0.44–0.59 t/capita (efficiency improvements, CCS and/or replacement of coal by gas, renewable energy and nuclear power)
- in the **heating sector** (industry, residential, services etc.) through efficiency improvements, reduction of coal consumption and the use of renewable energy sources (heat pump, waste management, solar energy, geothermal energy), are reduced by 14 % (*a*) or remain unchanged (*b*), from 0.21 t/capita to 0.18–0.21 t/capita
- in the **transport sector** increase by 44 % (*a*) or not more than 59 % (*b*), from 0.22 to 0.31–0.35 t/capita, through efficiency improvements, gas fuels and biofuels.

# Chapter 11
# G-20

## 11.1 Population and Gross Domestic Product 2011

The G-20 group includes the G-8 countries, emerging economies and the EU-27. In 2011 the group had a population of 4,527 million or 65 % of the world's population (Fig. 11.1). The group generates a GDP (PPP) of around $61,000 billion ($ of 2007), representing 83 % of the world GDP. Among the emerging countries, China, India and Indonesia together have 63 % of the population but only 26 % of GDP.

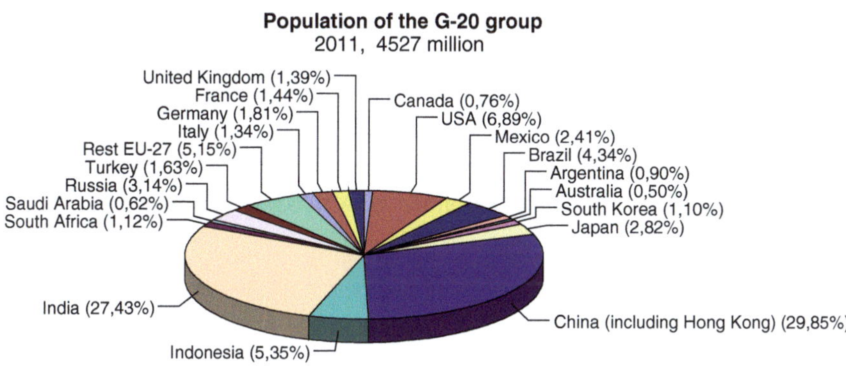


Fig. 11.1 Percent partition of the population of the G-20



V. Crastan, *Global Energy Demand and 2-degree Target, Report 2014*,
DOI 10.1007/978-3-319-12310-3_11

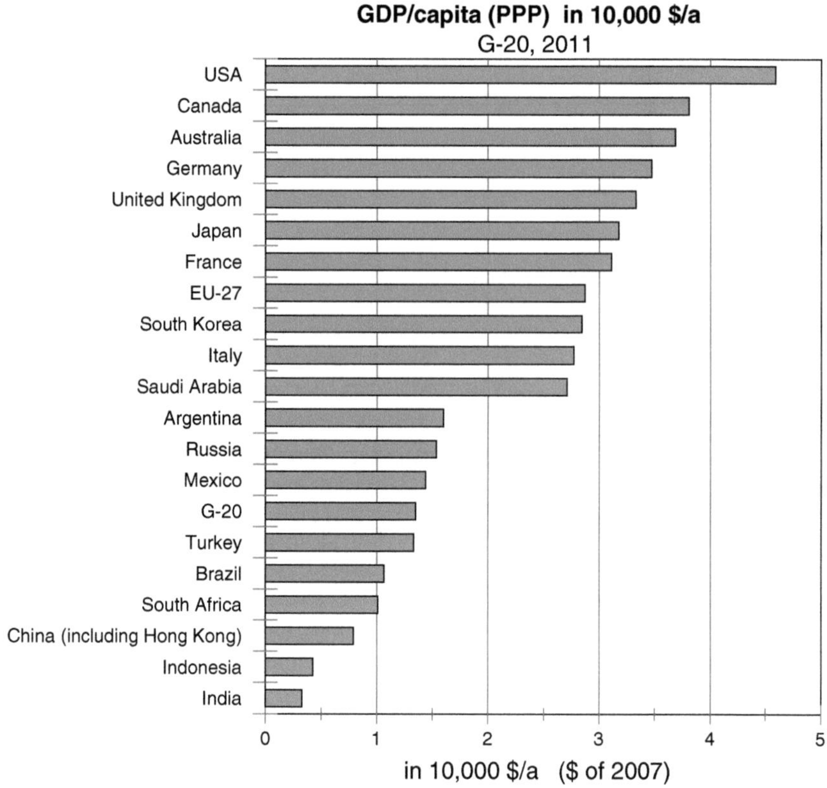


**Fig. 11.2** GDP (PPP) per capita in 2011 of the members of the G-20

The GDP (PPP) per capita in 2011 (Fig. 11.2) in $ of 2007 has an average value of $13,500 and ranges from $45,900 in the USA to $3,300 in India.

## 11.2 Energy Intensity 2011

The average energy intensity (gross energy, only energy use) in the G-20 is 1.85 kWh/$, and thus slightly lower than the world mean value. Insufficient energy intensities are those of Russia (despite undeniable progress), South Africa and China with values of nearly 3–3.5 kWh/$. The G-20 target 2030 required for climate protection is 1.12–1.19 kWh/$ (variants *a* and *b*). A significant improvement is essential (Fig. 11.3).

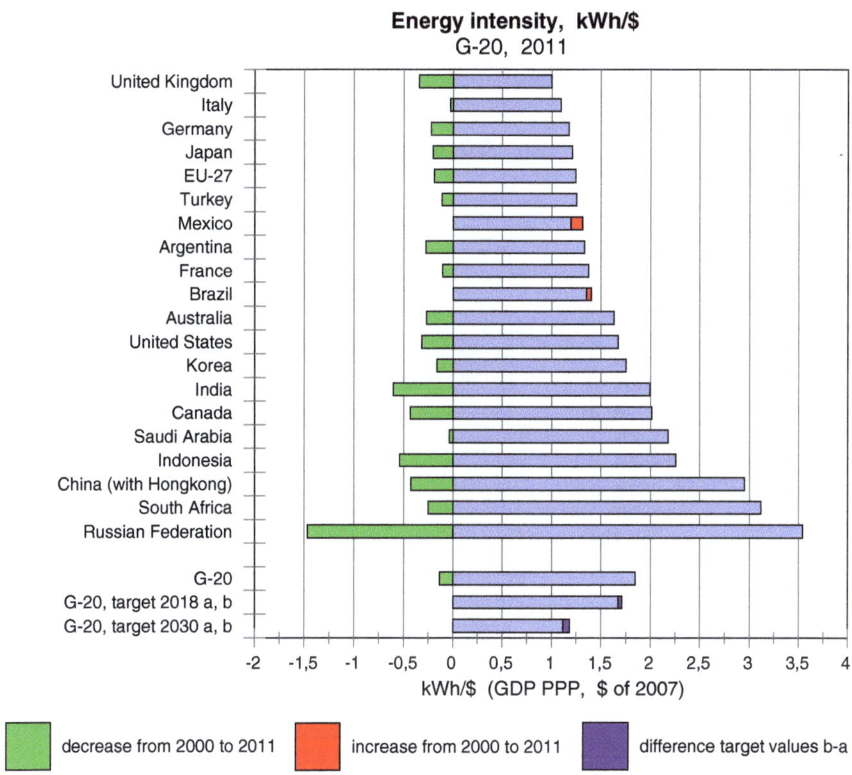


**Fig. 11.3** Energy intensity of the G-20 countries and changes from 2000 to 2011

## 11.3 $CO_2$ Intensity of Energy 2011

In 2011, the total emission of $CO_2$ from the G-20 countries was 25,490 Mt. The mean value of the $CO_2$ intensity was 226 g $CO_2$/kWh, thus a little above the world average. The $CO_2$ intensity is low in France and Brazil (<150 g $CO_2$/kWh): in France thanks to the production of electricity by water and nuclear power and in Brazil through hydroelectricity and bio-fuels. A very high and increasing $CO_2$ intensity (>250 g $CO_2$/kWh) is the case in China, Australia and Saudi Arabia, whose electricity generation is based almost exclusively on coal or oil. In the last 11 years, unfortunately, overall stagnation is seen or even a slight increase in the $CO_2$ intensity of the G-20 group. A reversal is required. The desirable G-20 climate target for 2030 is 183–195 g $CO_2$/kWh (variants *a* and *b*) (Fig. 11.4).

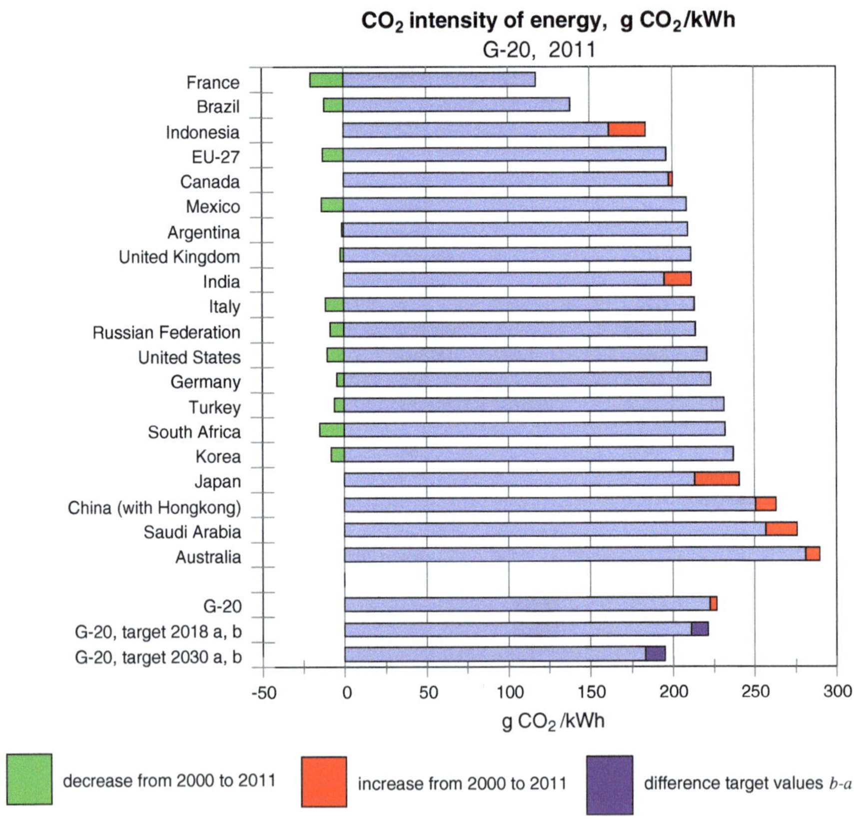


**Fig. 11.4** $CO_2$ intensity of energy of the G-20 countries and changes from 2000 to 2011

## 11.4 Indicator of $CO_2$ Sustainability 2011

Recall again, that the sustainability of energy consumption in terms of climate change can be characterised by the product of energy intensity and $CO_2$ intensity of energy. The $CO_2$ indicator calculated this way is 418 g $CO_2$/$ for the G-20, thus slightly lower than the global mean value.

Only France and Brazil have values that are already lower than the value of 204 to 232 g $CO_2$/$, the aim for 2030 (variants *a* and *b*). At the other end of the scale, with 600 to more than 750 g $CO_2$/$, are countries with poor efficiency (Russia), or that produce electricity almost exclusively with coal (China, South Africa) or oil (Saudi Arabia), see also the Appendix.

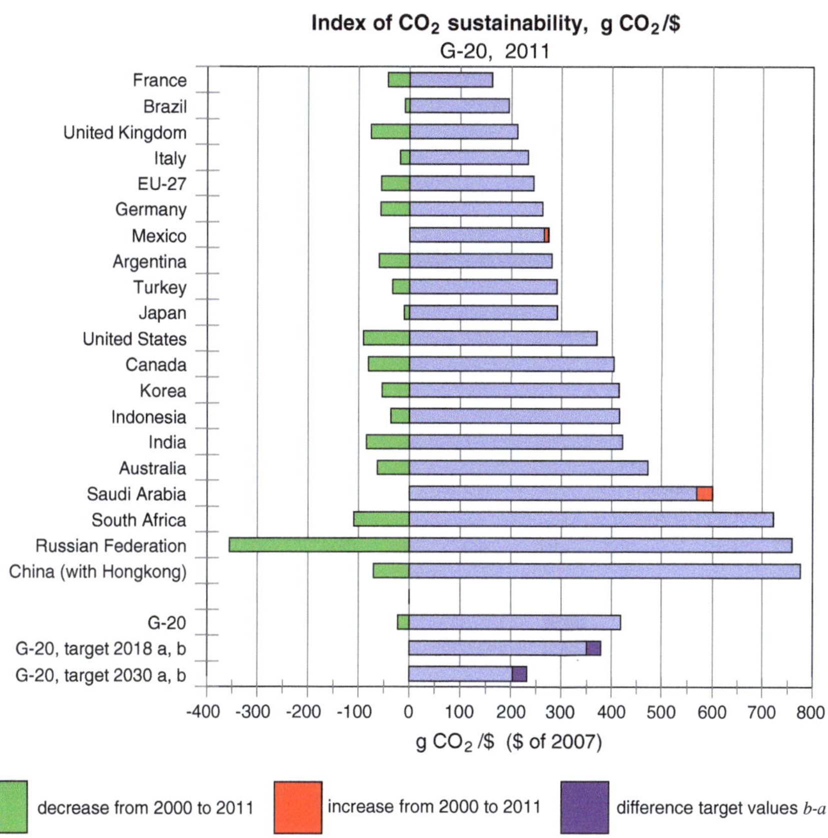


**Fig. 11.5** Indicator of $CO_2$ sustainability of the G-20 countries and changes from 2000 to 2010

The change in the $CO_2$ sustainability index is shown in more detail in Fig. 11.6 for the period 2000–2008 and for 2008–2011. Overall, a reduction of about 20 g $CO_2$/$ from 2000 to 2011 has been made, which in the medium term seems highly insufficient. By 2030, an overall reduction of almost 200 g CO2/$ is necessary (see Fig. 11.5) what requires increased efforts.

In the last 3 years, the development in key countries such as China, USA, EU-27, Canada, and South Africa is likely to be assessed as positive, which is encouraging. In several countries however the situation has worsened: Russia, South Korea, Japan (because of Fukushima), Saudi Arabia, Mexico and Brazil (Fig. 11.6).

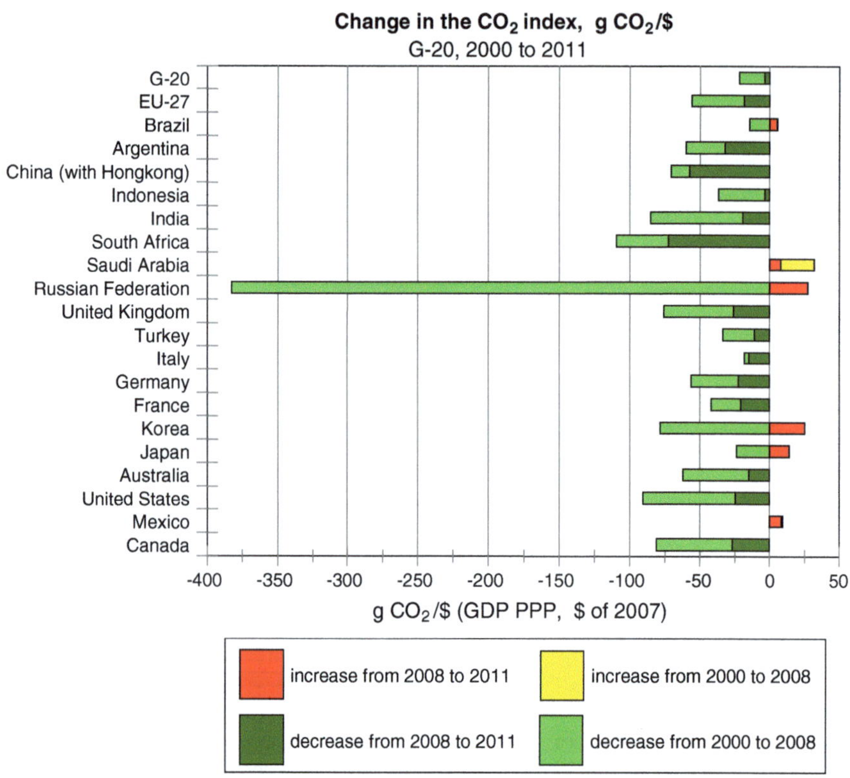


**Fig. 11.6** Change in the $CO_2$ index from 2000 to 2008 and 2008 to 2011 for the G-20 countries

## 11.5 Actual Indicators and $CO_2$ Emissions from 1990 to 2011 and Values Required from 2011 to meet the 2 °C Target

### *11.5.1 Energy Intensity, $CO_2$ Intensity and $CO_2$ Sustainability up to 2030*

The G-20 group had in 2011 an overall $CO_2$ sustainability indicator of 418 g $CO_2$/\$, lying 55 % above the OECD average (269 g $CO_2$/\$). This is also true for both components (+43 % for energy intensity and +8 % for $CO_2$ intensity of energy). For the composition of the contributing energy sources see the Appendix. The actual value of the indicators from 1980 to 2011 and the values required for the 2-degree target by 2030, for the variants *a* and *b* are shown in Fig. 11.7.

The emission share of the G-20 Group in 2030, to meet the 2-degree target, follows as the sum of the contributions of the members in accordance with Chap. 12 and is 22,100 Mt (variant *a*) or 25,200 Mt (variant *b*). The GDP (PPP) of 108,500 billion dollars (\$ 2007), as the sum of the members, leads to a necessary $CO_2$ sustainability indicator of $\eta$ = 204–232 g $CO_2$/\$.

The values for 2018 correspond to the data of the IMF (International Monetary Fund), with a GDP (PPP) forecast level for the G-20 group of \$76,500 billion (\$ of 2007).

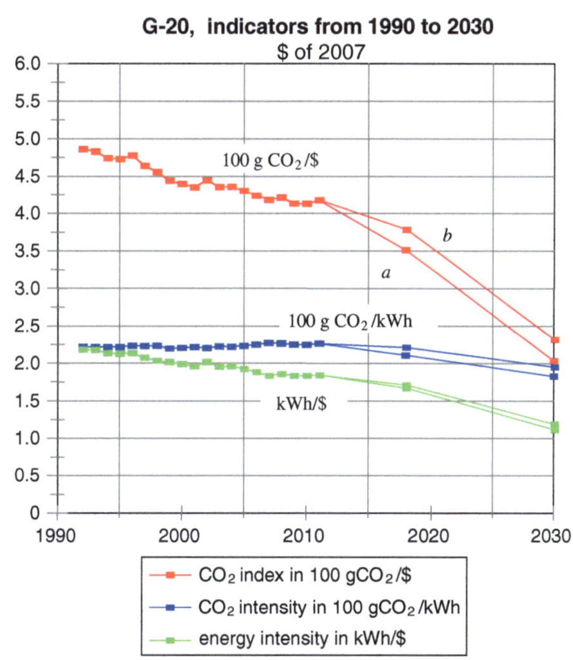


**Fig. 11.7** G-20 indicators from 1990 to 2011 and climate protection scenario until 2030

The gross energy demand for 2030 (only energy use and without marine and aviation bunkers) is in this scenario 10,400–11,100 Mtoe. The global per capita indicators for energy and $CO_2$ emissions can then be estimated as $e$ = 2.7–2.9 kW/capita (decrease of 10 % to increase of 2 % relative to 2011), and α = 4.4–5.0 t $CO_2$/capita (a decrease of 23 or 12 % relative to 2011).

The necessary change in the trend of the indicators to meet the 2-degree target is illustrated in Fig. 11.8 for variant *a*. A significant improvement of the energy efficiency trends and a turnaround in the $CO_2$ intensity of energy (−1.1 %/a) are necessary. They can only be achieved if all countries and especially those with heavy $CO_2$ emissions participate. For this, see the contributions of each country in Chap. 12.

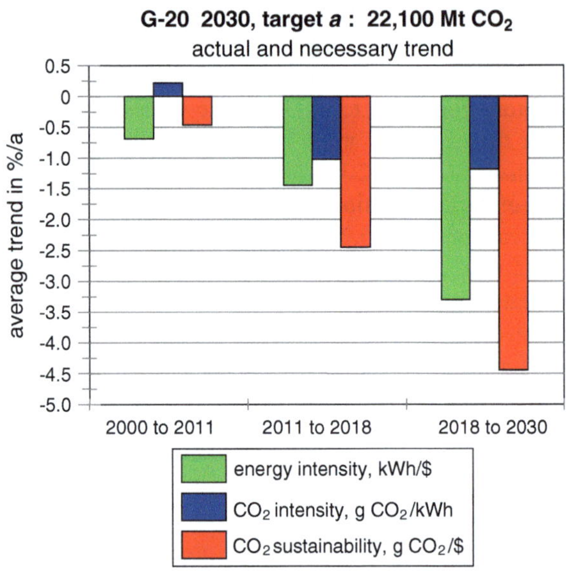


**Fig. 11.8** Trend from 2000 to 2011 of the G-20 group and necessary change in the trend from 2011 to achieve the 2-degree target, variant *a*

With variant *b* (Fig. 11.9) the trend reversal of the $CO_2$ intensity of energy in particular, is before 2018 more moderate.

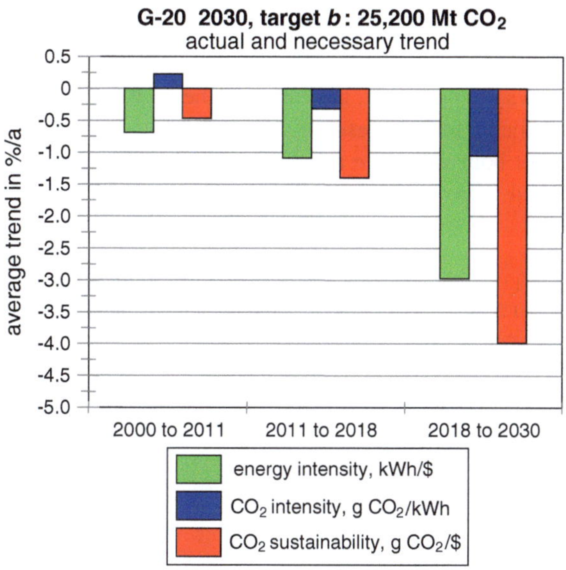


**Fig. 11.9** Trend from 2000 to 2011 of the G-20 group and necessary change in the trend from 2011 to achieve the 2-degree target, variant *b*

## 11.5.2 $CO_2$ Emissions until 2050

Figure 11.10 shows the actual $CO_2$ emissions from 1990 to 2011, and those corresponding to the indicators of Fig. 11.7 and permissible for meeting the 2-degree climate target up to 2050, for variants *a* and *b*.

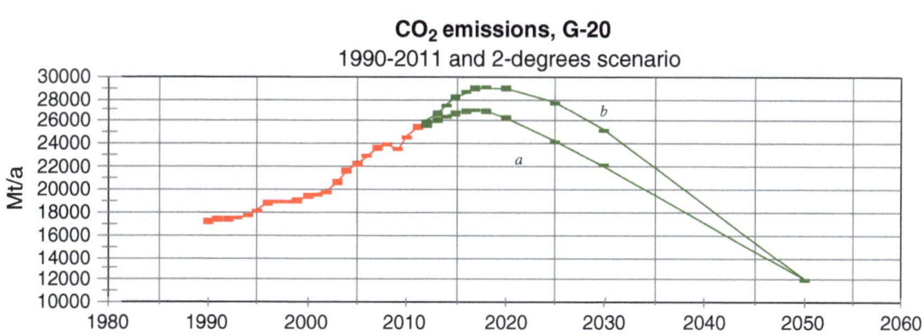


**Fig. 11.10** $CO_2$ emissions of the G-20, from 1990 to 2011 and climate protection scenario until 2050

### *11.5.3 Per Capita Indicators up to 2030*

Figure 11.11 shows the per capita indicators of the G-20 group from 1990 to 2011 as well as the profile up to 2030 resulting from the above considerations to meet the 2-degree target, for both variants *a* and *b*. The GDP (PPP) data of 2018 correspond to the statistics and forecasts of the IMF.

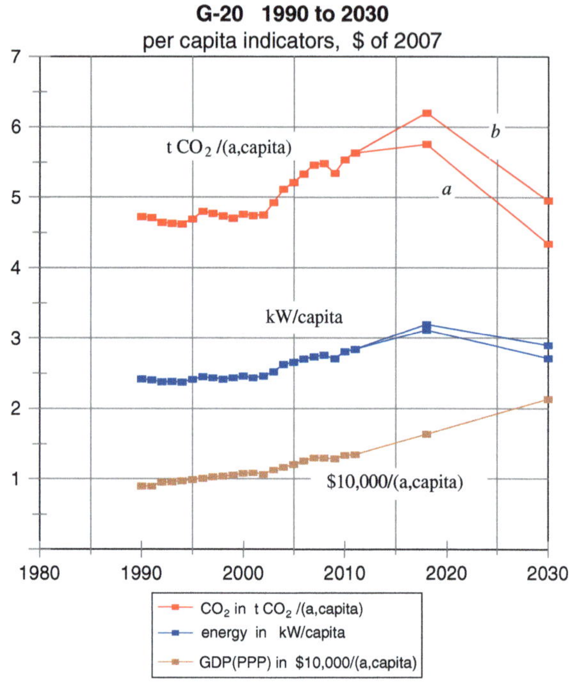


**Fig. 11.11** Per capita indicators of the G-20 from 1990 to 2011 and 2-degrees-scenario until 2030

Figure 11.12 shows, in more detail **for each consumption sector,** the per capita emissions of the resident population in 2011, and those according to the 2-degree scenario for 2030 (variant *a* and *b*).

The four consumption sectors are (total specific emissions of $CO_2$ in 2011: 5.6 t/capita):

- industry (1.2 t/capita)
- transport (1.1 t/capita)
- residential, services, agriculture, etc. (1.0 t/capita)
- energy-sector losses (2.3 t/capita).

The energy and emissions data for 2011 correspond to the G-20-Annex A5.

Final-energy carriers emit $CO_2$ directly or indirectly (via electricity and district heating). The assumed distribution of these energy carriers within the consumer sectors, for both variants in 2030, is ***one way among others*** to meet the 2-degree condition.

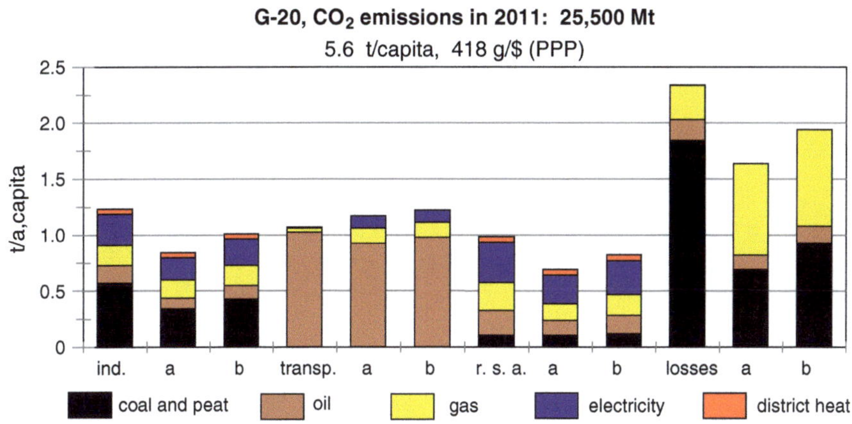


**Fig. 11.12** $CO_2$ emissions per capita of the 4 sectors: industry, transport, residential + services + agriculture, losses of the energy sector (definitions see Appendix), each sector represented by three bars: *first bar* 2011 (data in the text and the diagram title, according to the G-20-Annex A5), *second bar* possible 2 °C scenario **variant *a*** for **2030**, 22,100 Mt, 4.4 t/capita, 204 g/$ (PPP), *third bar* possible 2 °C scenario **variant *b*** for **2030**, 25,200 Mt, 5.0 t/capita, 232 g/$ (PPP)

The target value 2030 is, for example, achieved when the per capita emissions:

- in the **electricity and district heat production sector, including energy-sector losses**, are reduced by 26 % (variant *a*) or 13 % (variant *b*) from 3.1 t/capita in 2011 to 2.3–2.7 t/capita (efficiency improvements, replacement of coal by gas, CCS, renewable energy and nuclear power)
- in the **heating sector** (industry, residential, services etc.) through efficiency improvements (better insulation), reduction of coal consumption and the use of renewable energy sources (heat pump, waste management, solar energy, geo-thermal energy), are reduced by 34 % (*a*) or 19 % (*b*) (from 1.5 t/capita in 2011 to 1.0–1.2 t/capita)
- in the **transport sector** increase by 9 % (*a*) or not more than 14 % (*b*), from 1.07 in 2011 to 1.17–1.22 t/capita), through efficiency improvements, gas fuels, bio-fuels and electric mobility.

# Chapter 12
# G-20 Countries: Indicators, 2-degree Scenario

The G-20 members are ordered by the size of their $CO_2$ emissions in 2011, indicated in the table below, together with the respective indicator of $CO_2$ sustainability:

| | Mt $CO_2$ | g $CO_2$/$ | |
|---|---|---|---|
| China | 8,000 | 744 | |
| USA | 5,287 | 369 | |
| EU-27 | 3,543 | 245 | |
| India | 1,745 | 422 | |
| Russia | 1,653 | 759 | |
| Japan | 1,186 | 292 | |
| Germany | 748 | 263 | |
| South Korea | 588 | 414 | |
| Canada | 530 | 404 | |
| Saudi Arabia | 457 | 601 | |
| United Kingdom | 443 | 212 | |
| Mexico | 432 | 275 | |
| Indonesia | 426 | 415 | |
| Brazil | 408 | 195 | |
| Australia | 397 | 472 | |
| Italy | 393 | 234 | |
| South Africa | 368 | 722 | |
| France | 328 | 162 | |
| Turkey | 286 | 291 | |
| Argentina | 184 | 281 | |
| G-20 | 25,489 | 418 | G-20 target for 2030: 204–232 g $CO_2$/$ |

To achieve the climate target it is appropriate to set the percentage reduction required by 2030 to a higher value, the higher is the indicator of 2011, basically like the **green curve** in Fig. 12.1a (variant *a*) or Fig. 12.1b (variant *b*).



V. Crastan, *Global Energy Demand and 2-degree Target, Report 2014*,
DOI 10.1007/978-3-319-12310-3_12

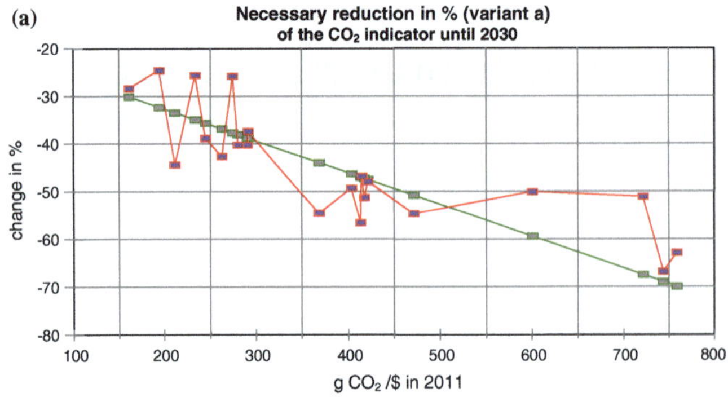


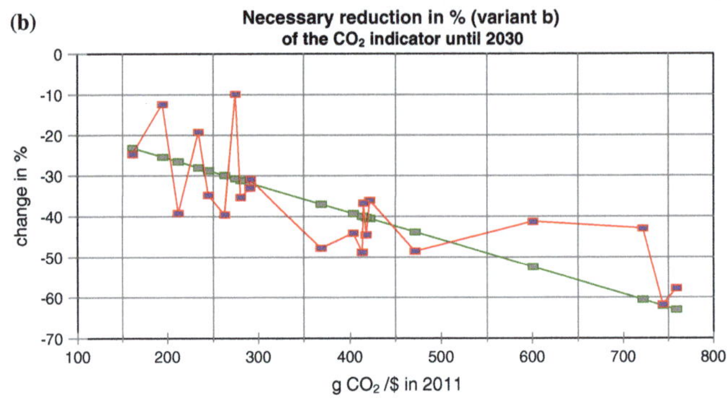


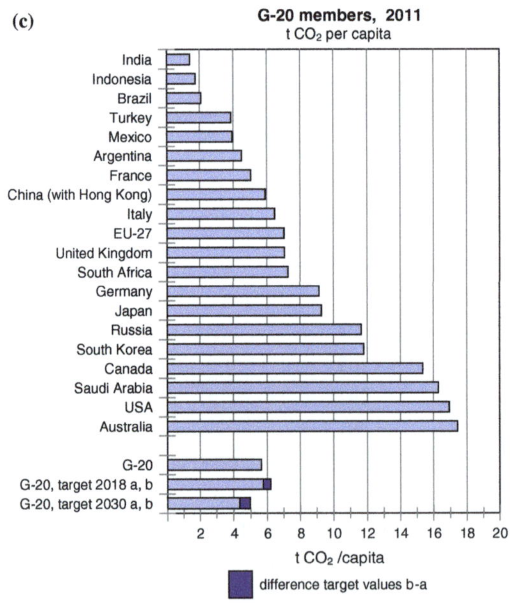


**Fig. 12.1** **a** Necessary reduction in percentage of the $CO_2$ indicator up to 2030 relative to 2011 for the 2-degree scenario variant *a*. **b** Necessary reduction in percentage of the $CO_2$ indicator up to 2030 relative to 2011 for the 2-degree scenario variant *b*. **c** Per capita emissions of the G-20 members in 2011 and 2 degree target values of the G-20 group for 2018 and 2030 according to the variants *a* and *b* of Chap. 11

The **red curve** in Fig. 12.1a (variant *a*) or Fig. 12.1b (variant *b*) shows the effective reduction assumed in the following sections for the separate G-20 members, taking account of local trends and factors, but giving the wished-for overall result.

More details on the individual member countries are given in Sects. 12.1–12.19. The EU-27 has already been analysed in Chap. 5.

The comparison of the **per capita emissions** of the G-20 members in 2011 is illustrated in Fig. 12.1c. They are between 1.4 in India and 17.4 t $CO_2$/capita in Australia, and the mean value is about 5.6 t $CO_2$/capita.

The target values for 2018 and 2030 (variants *a* and *b*) of the G-20 group, compatible with the 2 degrees scenario from Chap. 11 are also listed. An overall increase to around 6 t $CO_2$/capita by 2018 is still allowed, but after that a reduction to well below 5 t $CO_2$/capita should be sought by 2030.

## 12.1 China (with Hong Kong)

### *12.1.1 Actual Indicators (Energy Intensity, $CO_2$ Intensity and $CO_2$ Sustainability) from 1990 to 2011 and Required up to 2030 to meet the 2 °C Target*

In 2011, China had a $CO_2$ sustainability indicator of 744 g $CO_2$/\$, putting it in the second-last position of the G-20 group. The reasons are the mainly coal-based electricity production and the poor energy efficiency (see also the appendix). The trend of the indicators from 1990 to 2011 and the necessary one for the 2-degree target by 2030, for the variants *a* and *b* are shown in Fig. 12.2.

The climate target to be aimed at for 2030 is 247–285 g $CO_2$/\$ for a share of 7,400–8,550 Mt of the total G-20 $CO_2$ emissions of 22,100–25,200 Mt. The values for 2018 take into account the GDP (PPP) forecast by the International Monetary Fund (IMF) of around \$18,600 billion (\$ of 2007). The GDP for 2030 is estimated at \$30,000 billion.

**Fig. 12.2** China indicators 1990–2011 and 2-degrees scenario up to 2030

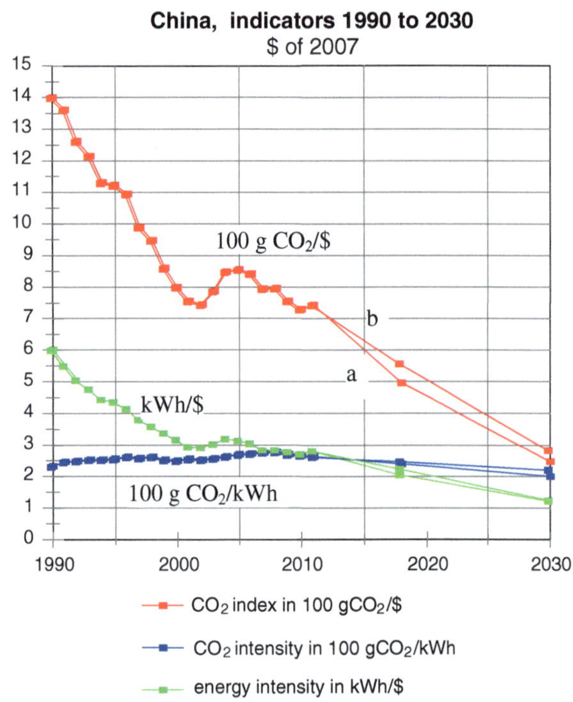


With this scenario, the gross energy demand in 2030 is 3,200–3,300 Mtoe. The per capita indicators for energy and $CO_2$ emissions would be: e ≈ 3.1–3.2 kW/capita (increase of 20–24 % relative to 2011) and $\alpha$ = 5.4–6.2 t $CO_2$ per capita (decrease of 10 % (var. *a*) to increase of 5 % (var. *b*) relative to 2011).

The necessary change in the trend of the indicators to meet the 2-degree target is illustrated in Fig. 12.3 for variant *a*. A significant and rapid improvement of energy efficiency trends and a decided turnaround in the $CO_2$ intensity of energy are necessary.

**Fig. 12.3** Necessary trend change to meet the 2-degrees target, variant *a*

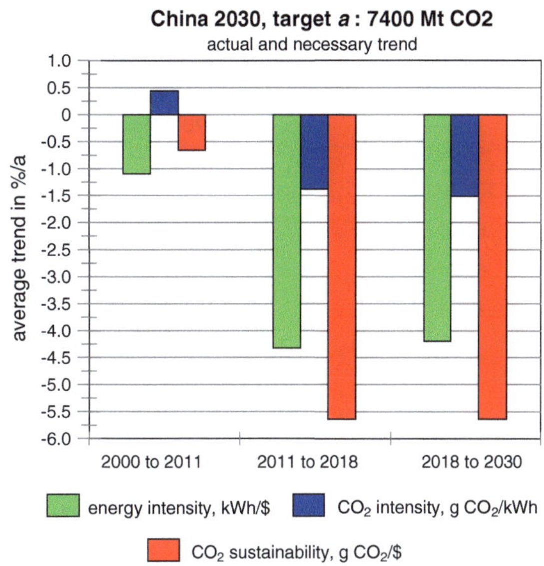


With the less strict variant *b* (Fig. 12.4), the stated trend changes are also necessary, but smaller changes are required both for energy intensity (only until 2018) and for $CO_2$ intensity of energy.

**Fig. 12.4** Necessary trend change to meet the 2-degrees target, variant *b*

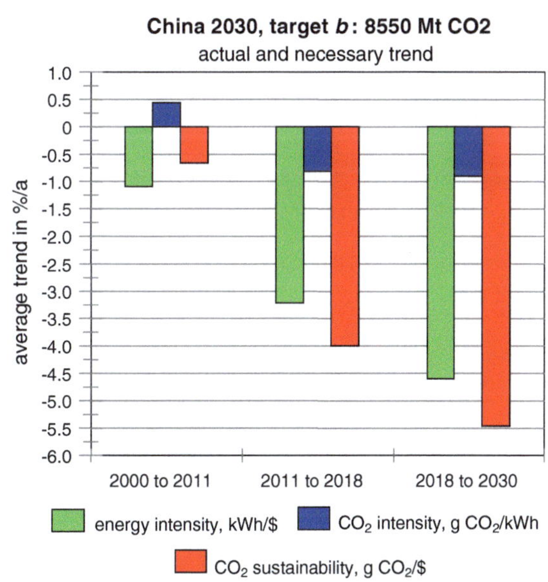

### 12.1.2 $CO_2$ Emissions until 2050

The actual $CO_2$ emissions in China from 1970 to 2011, and those allowed to 2050 for compliance with the 2 °C climate target (according to the indicators of Figs. 12.2, 12.3, 12.4) are shown in Fig. 12.5.

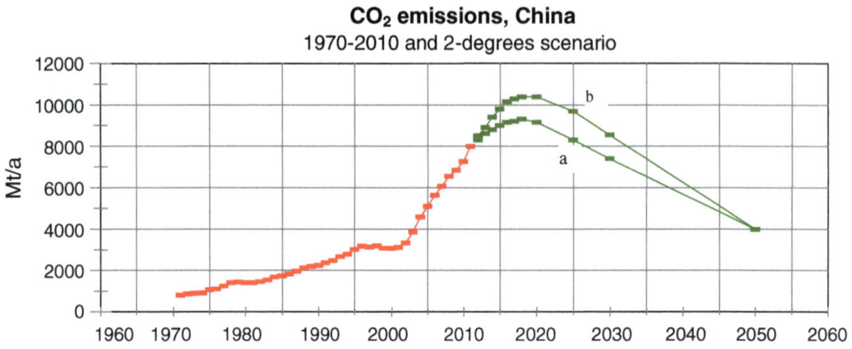


**Fig. 12.5** $CO_2$ emissions of China from 1970 to 2011 and climate protection scenario until 2050

### 12.1.3 Per Capita Indicators up to 2030

Figure 12.6 shows the per capita indicators of China from 1980 to 2011 as well as the profile up to 2030 resulting from the above considerations to meet the 2-degree target, for both variants *a* and *b*. The GDP (PPP) data up to 2018 correspond to the statistics and forecasts of the IMF.

**Fig. 12.6** Per capita indicators of the China from 1980 to 2011 and 2-degrees scenario until 2030

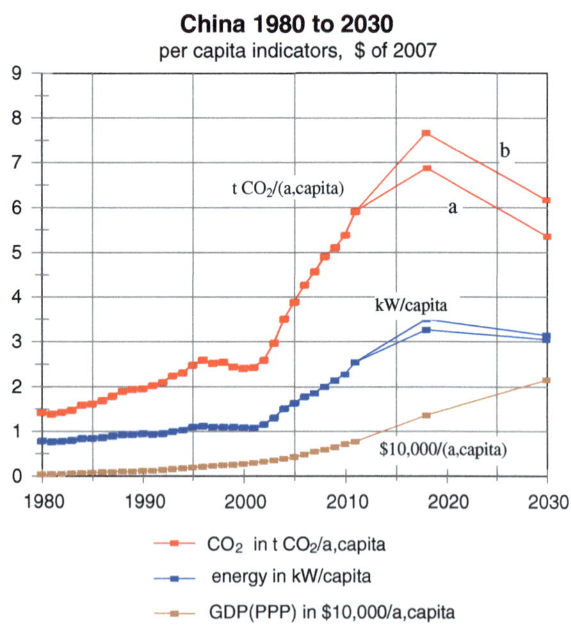

The following Fig. 12.7 shows, in more detail **per consumer sector**, the per capita emissions of the resident population in 2011, and those corresponding to the 2-degree scenario for 2030 (variant *a* and *b*).

The four consumer sectors are (total specific emissions of $CO_2$ in 2011: 5.9 t/capita):

- industry (2.0 t/capita)
- transport (0.4 t/capita)
- residential, services, agriculture, etc. (0.7 t/capita)
- energy-sector losses (2.8 t/capita).

The energy and emissions data for 2011 correspond to the China-Annex A5.

Final-energy carriers emit $CO_2$ directly or indirectly (via electricity and district heating). The assumed distribution of these energy carriers within the consumer sectors, for both variants in 2030, is ***one way among others*** to meet the 2-degree condition.

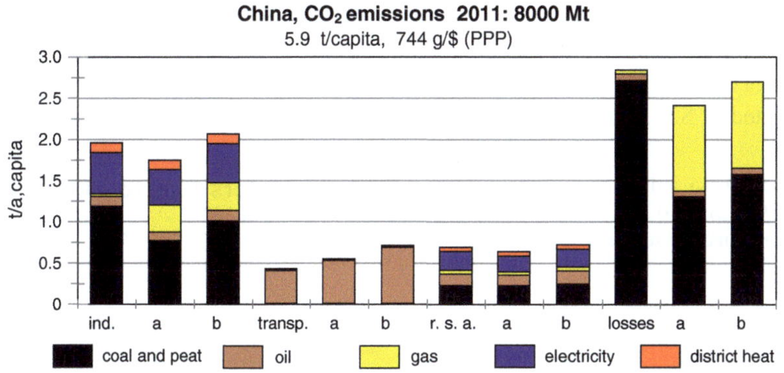


**Fig. 12.7** $CO_2$ emissions per capita of the 4 sectors: industry, transport, residential + services + agriculture, energy-sector losses (definitions see Appendix), each sector represented by three bars: *first bar* 2011 (data in the text and the diagram title, according to the China-Annex A5), *second bar* possible 2 °C scenario **variant *a*** for **2030**, 7,400 Mt, 5.4 t /capita, 247 g /$ (PPP), *third bar* possible 2 °C scenario **variant *b*** for **2030**, 8,550 Mt, 6.2 t/capita, 285 g/$ (PPP)

The target value for 2030 can, for example, be achieved if the per capita emissions:

- in the **electricity and district heat production sector, including energy-sector losses**, are reduced by 15 % (variant *a*) or 5 % (variant *b*) from 3.75 t/capita in 2011 to 3.2–3.6 t/capita (efficiency improvements, replacement of coal by gas, CCS, renewable energy and nuclear power)
- in the **heating sector** (industry, residential, services etc.) through efficiency improvements (better insulation), reduction of coal consumption, gas instead of coal, and the use of renewable energy sources (heat pump, waste management, solar energy, geothermal energy), are reduced by 10 % (*a*) or increase by not more than 10 % (*b*) (from 1.75 t/capita to 1.6–1.9 t/capita)
- in the **transport sector** increase by 30 % (*a*) but not more than 70 % (*b*), from 0.4 to 0.55–0.7 t/capita, through efficiency improvements, gas fuels, bio-fuels and electric mobility.

**The contribution of China, together with that of the United States, India and Russia is likely to be of crucial importance for achieving the climate goals.**

## 12.2 USA

### *12.2.1 Actual Indicators (Energy Intensity, $CO_2$ Intensity and $CO_2$ Sustainability) from 1980 to 2011 and Required up to 2030 to meet the 2 °C Target*

The $CO_2$ sustainability of the United States of America, at 369 g $CO_2$/\$, rates as rather bad in the OECD context. The reasons are the strong coal base of electricity production and the high $CO_2$ emissions per capita in the transport sector (see Appendix and Fig. 12.13). The actual figure of the indicators from 1980 to 2011 and the necessary figure for the 2-degree target up to 2030, for the variants *a* and *b* are shown in Fig. 12.8.

The climate target to be aimed at for 2030 is 168–193 g $CO_2$/\$ for a $CO_2$ share of the USA of 3,700–4,250 Mt of total $CO_2$ emissions of the G-20 group. The values for 2018 take into account the GDP (PPP forecast by the International Monetary Fund IMF of around \$17,460 billion (\$ of 2007). The GDP for 2030 is estimated at \$22,000 billion.

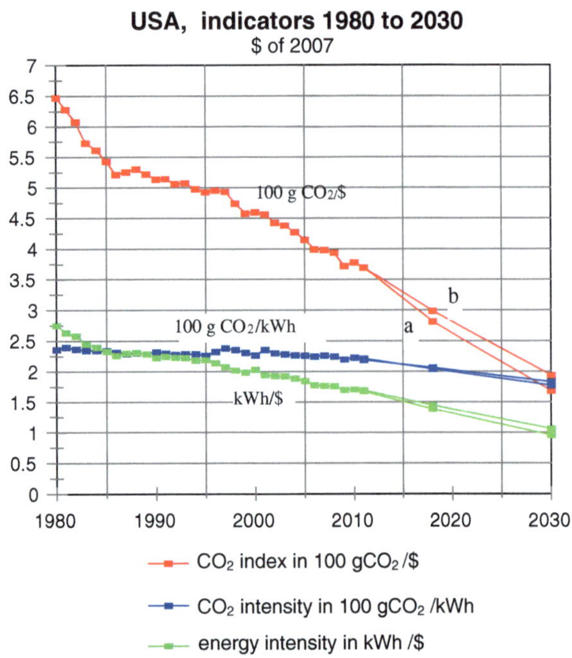


**Fig. 12.8** USA indicators from 1980 to 2010 and climate protection scenario up to 2030

The gross energy demand in 2030 with this scenario is 1,800–2,000 Mtoe (only energy use). The per capita indicators for energy and $CO_2$ emissions would be: e ≈ 6.5–7.2 kW/capita (reduction of 25 % (var. *a*) to 19 % (var. *b*) relative to 2011) and $\alpha$ = 10.0 t $CO_2$ per capita (reduction of 41 % (*a*) or 32 % (*b*) relative to 2011).

The necessary change in the trend of the indicators, relative to the period 2000–2011, to meet the 2-degree target is illustrated in Fig. 12.9 for variant *a*. A significant improvement of the energy efficiency trend and a stronger decrease of the $CO_2$ intensity (about −1 %/a) are necessary.

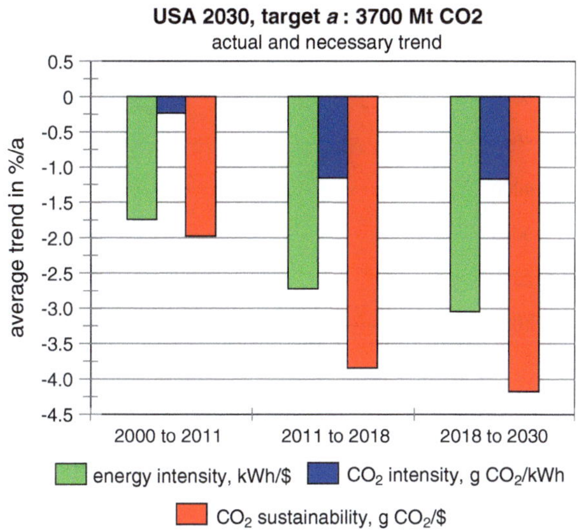


**Fig. 12.9** Necessary trend change to achieve the 2-degree target, variant *a*

With the less strict variant *b* (Fig. 12.10), the stated changes in trend are similar, but occur a little later and are correspondingly less severe.

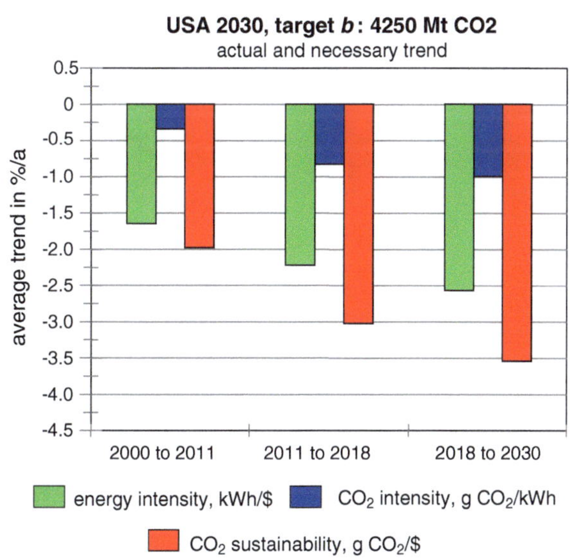


**Fig. 12.10** Necessary trend change to achieve the 2-degree target, variant *b*

### 12.2.2 $CO_2$ Emissions until 2050

The $CO_2$ emissions of the United States from 1970 to 2011, and up to 2050 that permit the 2-degree climate target (according to the indicators of Figs. 12.8, 12.9, 12.10), are shown in Fig. 12.11.

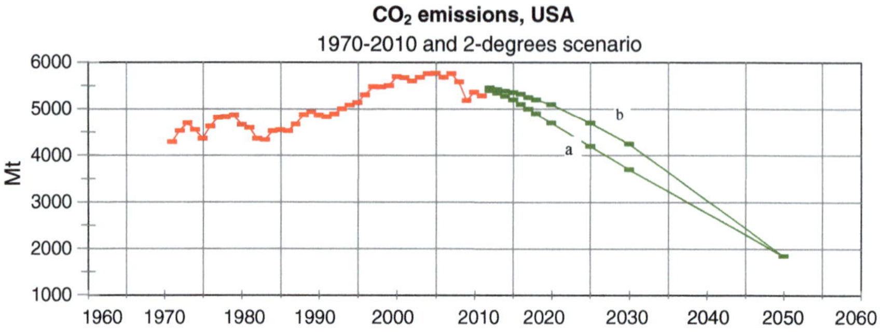


**Fig. 12.11** $CO_2$ emissions of the USA 1970 to 2011 and 2-degrees scenario until 2050

### 12.2.3 Per Capita Indicators up to 2030

Figure 12.12 shows the per capita indicators of the USA from 1980 to 2011 as well as the profile up to 2030 resulting from the above considerations to meet the 2-degree target, for both variants *a* and *b*. The GDP (PPP) data up to 2018 correspond to the statistics and forecasts by the IMF.

**Fig. 12.12** Per capita indicators of the USA from 1980 to 2011 and 2-degrees scenario until 2030

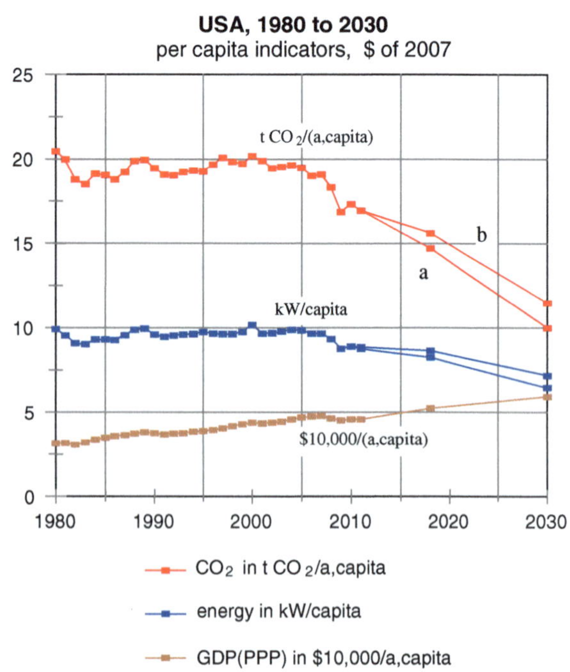

The following Fig. 12.13 shows, in more detail **per consumer sector,** the per capita emissions of the resident population in 2011, and those corresponding to the 2-degree scenario for 2030 (variant *a* and *b*).

The four consumer sectors are (total specific emissions of $CO_2$ in 2011: 16.9 t/capita):

- industry (2.1 t/capita)
- transport (5.5 t/capita)
- residential, services, agriculture, etc. (3.7 t/capita)
- energy-sector losses (5.6 t/capita).

The energy and emissions data for 2011 correspond to the USA Annex A5.

Final-energy carriers emit $CO_2$ directly or indirectly (via electricity and district heating). The assumed distribution of these energy carriers within the consumer sectors, for both variants in 2030, is ***one way among others*** to meet the 2-degree condition.

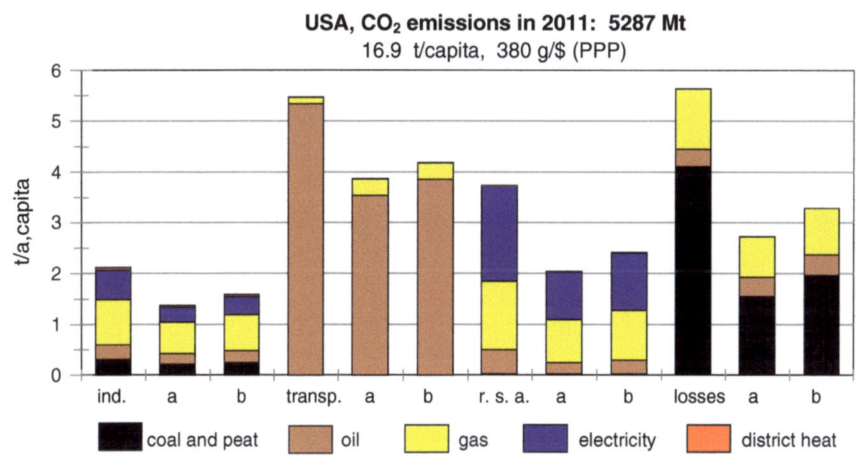


**Fig. 12.13** $CO_2$ emissions per capita of the 4 sectors: industry, transport, residential + services + agriculture, energy-sector losses (definitions see Appendix), each sector represented by three bars: *first bar* 2011 (data in the text and the diagram title, according to the USA-Annex A5), *second bar* possible 2 °C scenario **variant *a*** for **2030**, 3,700 Mt, 10.0 t/capita, 168 g/$ (PPP), *third bar* possible 2 °C scenario **variant *b*** for **2030**, 4,250 Mt, 11.5 t/capita, 193 g/$ (PPP)

The target value for 2030 can, for example, be achieved if the per capita emissions:

- in the **electricity and district heat production sector, including energy-sector losses**, are reduced by 50 % (variant *a*) or 40 % (variant *b*) from 8.2 t/capita in 2011 to 3.2–4.8 t/capita (efficiency improvements, replacement of coal by gas, CCS, renewable energy and nuclear power)

- in the **heating sector** (industry, residential, services etc.) through efficiency improvements (better insulation), reduction of coal consumption and the use of renewable energy sources (heat pump, waste management, solar energy, geothermal energy), are reduced by about 35 % (*a*) or 30 % (*b*) (from 3.3 t/capita to 2.1–2.5 t/capita)
- in the **transport sector** are reduced by 30 % (*a*) or 24 % (*b*), from 5.4 to 3.9–4.2 t/capita), through efficiency improvements, gas fuels, bio-fuels and electric mobility.

## 12.3 India

### 12.3.1 Actual Indicators (Energy Intensity, $CO_2$ Intensity and $CO_2$ Sustainability) from 1980 to 2011 and Required up to 2030 to meet the 2 °C Target

India is in 2011, with a $CO_2$ indicator of sustainability of 422 g $CO_2$/$ in the middle of the G-20 group (about the same as the world average). A negative point is the mainly coal-based electricity production (see Appendix). The actual graph of the indicators from 1980 to 2011 and the necessary one for the 2-degree target by 2030, for the variants *a* and *b*, are shown in Fig. 12.14.

The climate target to be aimed at for 2030 is 220–270 g $CO_2$/$ for India's share of 2,200–2,700 Mt of total $CO_2$ emissions of the G-20 group. The values for 2018 take into account the GDP (PPP) forecast by the International Monetary Fund IMF of around $6,000 billion ($ of 2007). The GDP for 2030 is estimated at $10,000 billion.

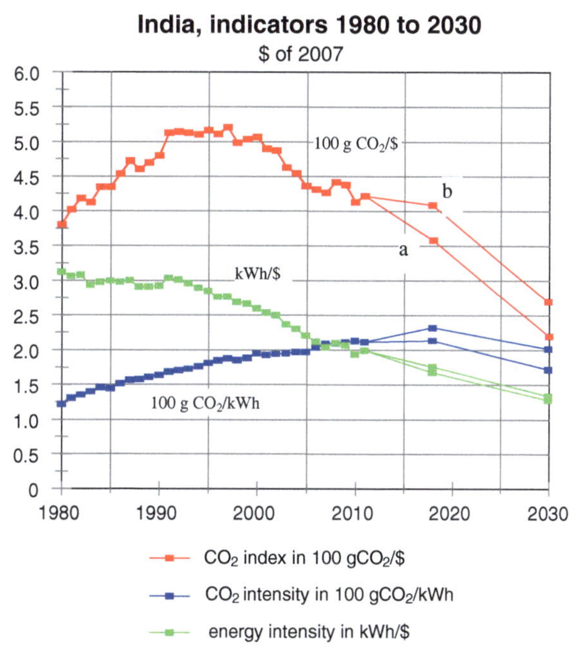


**Fig. 12.14** Indicators of India from 1980 to 2011 and 2-degrees scenario until 2030

The gross energy demand in 2030 with this scenario is 1,100–1,150 Mtoe (only energy use). The per capita indicators for energy and $CO_2$ emissions would be: e ≈ 1.0–1.1 kW/capita (increase of 34 % (var. *a*) to 40 % (var. *b*) relative to 2011) and $\alpha$ = 1.5–1.9 t $CO_2$ per capita (increase of 9 % (*a*) or 33 % (*b*) relative to 2011).

The necessary change in the trend of the indicators, relative to the period 2000 to 2011, to meet the 2-degree target is illustrated in Fig. 12.15 for variant *a*. The maintenance of the good energy efficiency trend from 2000 to 2011 (that in recent years rather tails off) and a stabilization and then a clear turnaround in the $CO_2$ intensity of energy after 2018, are required.

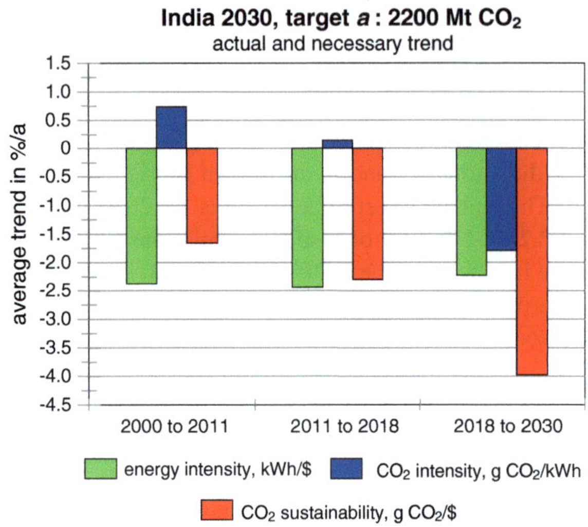


**Fig. 12.15** Necessary trend change to achieve the 2-degrees target, variant *a*

With the initially less strict variant *b* (Fig. 12.16), that perhaps takes better account of India's lagging state of development, the change of trend in $CO_2$ intensity in particular starts later.

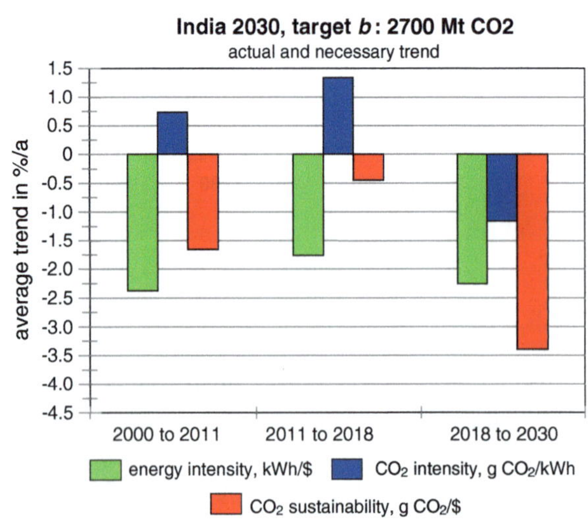


**Fig. 12.16** Necessary trend change to achieve the 2-degrees target, variant *b*

### 12.3.2 $CO_2$ Emissions until 2050

The actual $CO_2$ emissions of India from 1970 to 2011, and the values permissible up to 2050 for compliance with the 2 °C climate target, (according to the indicators of Figs. 12.14, 12.15, 12.16) are shown in Fig. 12.17 for both variants *a* and *b*. Variant *b* requires a much bigger change after 2040.

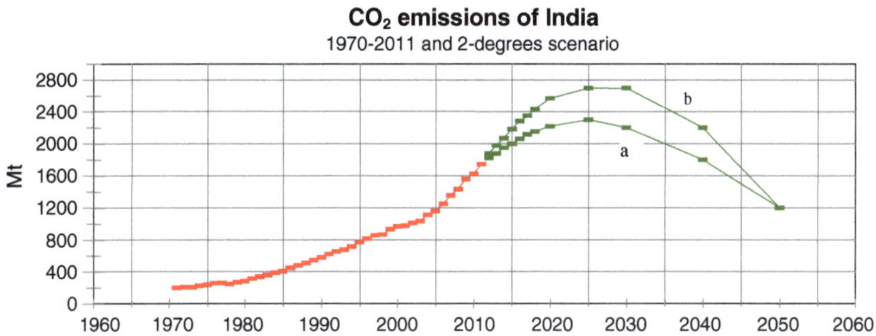


**Fig. 12.17** $CO_2$ emissions of India 1970 to 2011 and 2-degrees scenario until 2050

### 12.3.3 Per Capita Indicators up to 2030

Figure 12.18 shows the per capita indicators of India from 1980 to 2011 as well as the profile up to 2030 resulting from the above considerations to meet the 2-degree target, for both variants *a* and *b*. The GDP (PPP) data up to 2018 correspond to the statistics and forecasts by the IMF.

**Fig. 12.18** Per capita indicators of India from 1980 to 2011 and 2-degrees scenario until 2030

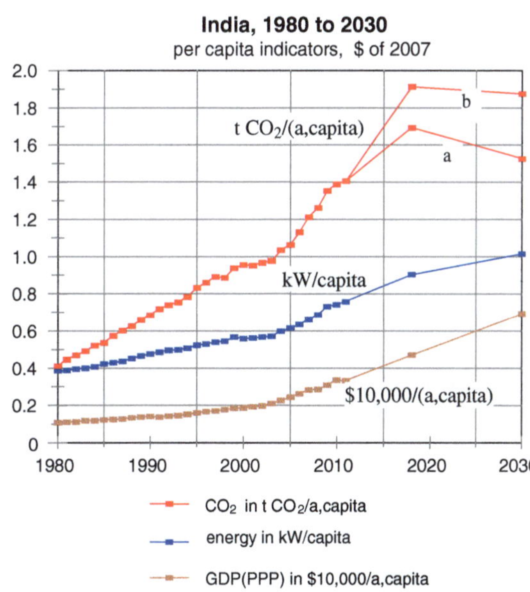

The following Fig. 12.19 shows the per capita emissions 2011 of the resident population, and those according to the 2-degree scenario 2030 (variant *a* and *b*) in more detail **per consumer sector**.

The four consumer sectors are (total specific $CO_2$ emissions in 2011: 1.4 t/capita):

- industry (0.38 t/capita)
- transport (0.13 t/capita)
- residential, services, agriculture etc. (0.20 t/capita)
- energy-sector losses (0.69 t/capita).

The energy and emissions data for 2011 correspond to the India-Annex A5.

Final-energy carriers emit $CO_2$ directly or indirectly (via electricity and district heating). The assumed distribution of these energy carriers within the consumer sectors, for both variants in 2030, is ***one way among others*** to meet the 2-degree condition.

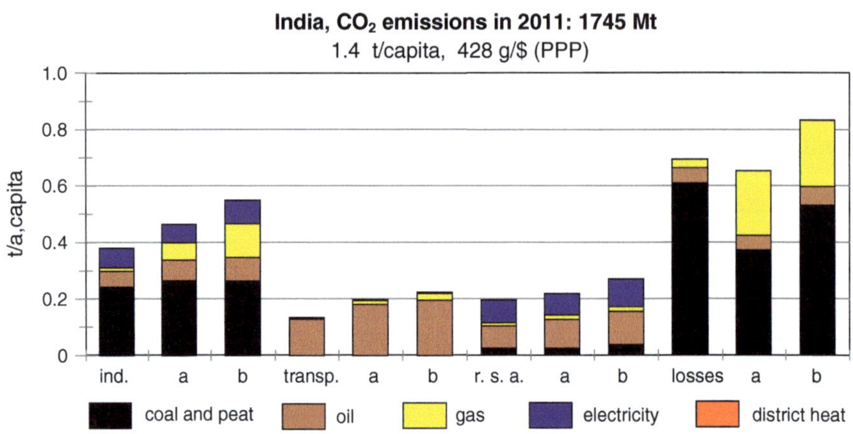


**Fig. 12.19** $CO_2$ emissions per capita of the 4 sectors: industry, transport, residential + services + agriculture, energy-sector losses (definitions see Appendix), each sector represented by three bars: *first bar* **2011** (data in the text and the diagram title, according to the India-Annex A5), *second bar* possible 2 °C scenario **variant *a*** for **2030**, 2,200 Mt, 1.53 t/capita, 220 g /$ (PPP), *third bar* possible 2 °C scenario **variant *b*** for **2030**, 2,700 Mt, 1.88 t/capita, 270 g /$ (PPP)

The target value for 2030 can, for example, be achieved if the per capita emissions:

- in the **electricity and district heat production sector, including energy-sector losses**, are reduced by 6 % (variant *a*) or increase not more than 20 % (variant *b*) from 0.85 t/capita in 2011 to 0.8–1.0 t/capita (efficiency improvements, replacement of coal by gas, CCS, renewable energy and nuclear power)
- in the **heating sector** (industry, residential, services etc.) through efficiency improvements, reduction of the percentage of coal consumption and the use of renewable energy sources (heat pump, waste management, solar energy, geothermal energy), increase by not more than 30 % (*a*) or 50 % (*b*) (from 0.42 t/capita in 2011 to 0.54–0.64 t/capita)
- in the **transport sector** increase by 50 % (*a*) to 70 % (*b*), from 0.13 in 2011 to 0.20–0.22 t/capita), through efficiency improvements, gas fuels, bio-fuels and electric mobility.

## 12.4 Russia

### *12.4.1 Actual Indicators (Energy Intensity, $CO_2$ Intensity and $CO_2$ Sustainability) from 1990 to 2011 and Required up to 2030 to meet the 2 °C Target*

Russia is, at 759 g $CO_2$/\$, at the bottom of the G-20 rankings. This is mainly due to extremely poor energy efficiency (3.6 kWh/\$, high losses in the energy sector; see Appendix). The actual graph of the indicators from 1990 to 2011 and the values necessary for meeting the 2-degree target by 2030, for the variants *a* and *b*, are shown in Fig. 12.20.

The $CO_2$ sustainability indicator to be aimed at for 2030 (variants *a* and *b*) is 282–321 g $CO_2$/\$ (−63 to −58 % relative to 2011) for Russia to have a share of 1,100–1,250 Mt (−33 to −24 % relative to 2011) of total $CO_2$ emissions of the G-20 group. The values for 2018 take into account the GDP (PPP) forecast by the International Monetary Fund IMF of around \$2,700 billion (\$ of 2007). The GDP for 2030 is estimated at \$3,900 billion.

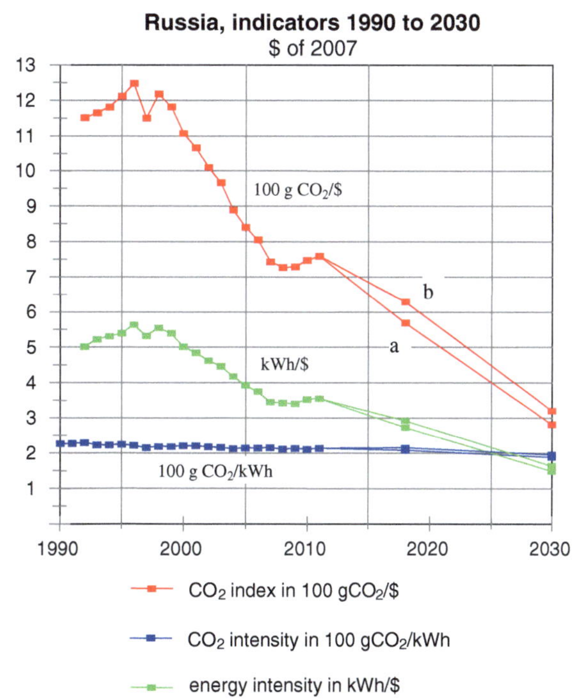


**Fig. 12.20** Indicators of Russia from 1990 to 2011 and 2-degrees scenario up to 2030

With this scenario, the gross energy demand in 2030 is 500–550 Mtoe. The per capita indicators for energy and $CO_2$ emissions would be (variants *a* and *b*): e ≈ 5.0–5.5 kW/capita) and α = 8.3–9.4 t $CO_2$ per capita (both about 30 % (*a*) or 20 % (*b*) lower than in 2011).

The change in the trend of the indicators required to meet the 2-degree target is illustrated in Fig. 12.21 for variant *a*:

- The good trend in energy intensity must be maintained and strengthened.
- The trend in $CO_2$ intensity has to be improved somewhat, especially after 2018.

But the trend development of energy intensity in the last three years (Fig. 12.20) is alarming, and should be reversed quickly.

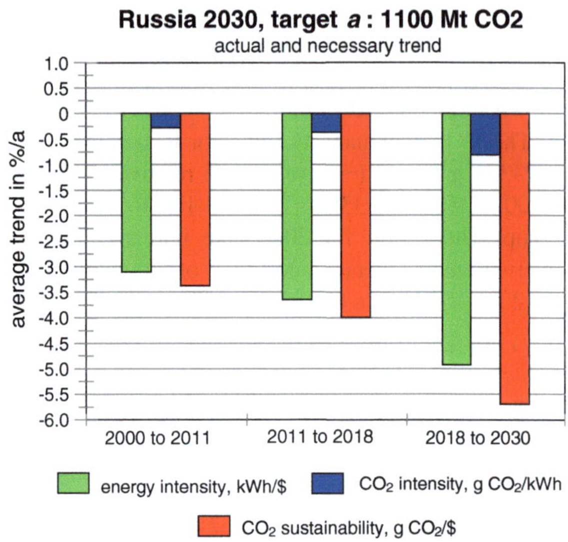


**Fig. 12.21** Necessary trend change to achieve the 2-degree target, variant *a*

Variant *b* (Fig. 12.22) clearly allows more time for that until 2018, but a big change of trend must occur after this date.

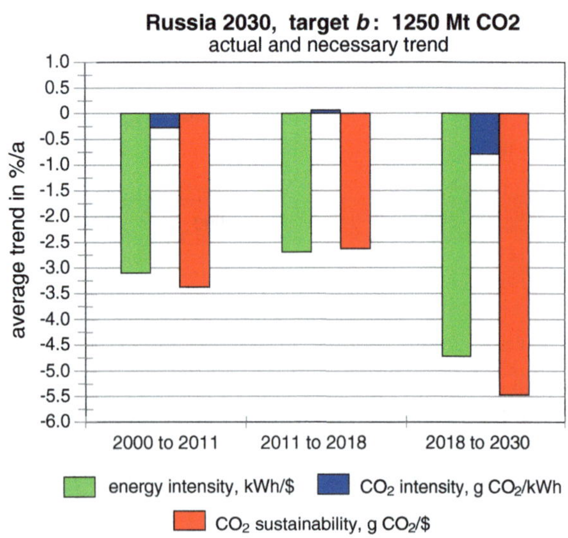


**Fig. 12.22** Necessary trend change to achieve the 2-degree target, variant *b*

## 12.4.2 $CO_2$ Emissions until 2050

The actual $CO_2$ emissions of Russia from 1990 to 2011, and those permissible up to 2050 to achieve the 2 °C climate target (according to the indicators of Figs. 12.20, 12.21, 12.22) are shown in Fig. 12.23 for both variants *a* and *b*.

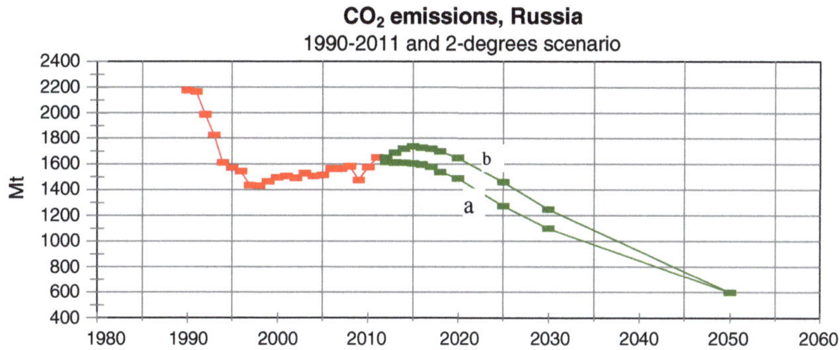


**Fig. 12.23** $CO_2$ emissions of Russia from 1990 to 2011 and 2-degrees scenario until 2050

## 12.4.3 Per Capita Indicators up to 2030

Figure 12.24 shows the per capita indicators of Russia from 1990 to 2011 as well as the profile up to 2030 resulting from the above considerations to meet the 2-degree target, for both variants *a* and *b*. The GDP (PPP) data up to 2018 correspond to the statistics and forecasts by the IMF.

**Fig. 12.24** Per capita indicators of Russia from 1990 to 2011 and 2-degrees scenario until 2030

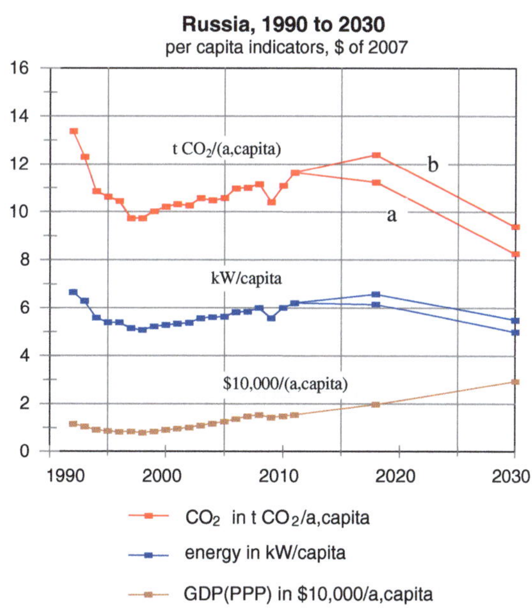

Figure 12.25 shows the per capita emissions of 2011 of the resident population, and those according to the 2-degree scenario 2030 (variant *a* and *b*) in more detail **per consumer sector**.

The four consumer sectors are (total specific $CO_2$ emissions in 2011: 11.6 t/capita):

- industry (2.2 t/capita)
- transport (1.8 t/capita)
- residential, services, agriculture etc. (2.8 t/capita)
- energy-sector losses (4.8 t/capita).

The energy and emissions data for 2011 correspond to the Russia-Annex A5.

Final-energy carriers emit $CO_2$ directly or indirectly (via electricity and district heating). The assumed distribution of these energy carriers within the consumer sectors, for both variants in 2030, is ***one way among others*** to meet the 2-degree condition.

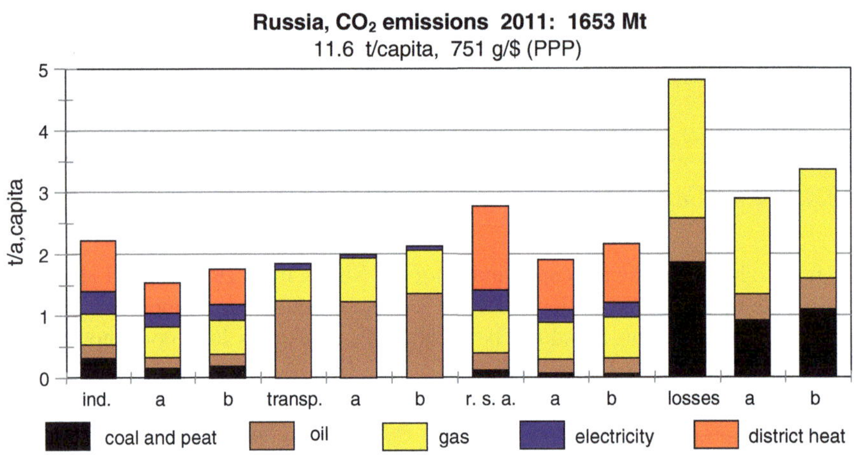


**Fig. 12.25** $CO_2$ emissions per capita of the 4 sectors: industry, transport, residential + services + agriculture, energy-sector losses (definitions see Appendix), each sector represented by three bars: *first bar* **2011** (data in the text and the diagram title, according to the Russia-Annex A5), *second bar* possible 2 °C scenario **variant *a*** for **2030**, 1,100 Mt, 8.3 t /capita, 282 g /$ (PPP), *third bar* possible 2 °C scenario **variant *b*** for **2030**, 1,250 Mt, 9.4 t/capita, 321 g/$ (PPP)

The target value for 2030 can, for example, be achieved if the per capita emissions:

- in the **electricity and district heat production sector, including energy-sector losses**, are reduced by 40 % (variant *a*) or 30 % (variant *b*) from 7.8 t/capita in 2011 to 4.4–5.5 t/capita (efficiency improvements, replacement of coal by gas, CCS, renewable energy and nuclear power)
- in the **heating sector** (industry, residential, services etc.) through efficiency improvements, reduction of coal and oil consumption and the use of renewable energy sources (heat pump, waste management, solar energy, geothermal energy), are reduced by 20 % (*a*) or 10 % (*b*) (from 2.1 t/capita in 2011 to 1.7–1.9 t/capita)
- in the **transport sector** increase by not more than 15 % (*a*) or 25 % (*b*), from 1.85 in 2011 to 2.0–2.13 t/capita, through efficiency improvements, gas fuels, bio-fuels and, as a more distant goal, electric mobility.

## 12.5 Japan

### *12.5.1 Actual Indicators (Energy Intensity, $CO_2$ Intensity and $CO_2$ Sustainability) from 1990 to 2011 and Required up to 2030 to meet the 2 °C Target*

Japan in 2011, with 292 g $CO_2$/\$, has, in the context of the G-20 group, a fairly good $CO_2$ sustainability indicator. The actual profile of the indicators from 1980 to 2011 and the necessary one for the 2-degree target by 2030, for the variants *a* and *b*, is shown in Fig. 12.26. The significant deterioration of $CO_2$ intensity of energy in 2011 as a result of the Fukushima accident is striking.

The climate target to be aimed at for 2030 is 183–202 g $CO_2$/\$ f (−37 to −31 % relative to 2011) for a share of Japan of 950–1,050 Mt of total $CO_2$ emissions of the G-20 group. The values for 2018 take into account the GDP (PPP) forecast by the International Monetary Fund IMF of around \$4,470 billion (\$ of 2007). The GDP for 2030 is estimated at \$5,200 billion.

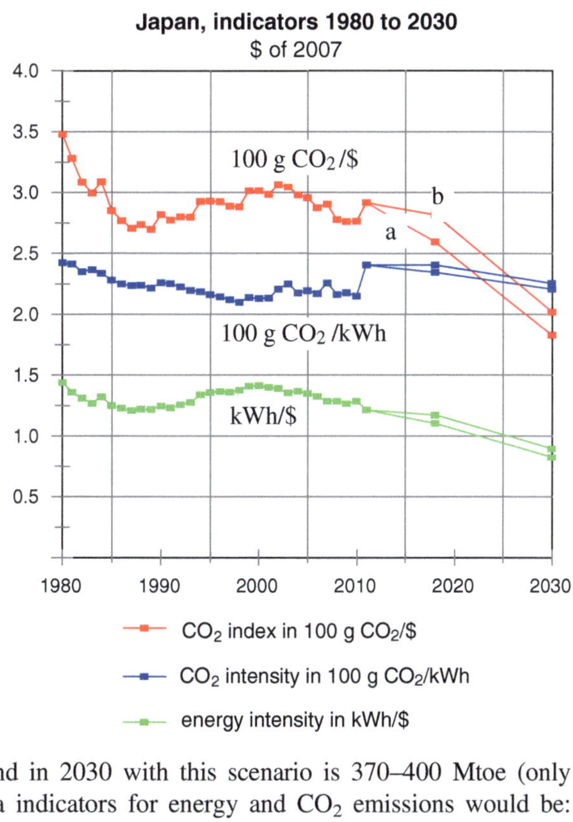


**Fig. 12.26** Indicators of Japan from 1980 to 2011 and 2-degrees scenario up to 2030 (significant deterioration in 2011 due to Fukushima)

The gross energy demand in 2030 with this scenario is 370–400 Mtoe (only energy use). The per capita indicators for energy and $CO_2$ emissions would be: e ≈ 3.9–4.2 kW/capita (decrease of 12 % (var. *a*) or 5 % (var. *b*) relative to 2011) and α = 7.5–8.3 t $CO_2$ per capita (−19 % (*a*) to −11 % (*b*) relative to 2011).

The necessary change in the trend of the indicators, relative to the period 2000 to 2011, to meet the 2-degree target is illustrated in in Fig. 12.27 for variant *a*. An improvement in the energy efficiency trend and an effort in $CO_2$ intensity of energy (−0.5 %/a after 2018) are required, despite Fukushima.

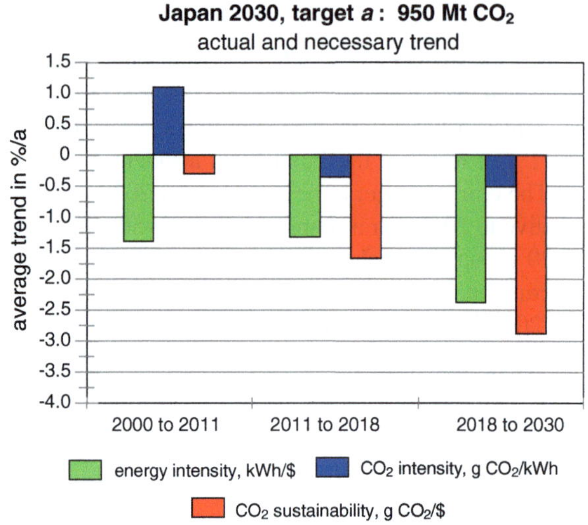


**Fig. 12.27** Necessary trend change to meet the 2-degrees target, variant *a*

Variant *b* (Fig. 12.28) is less strict until 2018, but requires more effort after 2030 to meet the 2-degree-goal (see also Fig. 12.29).

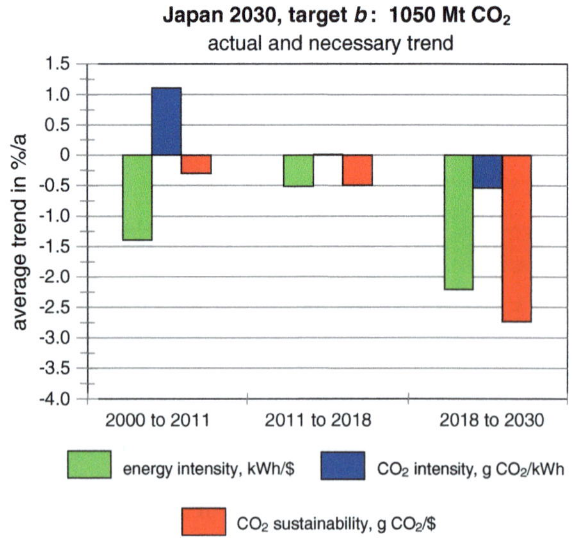


**Fig. 12.28** Necessary trend change to meet the 2-degrees target, variant *b*

## 12.5.2 $CO_2$ Emissions until 2050

The actual $CO_2$ emissions in Japan from 1970 to 2011 and those allowed to 2050 for compliance with the 2 °C climate target (according to the indicators of Figs. 12.26, 12.27, 12.28) are shown in Fig. 12.29.

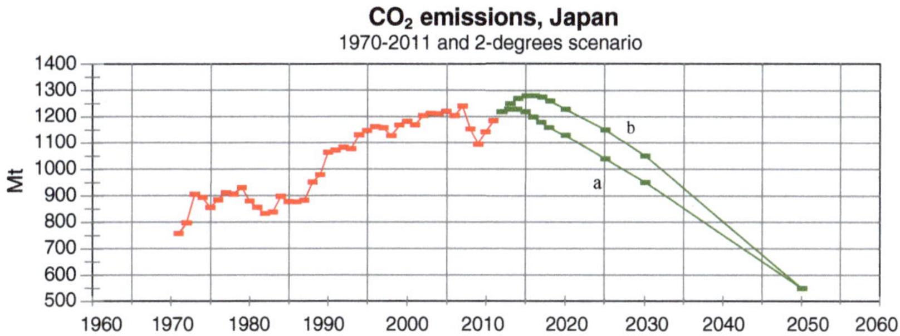


**Fig. 12.29** $CO_2$ emissions of Japan from 1970 to 2011 and climate protection scenario until 2050

## 12.5.3 Per Capita Indicators up to 2030

Figure 12.30 shows the **per capita indicators** of Japan from 1980 to 2011 as well as the profile up to 2030 resulting from the above considerations to meet the 2-degree target, for both variants *a* and *b*. The GDP (PPP) data up to 2018 correspond to the statistics and forecasts of the IMF.

**Fig. 12.30** Per capita indicators of Japan from 1980 to 2011 and 2-degrees scenario until 2030

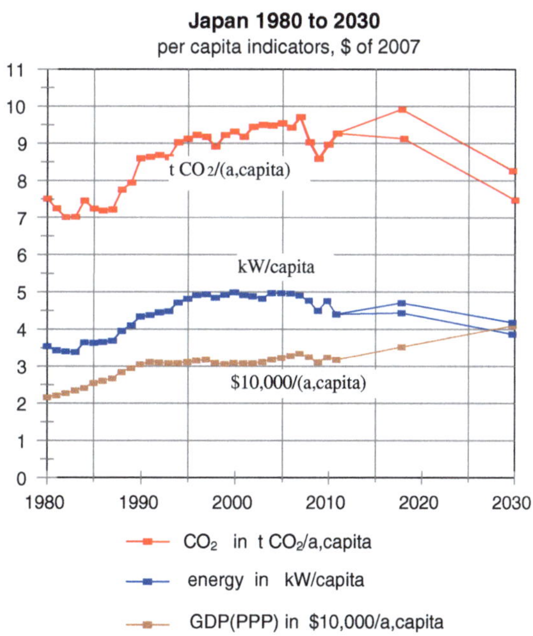

Figure 12.31 shows, in more detail **per consumer sector**, the per capita emissions of the resident population in 2011, and those corresponding to the 2-degree scenario for 2030 (variant *a* and *b*).

The four consumer sectors are (total specific emissions of $CO_2$ in 2011: 9.3 t/capita):

- industry (2.0 t/capita)
- transport (1.8 t/capita)
- residential, services, agriculture, etc. (2.3 t/capita)
- energy-sector losses (3.2 t/capita).

The energy and emissions data for 2011 correspond to the Japan Annex A5.

Final-energy carriers emit $CO_2$ directly or indirectly (via electricity and district heating). The assumed distribution of these energy carriers within the consumer sectors, for both variants in 2030, is ***one way among others*** to meet the 2-degree condition.

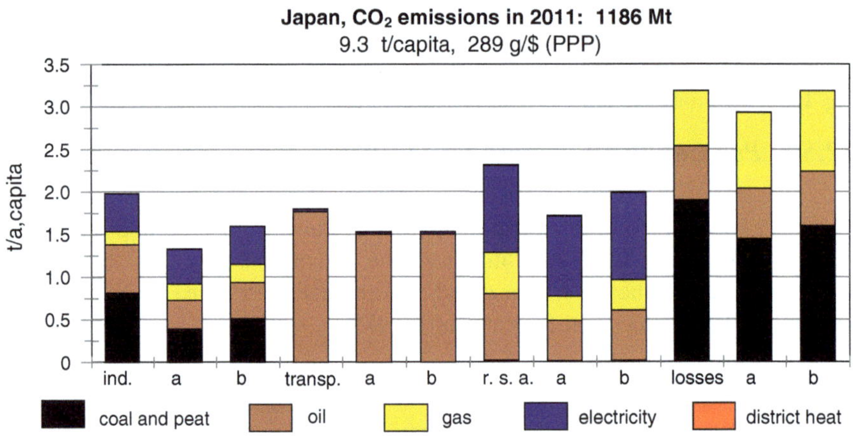


**Fig. 12.31** $CO_2$ emissions per capita of the 4 sectors: industry, transport, residential + services + agriculture, energy-sector losses (definitions see Appendix), each sector represented by three bars: *first bar* 2011 (data in the text and the diagram title, according to the Japan-Annex A5), *second bar* possible 2 °C scenario **variant *a*** for **2030**, 950 Mt, 7.5 t/capita, 183 g/$ (PPP), *third bar* possible 2 °C scenario **variant *b*** for **2030**, 1,050 Mt, 8.3 t/capita, 202 g/$ (PPP)

The target value for 2030 can, for example, be achieved if the per capita emissions:

- in the **electricity and district heat production sector, including energy-sector losses**, are reduced by 8 % (variant *a*) or kept constant (variant *b*) from 4.7 t/capita in 2011 to 4.3–4.7 t/capita (efficiency improvements, replacement of coal by gas, CCS, renewable energy and nuclear power)
- in the **heating sector** (industry, residential, services etc.) through efficiency improvements, reduction of coal consumption, gas instead of coal, and the use of renewable energy sources (heat pump, waste management, solar energy, geothermal energy), are reduced by 20 % (*a*) or by 10 % (*b*) (from 2.1 t/capita to 1.7–1.9 t/capita)
- in the **transport sector** are reduced by 15 % (*a* and *b*) from 1.8 to 1.53 t/capita, through efficiency improvements, gas fuels, bio-fuels and electric mobility.

## 12.6 Germany

### *12.6.1 Actual Indicators (Energy Intensity, $CO_2$ Intensity and $CO_2$ Sustainability) from 1990 to 2011 and Required up to 2030 to meet the 2 °C Target*

The indicator of $CO_2$ sustainability of Germany, at 263 g $CO_2$/$, rates quite well in the context of the G-20, but is above the average of the EU-27. The reason is the electricity production, still too heavily coal-based (see Appendix). The actual figure of the indicators from 1980 to 2011 and the necessary figure for the 2-degree target up to 2030, for the variant *a* and the very similar variant *b* are shown in Fig. 12.32.

The climate target to be aimed at for 2030 is 151–159 g $CO_2$/$ (about −40 % relative to 2011) for Germany's share of 560–590 Mt of total $CO_2$ emissions of the G-20 group of 22,000–25,000 Mt. The values for 2018 take into account the GDP (PPP) forecast by the International Monetary Fund IMF of around 3,100 Bn $ (of 2007). The GDP for 2030 is estimated at 3,700 billion.

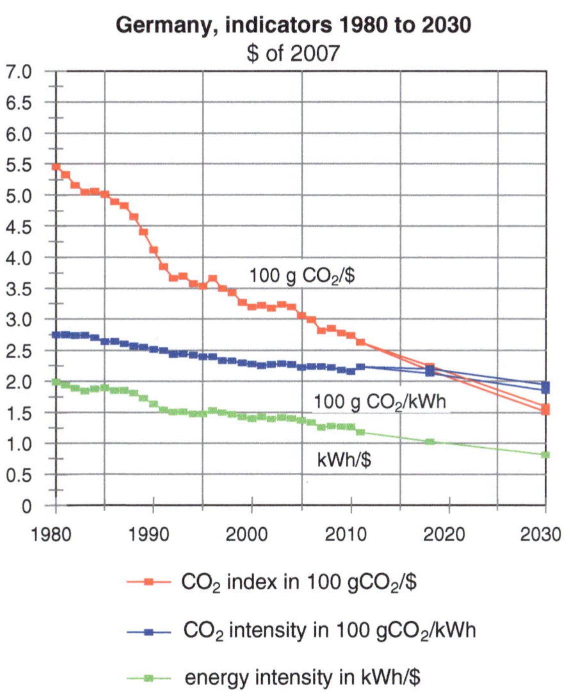


**Fig. 12.32** Indicators of Germany from 1980 to 2011 and 2-degrees scenario up to 2030

The gross energy demand in 2030 with this scenario is 260 Mtoe (only energy use). The per capita indicators for energy and $CO_2$ emissions would then be: e ≈ 4.3 kW/capita (−8 %) and $\alpha$ = 7.0–7.1 t $CO_2$ per capita (on average around −22 % relative to 2011).

The necessary change in the trend of the indicators, relative to the period 2000 to 2011, to meet the 2-degree target is illustrated in Fig. 12.33 for variant *a*. Maintaining and slightly improving the trend of energy efficiency, and a big improvement in the trend of $CO_2$ intensity of energy (−1.2 %/a from 2018) are necessary.

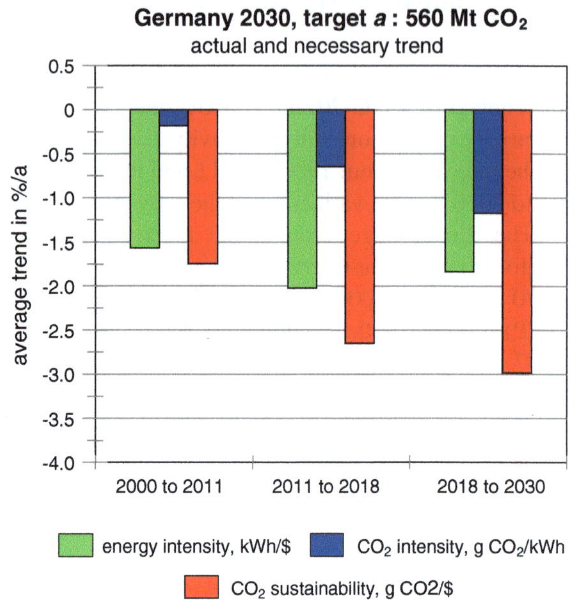


**Fig. 12.33** Necessary trend change to achieve the 2-degrees target, variant *a*

With variant *b* (Fig. 12.34), the trend in $CO_2$ intensity is less strict up to 2018.

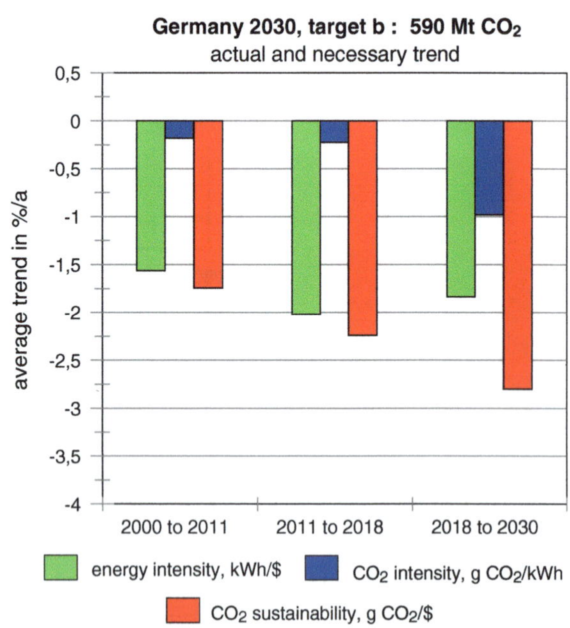


**Fig. 12.34** Necessary trend change to achieve the 2-degrees target, variant *b*

## 12.6.2 $CO_2$ Emissions until 2050

The actual $CO_2$ emissions in Germany from 1970 to 2011, and those allowed to 2050 for compliance with the 2 °C climate target (according to the indicators of Figs. 12.32, 12.33, 12.34) are shown in Fig. 12.35.

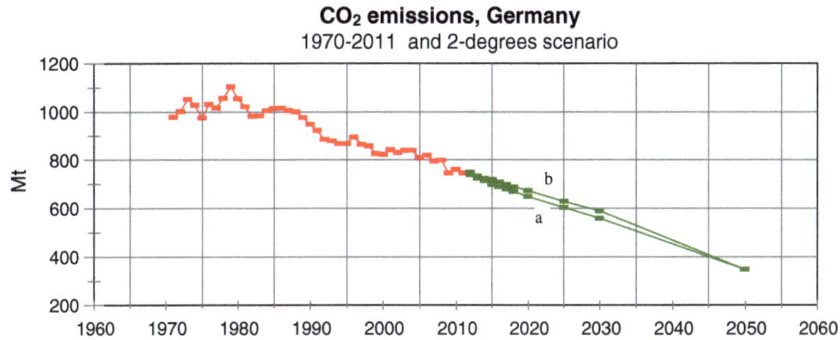


**Fig. 12.35** $CO_2$ emissions of Germany 1970 to 2011 and climate protection scenario up to 2050

## 12.6.3 Per Capita Indicators up to 2030

Figure 12.36 shows the per capita indicators of Germany from 1980 to 2011 as well as the profile up to 2030 resulting from the above considerations to meet the 2-degree target, for both variants *a* and *b*. The GDP (PPP) data up to 2018 correspond to the statistics and forecasts of the IMF.

**Fig. 12.36** Per capita indicators of Germany from 1980 to 2011 and 2-degrees scenario until 2030

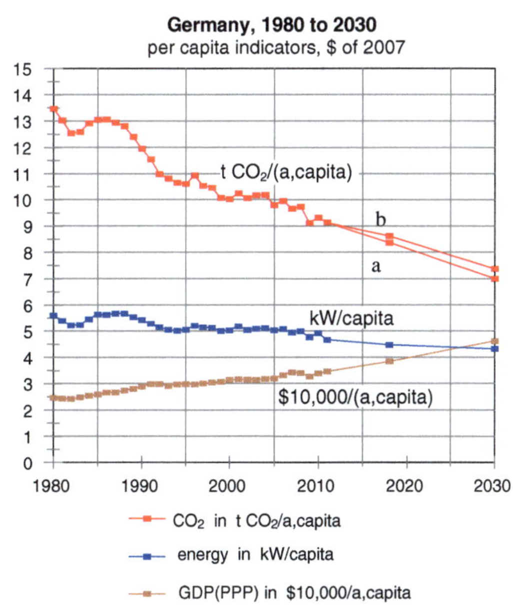

The following Fig. 12.37 shows, in more detail **per consumer sector**, the per capita emissions of the resident population in 2011, and those corresponding to the 2-degree scenario for 2030 (variant *a* and *b*).

The four consumer sectors are (total specific emissions of $CO_2$ in 2011: 9.1 t/capita):

- industry (1.6 t/capita)
- transport (2.0 t/capita)
- residential, services, agriculture, etc. (2.5 t/capita)
- energy-sector losses (3.0 t/capita).

The energy and emissions data for 2011 correspond to the Germany-Annex A5.

Final-energy carriers emit $CO_2$ directly or indirectly (via electricity and district heating). The assumed distribution of these energy carriers within the consumer sectors, for both variants in 2030, is ***one way among others*** to meet the 2-degree condition.

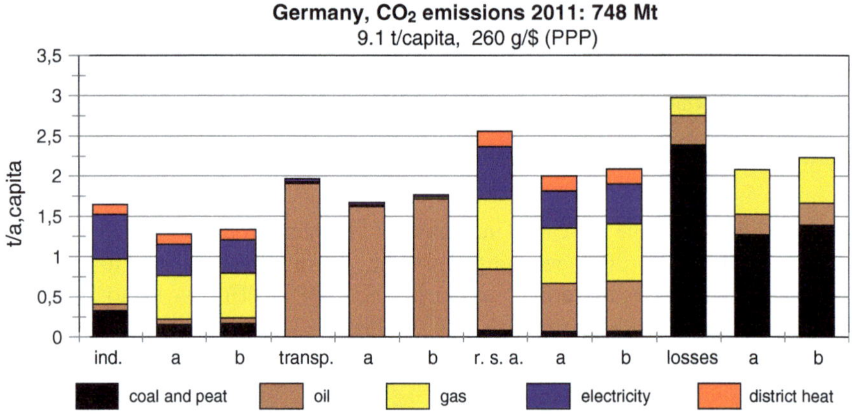


**Fig. 12.37** $CO_2$ emissions per capita of the 4 sectors: industry, transport, residential + services + agriculture, energy-sector losses (definitions see Appendix), each sector represented by three bars: *first bar* 2011 (data in the text and the diagram title, according to the Germany-Annex A5), *second bar* possible 2 °C scenario **variant *a*** for **2030**, 560 Mt, 7.0 t/capita, 151 g/$ (PPP), *third bar* possible 2 °C scenario **variant *b*** for **2030**, 590 Mt, 7.4, t/capita, 159 g/$ (PPP)

The target value for 2030 can, for example, be achieved if the per capita emissions:

- in the **electricity and district heat production sector, including energy-sector losses**, are reduced by 28 % (variant *a*) or 23 % (variant *b*) from 4.5 t/capita in 2011 to 3.3–3.5 t/capita: efficiency improvements, replacement of coal by gas, CCS and renewable energies (wind and solar energy)
- in the **heating sector** (industry, residential, services etc.) through efficiency improvements, reduction of coal consumption, gas instead of coal, and the use of renewable energy sources (heat pump, waste management, solar energy, geothermal energy), are reduced by 21 % (*a*) or by 18 % (*b*) (from 2.7 t/capita to 2.1–2.2 t/capita)
- in the **transport sector** are reduced by 15 % (*a*) or 10 % (*b*), from 2.0 to 1.7–1.8 t/capita, through efficiency improvements, gas fuels, bio-fuels and electric mobility.

## 12.7 South Korea

### *12.7.1 Actual Indicators (Energy Intensity, $CO_2$ Intensity and $CO_2$ Sustainability) from 1990 to 2011 and Required up to 2030 to meet the 2 °C Target*

South Korea had a relatively poor $CO_2$ sustainability indicator in 2011 as OECD member, with 414 g $CO_2$/\$. The reasons is the mainly coal-based electricity production (see also the appendix). The trend of the indicators from 1980 to 2011 and the necessary one for the 2-degree target by 2030, for the variants *a* and *b* are shown in Fig. 12.38.

The climate target to be aimed at for 2030 is 180–212 g $CO_2$/\$ (−56 to −48 % relative to 2011 for a share of 450–530 Mt of the total G-20 $CO_2$ emissions of 22,100–25,200 Mt. The values for 2018 take into account the GDP (PPP) forecast by the International Monetary Fund (IMF) of around \$1,800 billion (\$ of 2007). The GDP for 2030 is estimated at \$2,500 billion.

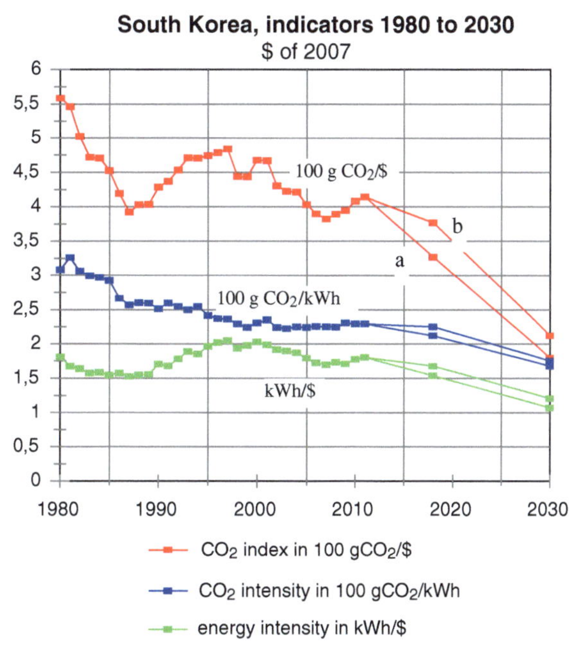


**Fig. 12.38** South Korea indicators 1980 to 2011 and 2-degrees scenario up to 2030

With this scenario, the gross energy demand in 2030 is 230–260 Mtoe. The per capita indicators for energy and $CO_2$ emissions would be: e ≈ 5.9–6.1 kW/capita (unchanged to +13 % relative to 2011) and α = 8.7–10.2 t $CO_2$ per capita (−27 % (variant *a*) to −16 % (variant *b*) relative to 2011).

The necessary change in the trend of the indicators to meet the 2-degree target is illustrated in Fig. 12.39 for variant *a*. A significant improvement of energy efficiency trends (especially compared to the past three years) and overcoming of the stagnation and a decided turnaround in the $CO_2$ intensity of energy are necessary.

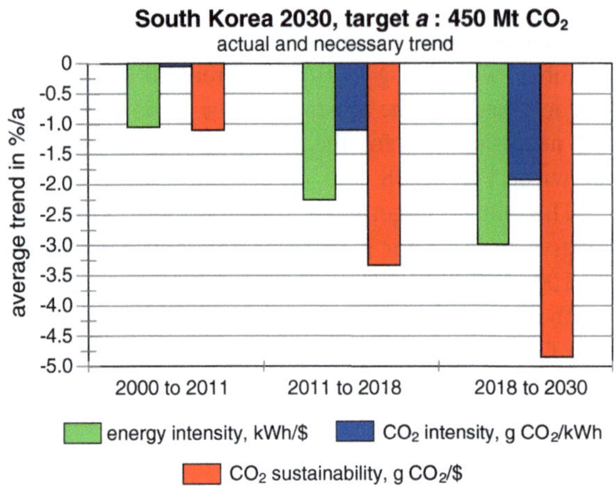


**Fig. 12.39** Necessary trend change to achieve the 2-degrees target, variant *a*

Variant *b* (Fig. 12.40) is, up to 2018, regarding both, the change of trend in energy intensity and $CO_2$ intensity, significantly milder.

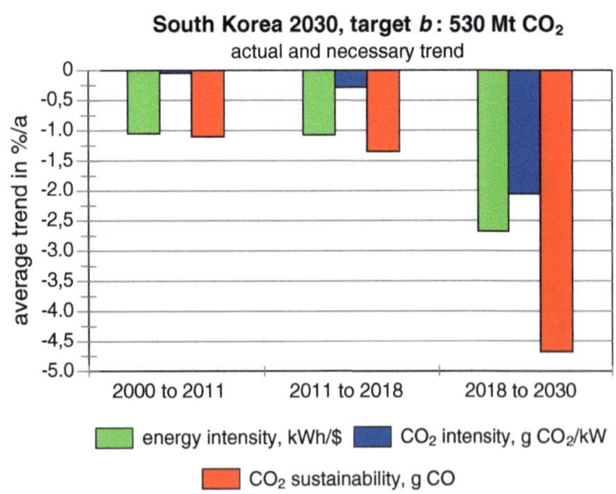


**Fig. 12.40** Necessary trend change to achieve the 2-degrees target, variant *b*

### 12.7.2 $CO_2$ Emissions until 2050

The $CO_2$ emissions of South Korea from 1970 to 2011, and up to 2050 that permit the 2-degree climate target (according to the indicators of Figs. 12.38, 12.39, 12.40), are shown in Fig. 12.41.

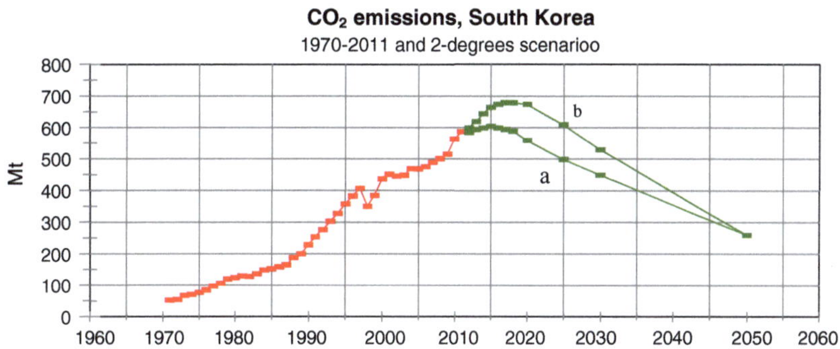


**Fig. 12.41** $CO_2$ emissions of South Korea 1970 to 2011 and 2-degrees scenario until 2050

### 12.7.3 Per Capita Indicators up to 2030

Figure 12.42 shows the per capita indicators of South Korea from 1980 to 2011 as well as the profile up to 2030 resulting from the above considerations to meet the 2-degree target, for both variants *a* and *b*. The GDP (PPP) data up to 2018 correspond to the statistics and forecasts of the IMF.

**Fig. 12.42** Per capita indicators of South Korea from 1980 to 2011 and 2-degrees scenario until 2030

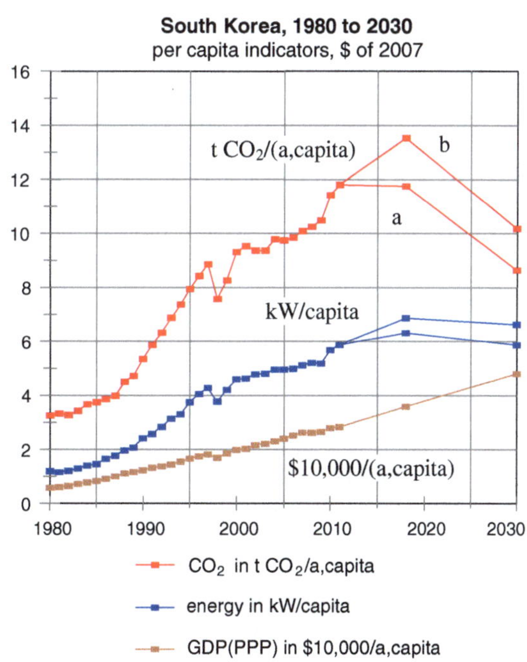

The following Fig. 12.43 shows, in more detail **per consumer sector**, the per capita emissions of the resident population in 2011, and those corresponding to the 2-degree scenario for 2030 (variant *a* and *b*).

The four consumer sectors are (total specific emissions of $CO_2$ in 2011: 11.8 t/capita):

- industry (2.5 t/capita)
- transport (1.8 t/capita)
- residential, services, agriculture, etc. (2.2 t/capita)
- energy-sector losses (5.3 t/capita).

The energy and emissions data for 2011 correspond to the South Korea-Annex A5.

Final-energy carriers emit $CO_2$ directly or indirectly (via electricity and district heating). The assumed distribution of these energy carriers within the consumer sectors, for both variants in 2030, is ***one way among others*** to meet the 2-degree condition.

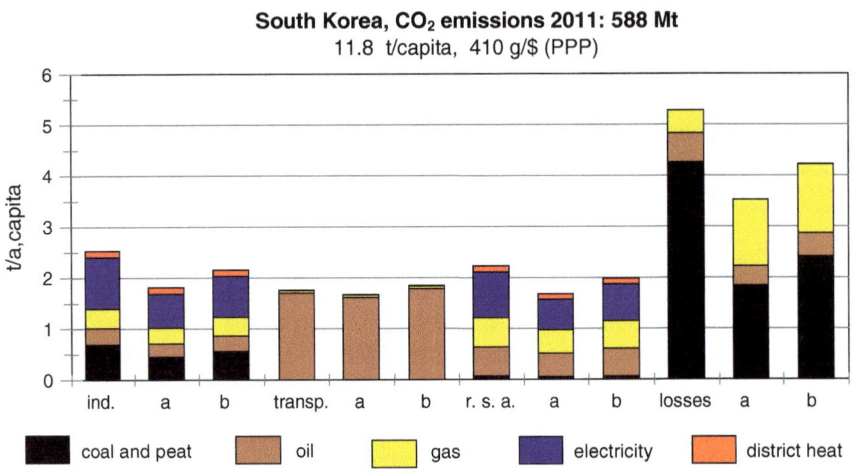


**Fig. 12.43** $CO_2$ emissions per capita of the 4 sectors: industry, transport, residential + services + agriculture, energy-sector losses (definitions see Appendix), each sector represented by three bars: *first bar* 2011 (data in the text and the diagram title, according to the South Korea-Annex A5), *second bar* possible 2 °C scenario **variant *a*** for **2030**, 450 Mt, 8.7 t/capita, 180 g/$ (PPP), *third bar* possible 2 °C scenario **variant *b*** for **2030**, 530 Mt, 10.2 t/capita, 212 g/$ (PPP)

The target value for 2030 can, for example, be achieved if the per capita emissions:

- in the **electricity and district heat production sector, including energy-sector losses**, are reduced by 32 % (variant *a*) or 19 % (variant *b*) from 7.45 t/capita in 2011 to 5.1–6.0 t/capita (efficiency improvements, replacement of coal by gas, CCS, renewable energy and nuclear power)
- in the **heating sector** (industry, residential, services etc.) through efficiency improvements, reduction of coal consumption, gas instead of coal, and the use of renewable energy sources (heat pump, waste management, solar energy, geothermal energy), are reduced by 24 % (*a*) or by 9 % (*b*) (from 2.6 t/capita to 2.0–2.4 t/capita)
- in the **transport sector** are reduced by 5 % (*a*) or increase not more than 5 % (*b*), from 1.75 to 1.65 to 1,85 t/capita, through efficiency improvements, gas fuels, bio-fuels and electric mobility.

## 12.8 Canada

### *12.8.1 Actual Indicators (Energy Intensity, $CO_2$ Intensity and $CO_2$ Sustainability) from 1980 to 2011 and Required up to 2030 to meet the 2 °C Target*

Canada is in 2011, with a $CO_2$ indicator of sustainability of 404 g $CO_2$/\$, slightly below the average of the G-20 countries. Particularly high are the $CO_2$ emissions of the transport sector (see Appendix). The actual graph of the indicators from 1980 to 2011 and the necessary one for the 2-degree target by 2030, for the variants *a* and *b*, are shown in Fig. 12.44.

The climate target to be aimed at for 2030 is 205–226 g $CO_2$/\$ for Canada's share of 390–430 Mt of total $CO_2$ emissions of the G-20 group. The values for 2018 take into account the GDP (PPP) forecast by the International Monetary Fund IMF of around \$1,520 billion (\$ of 2007). The GDP for 2030 is estimated at \$1,900 billion.

**Fig. 12.44** Indicators of Canada from 1980 to 2011 and 2-degrees scenario until 2030

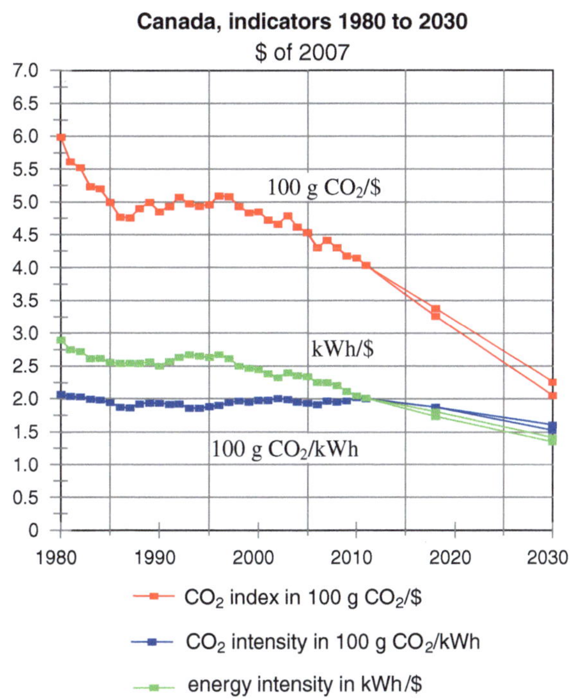


The gross energy demand in 2030 with this scenario is 220–230 Mtoe (only energy use). The per capita indicators for energy and $CO_2$ emissions would be: e ≈ 7.0–7.3 kW/capita: (−21 % (var. *a*) to −17 % (var. *b*) relative to 2011) and $\alpha$ = 9.3–10.2 t $CO_2$ per capita (−40 % (*a*) to −33 % (*b*) relative to 2011).

The necessary change in the trend of the indicators, relative to the period 2000–2011, to meet the 2-degree target is illustrated in Fig. 12.45 for variant *a*. The good energy efficiency trend must be maintained and slightly enhanced. Overcoming stagnation and a significant improvement in the trend in $CO_2$ intensity of energy are required.

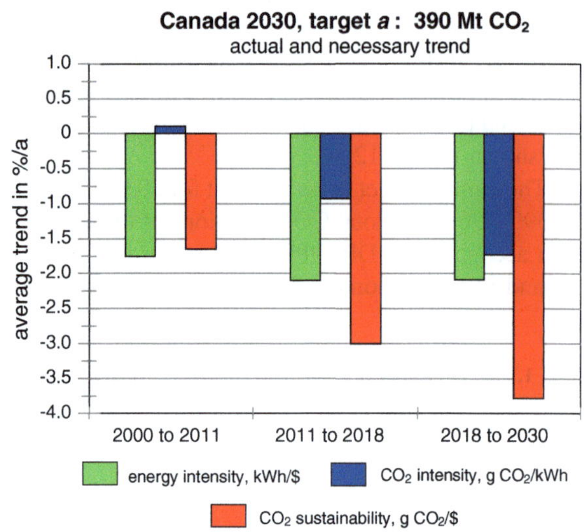


**Fig. 12.45** Necessary trend change to achieve the 2-degrees target, variant *a*

In variant *b* (Fig. 12.46), the trend changes are less strict and occur a little later.

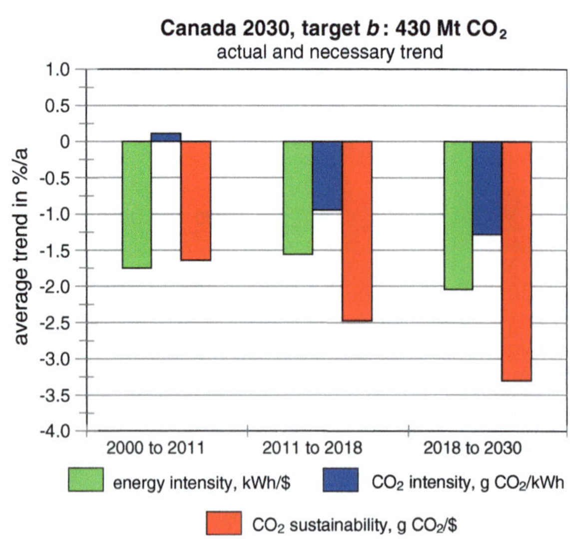


**Fig. 12.46** Necessary trend change to achieve the 2-degrees target, variant *b*

## 12.8.2 $CO_2$ Emissions until 2050

The actual $CO_2$ emissions in Canada from 1970 to 2011, and those allowed to 2050 for compliance with the 2 °C climate target (according to the indicators of Figs. 12.44, 12.45, 12.46) are shown in Fig. 12.47.

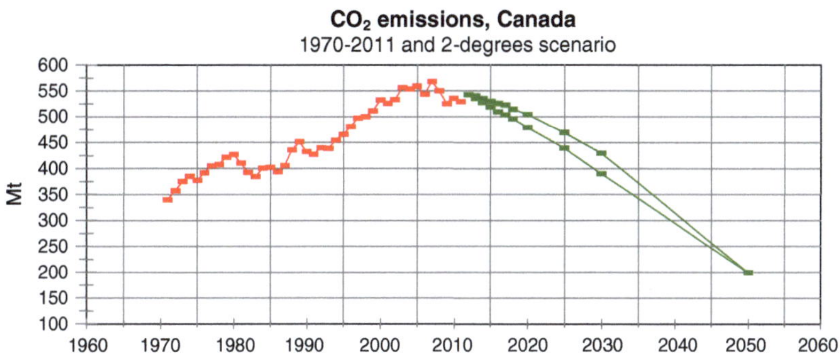


**Fig. 12.47** $CO_2$ emissions of Canada 1970 to 2011 and climate protection scenario to 2050

## 12.8.3 Per Capita Indicators up to 2030

Figure 12.48 shows the per capita indicators of Canada from 1980 to 2011 as well as the profile up to 2030 resulting from the above considerations to meet the 2-degree target, for both variants *a* and *b*. The GDP (PPP) data up to 2018 correspond to the statistics and forecasts of the IMF.

**Fig. 12.48** Per capita indicators of Canada from 1980 to 2011 and 2-degrees scenario until 2030

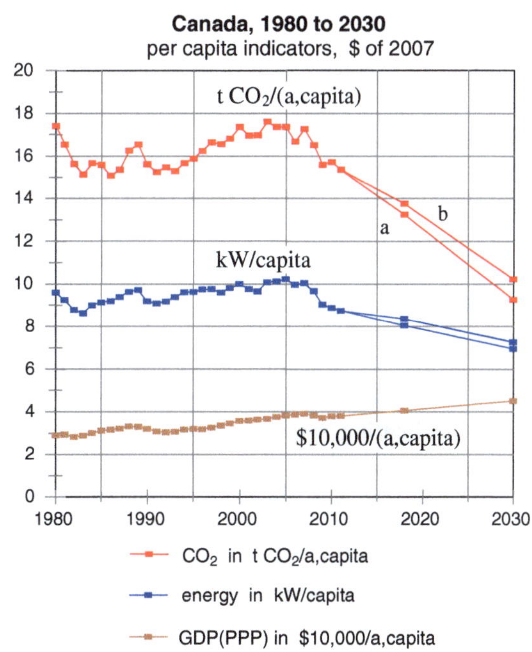

The following Fig. 12.49 shows, in more detail **per consumer sector,** the per capita emissions of the resident population in 2011, and those corresponding to the 2-degree scenario for 2030 (variant *a* and *b*).

The four consumer sectors are (total specific emissions of $CO_2$ in 2011: 15.4 t/capita):

- industry (3.4 t/capita)
- transport (5.9 t/capita)
- residential, services, agriculture, etc. (3.7 t/capita)
- energy-sector losses (2.4 t/capita).

The energy and emissions data for 2011 correspond to the Canada-Annex A5.

Final-energy carriers emit $CO_2$ directly or indirectly (via electricity and district heating). The assumed distribution of these energy carriers within the consumer sectors, for both variants in 2030, is ***one way among others*** to meet the 2-degree condition.

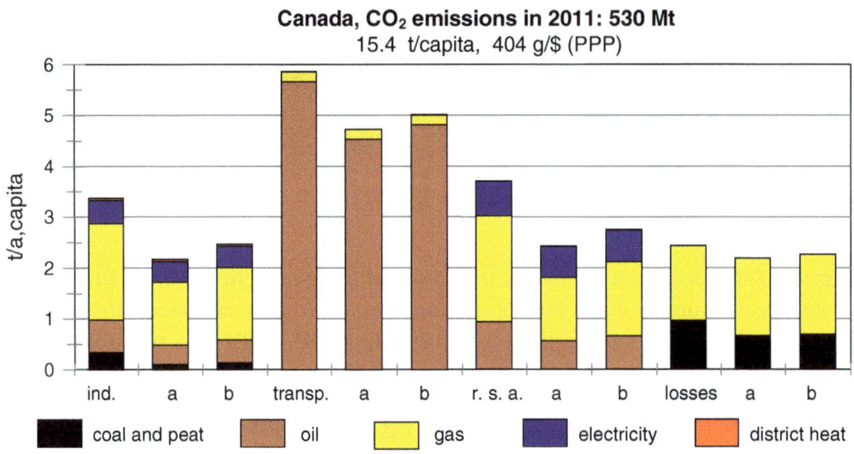


**Fig. 12.49** $CO_2$ emissions per capita of the 4 sectors: industry, transport, residential + services + agriculture, energy-sector losses (definitions see Appendix), each sector represented by three bars: *first bar* 2011 (data in the text and the diagram title, according to the Canada-Annex A5), *second bar* possible 2 °C scenario **variant *a*** for **2030**, 390 Mt, 9.3 t/capita, 205 g/$ (PPP), *third bar* possible 2 °C scenario **variant *b*** for **2030**, 430 Mt, 10.2 t/capita, 226 g/$ (PPP)

The target value for 2030 can, for example, be achieved if the per capita emissions:

- in the **electricity and district heat production sector, including energy-sector losses**, are reduced by 10 % (variant *a*) or 7 % (variant *b*) from 3.6 t/capita in 2011 to 3.3–3.4 t/capita (efficiency improvements, replacement of coal by gas, CCS, renewable energy and nuclear power)
- in the **heating sector** (industry, residential, services etc.) through efficiency improvements (better insulation), reduction of coal consumption, gas instead of coal and oil, and the use of renewable energy sources (heat pump, waste management, solar energy, geothermal energy), are reduced by 40 % (*a*) or 30 % (*b*) (from 5.9 t/capita to 3.5–4.1 t/capita)
- in the **transport sector** are reduced by 20 % (*a*) or 15 % (*b*), from 5.9 in 2011 to 4.7–5.0 t/capita, through efficiency improvements, gas fuels, bio-fuels and electric mobility.

## 12.9 Saudi Arabia

### *12.9.1 Actual Indicators (Energy Intensity, $CO_2$ Intensity and $CO_2$ Sustainability) from 1990 to 2011 and Required up to 2030 to meet the 2 °C Target*

In 2011, Saudi Arabia had a $CO_2$ sustainability indicator of 601 g $CO_2$/\$, putting it in the fourth-last position of the G-20 group. The reasons are the energy system based only on oil and the poor energy efficiency (see also the appendix). The trend of the indicators from 1990 to 2011 and the necessary one for the 2-degree target by 2030, for the variants *a* and *b* are shown in Fig. 12.50.

The climate target to be aimed at for 2030 is 300–353 g $CO_2$/\$ (−49 to −41 % relative to 2011) for a share of 450–530 Mt (−2 % (var. *a*) to +16 % (var. *b*) relative to 2011) of the total G-20 $CO_2$ emissions. The values for 2018 take into account the GDP (PPP) forecast by the International Monetary Fund (IMF) of around \$1,020 billion (\$ of 2007). The GDP for 2030 is estimated at \$1,500 billion.

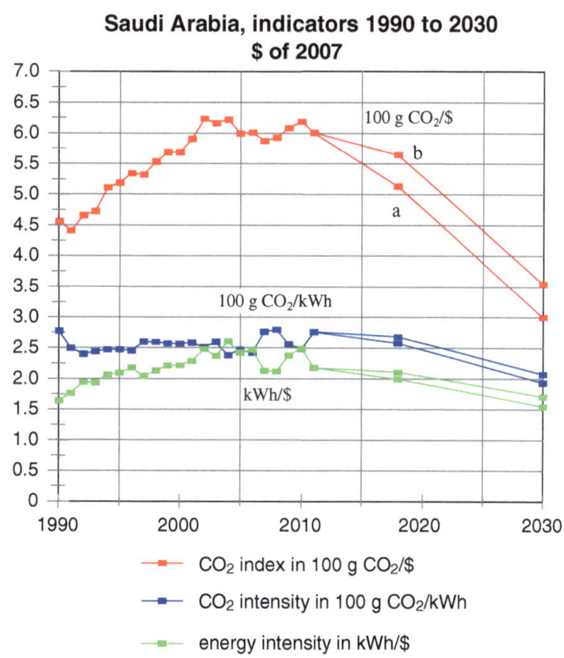


**Fig. 12.50** Saudi Arabia indicators 1990 to 2011 and 2-degrees scenario up to 2030

With this scenario, the gross energy demand in 2030 is 200–220 Mtoe. The per capita indicators for energy and $CO_2$ emissions would be: e ≈ 7.6–8.3 kW/capita (increase of 13–24 % relative to 2011) and $\alpha$ = 12.9–15.1 t $CO_2$ per capita (−21 % (var. *a*) to −7 % (var. *b*) relative to 2011).

The necessary change in the trend of the indicators relative to the period 2000–2011, to meet the 2-degree target is illustrated in Fig. 12.51 for variant *a*. A significant improvement of energy efficiency trends and a decided turnaround in the $CO_2$ intensity of energy are urgently needed.

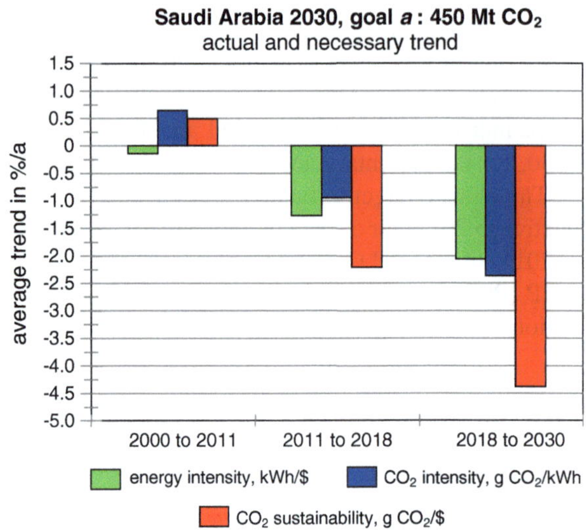


**Fig. 12.51** Necessary trend change to achieve the 2-degrees target, variant *a*

With variant *b* (Fig. 12.52) the trend improvement is significantly less strict, especially before 2018.

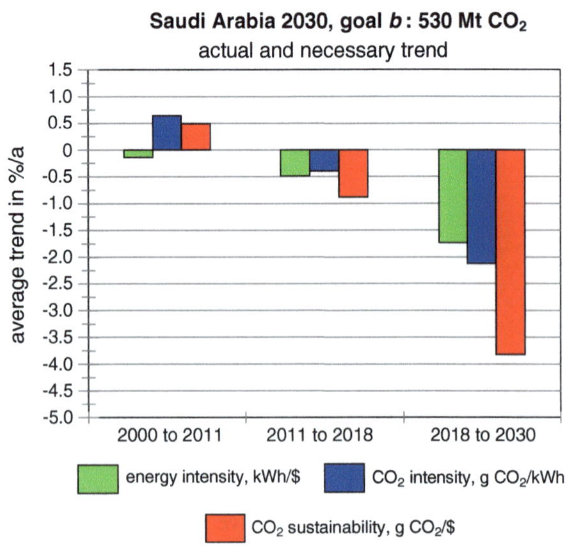


**Fig. 12.52** Necessary trend change to achieve the 2-degrees target, variant *b*

## 12.9.2 $CO_2$ Emissions until 2050

The actual $CO_2$ emissions in Saudi Arabia from 1980 to 2011, and those allowed to 2050 for compliance with the 2 °C climate target (according to the indicators of Figs. 12.50, 12.51, 12.52) are shown in Fig. 12.53.

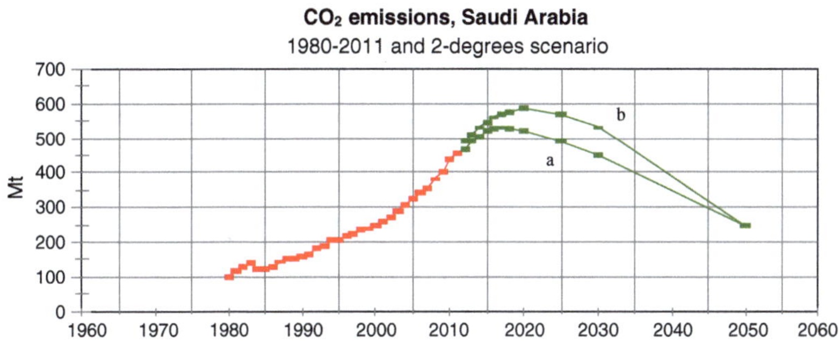


**Fig. 12.53** $CO_2$ emissions of Saudi Arabia 1980 to 2011 and 2-degrees scenario until 2050

## 12.9.3 Per Capita Indicators up to 2030

Figure 12.54 shows the per capita indicators of Saudi Arabia from 1980 to 2011 as well as the profile up to 2030 resulting from the above considerations to meet the 2-degree target, for both variants *a* and *b*. The GDP (PPP) data up to 2018 correspond to the statistics and forecasts of the IMF.

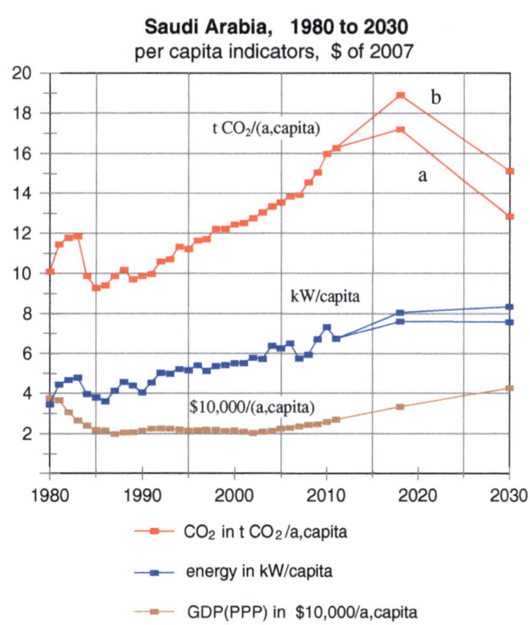


**Fig. 12.54** Per capita indicators of Saudi Arabia from 1980 to 2011 and 2-degrees scenario until 2030

The following Fig. 12.55 shows, in more detail **per consumer sector**, the per capita emissions of the resident population in 2011, and those corresponding to the 2-degree scenario for 2030 (variant *a* and *b*).

The four consumer sectors are (total specific emissions of $CO_2$ in 2011: 16.3 t/capita):

- industry (2.4 t/capita)
- transport (4.6 t/capita)
- residential, services, agriculture, etc. (1.9 t/capita)
- energy-sector losses (7.4 t/capita).

The energy and emissions data for 2011 correspond to the Saudi Arabia-Annex A5.

Final-energy carriers emit $CO_2$ directly or indirectly (via electricity and district heating). The assumed distribution of these energy carriers within the consumer sectors, for both variants in 2030, is ***one way among others*** to meet the 2-degree condition.

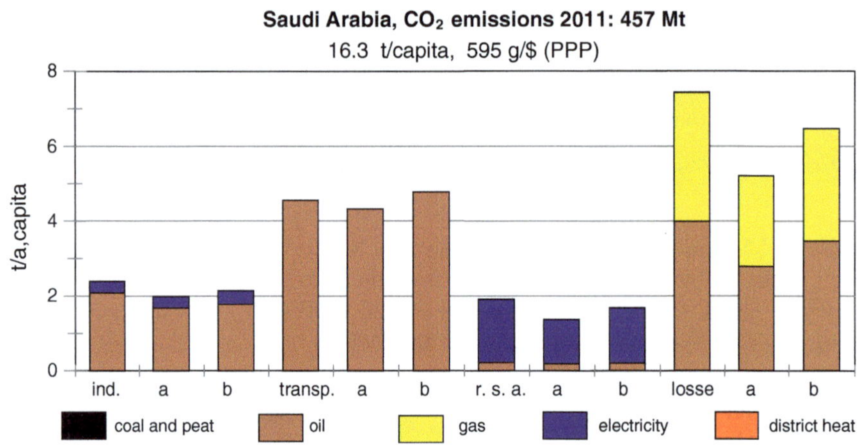


**Fig. 12.55** $CO_2$ emissions per capita of the 4 sectors: industry, transport, residential + services + agriculture, energy-sector losses (definitions see Appendix), each sector represented by three bars: *first bar* 2011 (data in the text and the diagram title, according to the Saudi Arabia-Annex A5), *second bar* possible 2 °C scenario **variant *a*** for **2030**, 450 Mt, 12.9 t/capita, 300 g/$ (PPP), *third bar* possible 2 °C scenario **variant *b*** for **2030**, 530 Mt, 15.1 t/capita, 353 g/$ (PPP)

The target value for 2030 can, for example, be achieved if the per capita emissions:

- in the **electricity and district heat production sector, including energy-sector losses**, are reduced by 45 % (variant *a*) or 14 % (variant *b*) from 9.4 t/capita in 2011 to 5.2–8.3 t/capita (efficiency improvements, replacement of oil by gas, CCS, renewable energy and nuclear power)
- in the **heating sector** (industry, residential, services etc.) through efficiency improvements (better insulation), gas instead of oil, use of renewable energy sources (heat pump, waste management, solar energy, geothermal energy), are reduced by 19 % (*a*) or 14 % (*b*) (from 2.3 t/capita to 1.9–2.0 t/capita)
- in the **transport sector** are reduced by 5 % (*a*) or increase but not more than 5 % (*b*), from 4.6 to 4.3–4.8 t/capita, through efficiency improvements, gas fuels, bio-fuels and electric mobility.

## 12.10 United Kingdom

### 12.10.1 Actual Indicators (Energy Intensity, $CO_2$ Intensity and $CO_2$ Sustainability) from 1990 to 2011 and Required up to 2030 to meet the 2 °C Target

In 2011 the United Kingdom had, with 212 g $CO_2$/\$, the third best ranking in the G-20 group, thanks above all to the good energy efficiency. The trend of the indicators from 1980 to 2011 and the necessary one for the 2-degree target by 2030, for variants *a* and *b* are shown in Fig. 12.56.

The climate target to be aimed at for 2030 is 118–129 g $CO_2$/\$ (−44 to −39 % relative to 2011) for a share of 330–360 Mt (−36 to −19 % relative to 2011) of the total G-20 $CO_2$ emissions. The values for 2018 take into account the GDP (PPP) forecast by the International Monetary Fund (IMF) of around \$2,350 billion (\$ of 2007). The GDP for 2030 is estimated at \$2,800 billion.

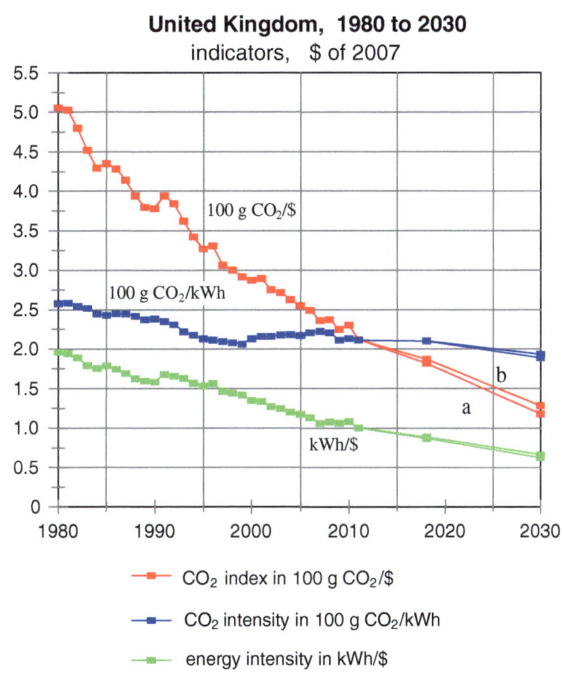


**Fig. 12.56** U.K. indicators 1980 to 2010 and climate protection scenario until 2030

With this scenario, the gross energy demand in 2030 is 150–160 Mtoe (only energy use). The per capita indicators for energy and $CO_2$ emissions would be: e ≈ 2.8–3.0 kW/capita (−26 to −22 % relative to 2011) and $\alpha$ = 4.7–5.1 t $CO_2$ per capita (−34 to −28 % relative to 2011).

The change in the trend of the indicators to be aimed at, to meet the 2-degree target is illustrated in Fig. 12.57 for variant *a*. It is necessary to maintain the good energy efficiency trend and from 2018 to overcome the stagnation in the trend of the $CO_2$ intensity of energy.

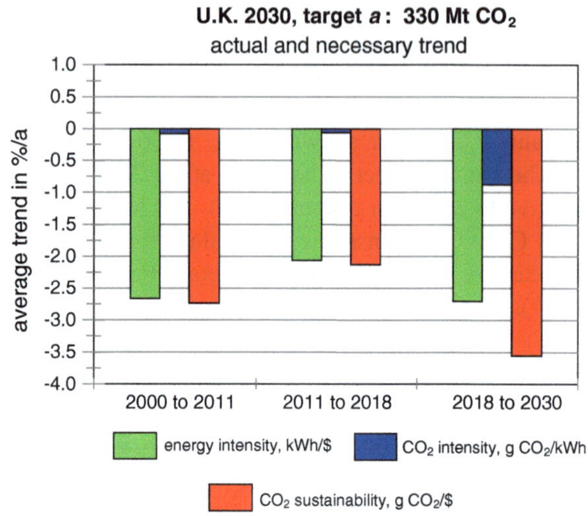


**Fig. 12.57** Necessary trend change to achieve the 2-degrees target, variant *a*

In variant *b* (Fig. 12.58) the change of trend both in terms of energy intensity and $CO_2$ intensity is slightly milder.

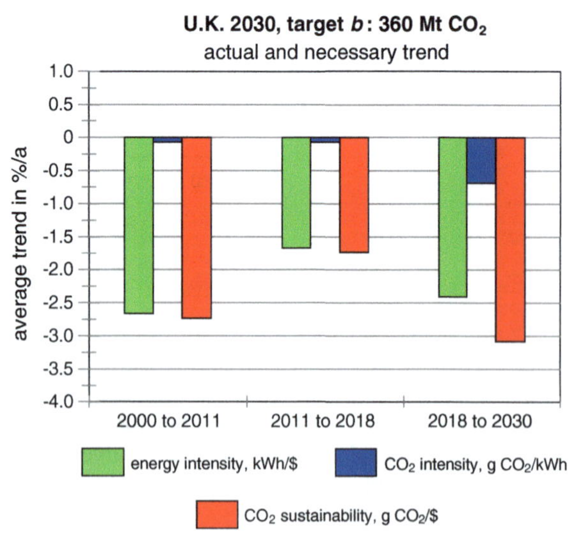


**Fig. 12.58** Necessary trend change to achieve the 2-degrees target, variant *b*

## 12.10.2 $CO_2$ Emissions until 2050

The actual $CO_2$ emissions in the United Kingdom from 1970 to 2011, and those allowed to 2050 for compliance with the 2 °C climate target (according to the indicators of Figs. 12.56, 12.57, 12.58) are shown in Fig. 12.59.

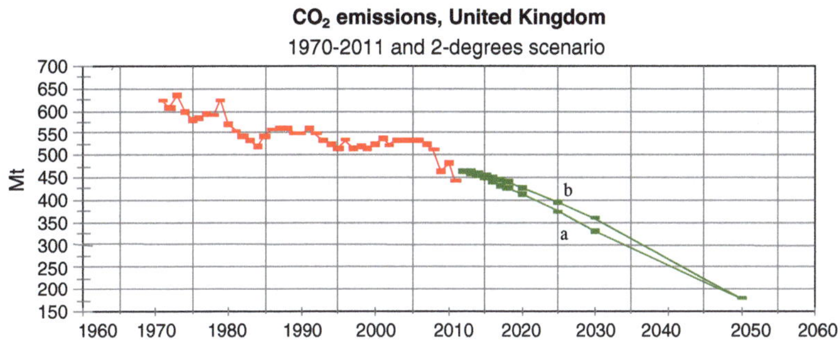


**Fig. 12.59** $CO_2$ emissions of the United Kingdom 1970 to 2011 and 2-degrees scenario until 2050

## 12.10.3 Per Capita Indicators up to 2030

Figure 12.60 shows the per capita indicators of the United Kingdom from 1980 to 2011 as well as the profile up to 2030 resulting from the above considerations to meet the 2-degree target, for both variants *a* and *b*. The GDP (PPP) data up to 2018 correspond to the statistics and forecasts of the IMF.

**Fig. 12.60** Per capita indicators of the U.K. from 1980 to 2011 and 2-degrees scenario until 2030

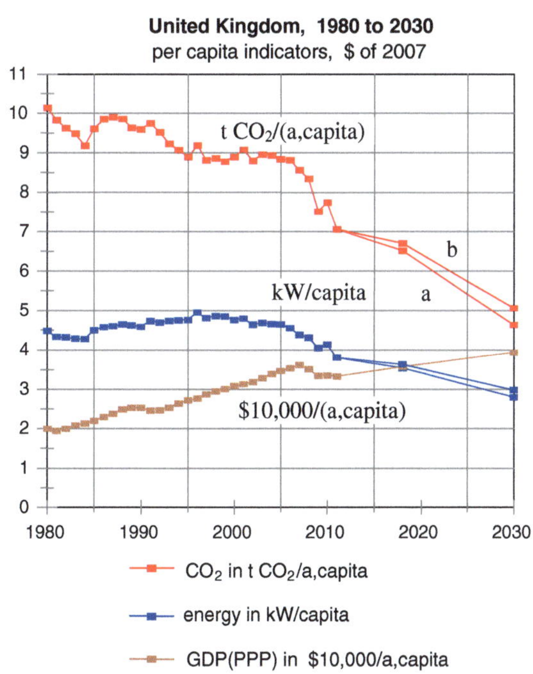

The following Fig. 12.61 shows, in more detail **per consumer sector**, the per capita emissions of the resident population in 2011, and those corresponding to the 2-degree scenario for 2030 (variant *a* and *b*).

The four consumer sectors are (total specific emissions of $CO_2$ in 2011: 7.1 t/capita):

- industry (1.0 t/capita)
- transport (1.9 t/capita)
- residential, services, agriculture, etc. (1.9 t/capita)
- energy-sector losses (2.3 t/capita).

The energy and emissions data for 2011 correspond to the U.K.-Annex A5.

Final-energy carriers emit $CO_2$ directly or indirectly (via electricity and district heating). The assumed distribution of these energy carriers within the consumer sectors, for both variants in 2030, is ***one way among others*** to meet the 2-degree condition.

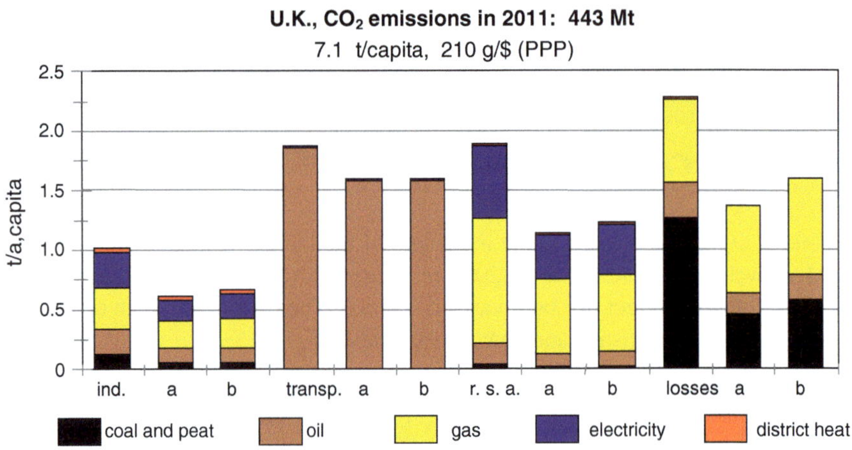


**Fig. 12.61** $CO_2$ emissions per capita of the 4 sectors: industry, transport, residential + services + agriculture, energy-sector losses (definitions see Appendix), each sector represented by three bars: *first bar* 2011 (data in the text and the diagram title, according to the U.K.-Annex A5), *second bar* possible 2 °C scenario **variant *a*** for **2030**, 330 Mt, 4.7 t/capita, 118 g/$ (PPP), *third bar* possible 2 °C scenario **variant *b*** for **2030**, 360 Mt, 5.1 t/capita, 129 g/$ (PPP)

The target value for 2030 can, for example, be achieved if the per capita emissions:

- in the **electricity and district heat production sector, including energy-sector losses**, are reduced by 40 % (variant *a*) or 30 % (variant *b*) from 3.3 t/capita in 2011 to 2.0–2.3 t/capita (efficiency improvements, replacement of coal by gas, CCS, renewable energy and nuclear power)
- in the **heating sector** (industry, residential, services etc.) through efficiency improvements (better insulation), reduction of coal consumption, gas instead of coal, and the use of renewable energy sources (heat pump, waste management, solar energy, geothermal energy), are reduced by 40 % (*a*) or 38 % (*b*) (from 1.9 t/capita to around 1.2 t/capita)
- in the **transport sector** are reduced by 15 % (*a* and *b*) from 1.9 to 1.6 t/capita, through efficiency improvements, gas fuels, bio-fuels and electric mobility.

# 12.11 Mexico

## *12.11.1 Actual Indicators (Energy Intensity, $CO_2$ Intensity and $CO_2$ Sustainability) from 1990 to 2011 and Required up to 2030 to meet the 2 °C Target*

Mexico has in 2011, with 275 g $CO_2$/$, a $CO_2$ sustainability indicator below the OECD average, thanks to the highly gas-oriented electricity production. The trend of the indicators from 1980 to 2011 and the necessary one for the 2-degree target by 2030, for the variants *a* and *b* are shown in Fig. 12.62.

The climate target to be aimed at for 2030 is 204–248 g $CO_2$/$ (−26 to −10 % relative to 2011) for a share of 510 to 620 Mt (+22 to +44 % relative to 2011) of the total G-20 $CO_2$ emissions. The values for 2018 take into account the GDP (PPP) forecast by the International Monetary Fund (IMF) of around $1,960 billion ($ of 2007). The GDP for 2030 is estimated at $2,500 billion.

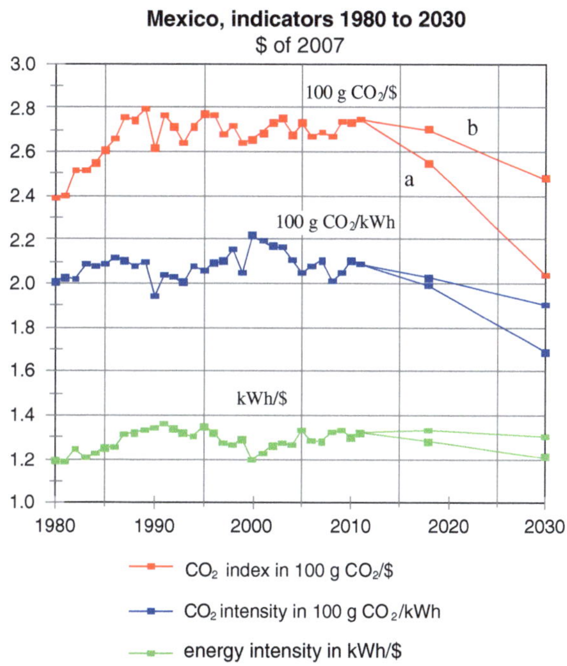


**Fig. 12.62** Mexico indicators 1990 to 2011 and 2-degrees scenario up to 2030

With this scenario, the gross energy demand in 2030 is 260–280 Mtoe. The per capita indicators for energy and $CO_2$ emissions would be: e ≈ 2.8–3.0 kW/capita (+28 to +37 % relative to 2011) and $\alpha$ = 4.1–5.0 t $CO_2$ per capita (+3 % (var. *a*) to +25 % (var. *b*) relative to 2011).

The necessary change in the trend of the indicators, relative to the period 2000–2011, to meet the 2-degree target is illustrated in Fig. 12.63 for variant *a*. A significant turnaround in energy intensity and the strengthening of the trend of the $CO_2$ intensity of energy (−1.3 %/a from 2018) are necessary.

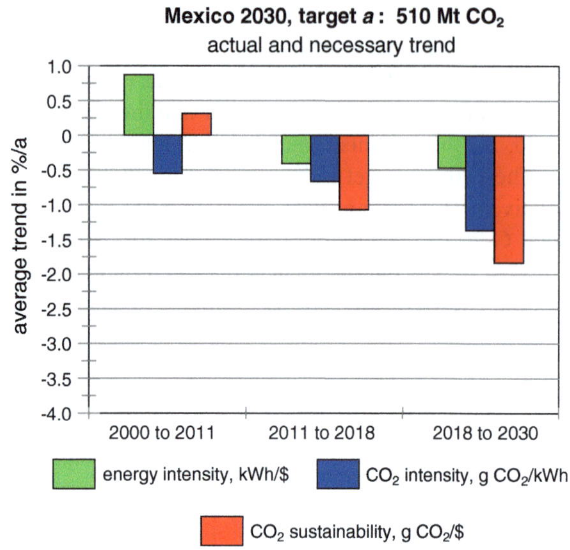


**Fig. 12.63** Necessary trend change to meet the 2-degrees target, variant *a*

With variant *b* (Fig. 12.64) the reversal of the trend in energy intensity is still necessary but more gentle, and maintaining the $CO_2$ intensity trend is sufficient.

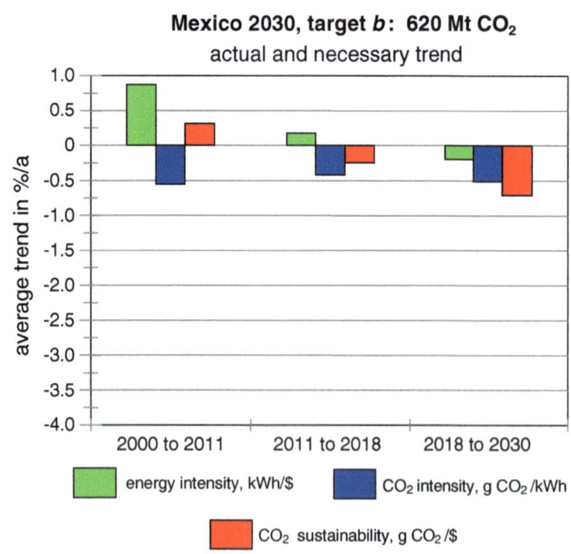


**Fig. 12.64** Necessary trend change to meet the 2-degrees target, variant *b*

## 12.11.2 $CO_2$ Emissions until 2050

The actual $CO_2$ emissions in Mexico from 1970 to 2011, and those allowed to 2050 for compliance with the 2 °C climate target (according to the indicators of Figs. 12.62, 12.63, 12.64) are shown in Fig. 12.65.

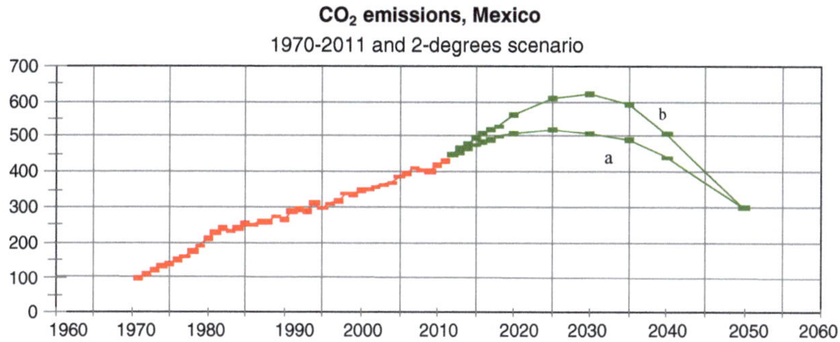


**Fig. 12.65** $CO_2$ emissions of Mexico from 1970 to 2011 and climate protection scenario until 2050

## 12.11.3 Per Capita Indicators up to 2030

Figure 12.66 shows the per capita indicators of Mexico from 1980 to 2011 as well as the profile up to 2030 resulting from the above considerations to meet the 2-degree target, for both variants *a* and *b*. The GDP (PPP) data up to 2018 correspond to the statistics and forecasts of the IMF.

**Fig. 12.66** Per capita indicators of Mexico from 1980 to 2011 and 2-degrees scenario until 2030

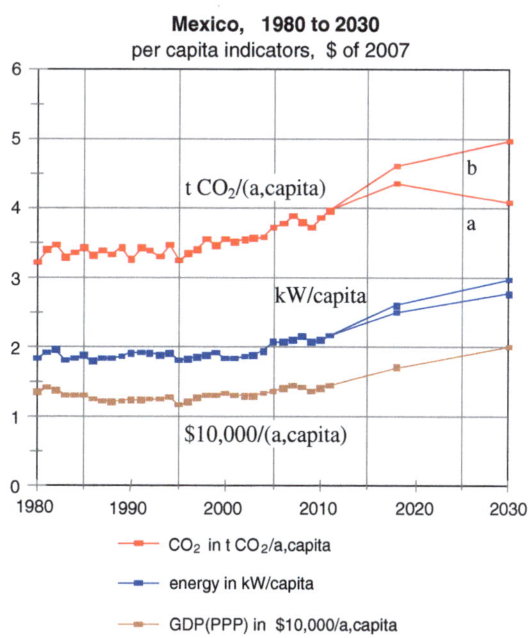

The following Fig. 12.67 shows, in more detail **per consumer sector**, the per capita emissions of the resident population in 2011, and those corresponding to the 2-degree scenario for 2030 (variant *a* and *b*).

The four consumer sectors are (total specific emissions of $CO_2$ in 2011: 4.0 t/capita):

- industry (0.6 t/capita)
- transport (1.4 t/capita)
- residential, services, agriculture, etc. (0.5 t/capita)
- energy-sector losses (1.5 t/capita).

The energy and emissions data for 2011 correspond to the Mexico-Annex A5.

Final-energy carriers emit $CO_2$ directly or indirectly (via electricity and district heating). The assumed distribution of these energy carriers within the consumer sectors, for both variants in 2030, is ***one way among others*** to meet the 2-degree condition.

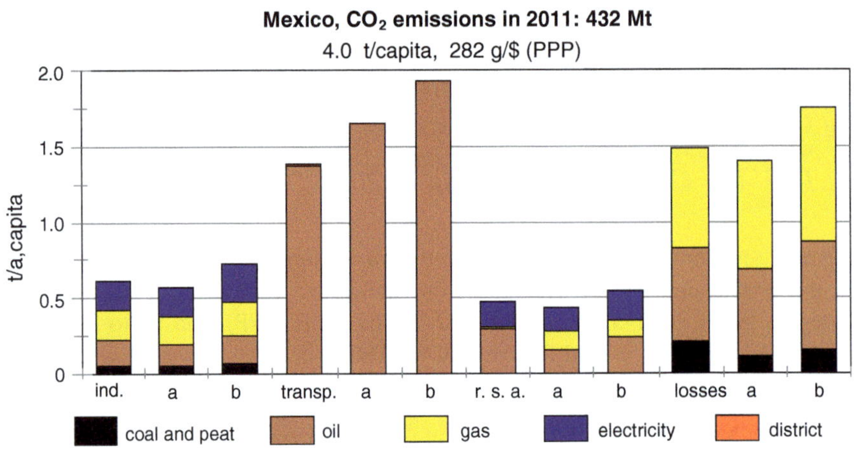


**Fig. 12.67** $CO_2$ emissions per capita of the 4 sectors: industry, transport, residential + services + agriculture, energy-sector losses (definitions see Appendix), each sector represented by three bars: *first bar* 2011 (data in the text and the diagram title, according to the Mexico-Annex A5), *second bar* possible 2 °C scenario **variant *a*** for **2030**, 510 Mt, 4.1 t/capita, 204 g/$ (PPP). Third bar: possible 2 °C scenario **variant *b*** for **2030**, 620 Mt, 5.0 t/capita, 248 g/$ (PPP)

The target value for 2030 can, for example, be achieved if the per capita emissions:

- in the **electricity and district heat production sector, including energy-sector losses**, are reduced by 5 % (variant *a*) or increase by not more than 30 % (variant *b*) from 1.9 t/capita in 2011 to 1.8–2.2 t/capita (efficiency improvements, replacement of coal by gas, CCS, renewable energy and nuclear power)
- in the **heating sector** (industry, residential, services etc.) through efficiency improvements, reduction of coal consumption, gas instead of coal, and the use of renewable energy sources (heat pump, waste management, solar energy, geothermal energy), are reduced by 11 % (*a*) or increase by not more than 15 % (*b*) (from 0.73 t/capita to 0.65–0.84 t/capita)
- in the **transport sector** increase by 22 % (*a*) but not more than 40 % (*b*), from 1.4 in 2011 to 1.7–1.9 t/capita, through efficiency improvements, gas fuels, biofuels and electric mobility.

## 12.12 Indonesia

### *12.12.1 Actual Indicators (Energy Intensity, $CO_2$ Intensity and $CO_2$ Sustainability) from 1990 to 2011 and Required up to 2030 to meet the 2 °C Target*

In 2011, Indonesia had a $CO_2$ sustainability indicator of 415 g $CO_2$/\$, corresponding to about the mean value of the G-20 group. This is despite a high consumption of fossil fuels for electricity production, but also thanks to geothermal energy and a high biomass share in the heat sector (see Appendix). The trend of the indicators from 1980 to 2011 and the necessary one for the 2-degree target by 2030, for variants *a* and *b* are shown in Fig. 12.68.

The climate target to be aimed at for 2030 is 221–263 g $CO_2$/\$ (−47 to −37 % relative to 2011) for Indonesia to have a share of 530–630 Mt (+24 to +48 % relative to 2011) of the total of G-20 $CO_2$ emissions. The values for 2018 take into account the GDP (PPP) forecast by the International Monetary Fund (IMF) of around \$1,530 billion (\$ of 2007). The GDP for 2030 is estimated at \$2,400 billion.

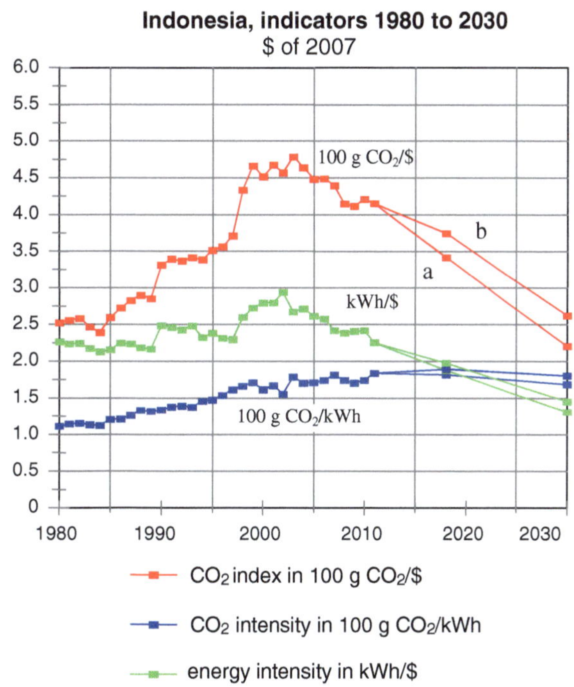


**Fig. 12.68** Indonesia indicators 1980 to 2011 and 2-degrees scenario up to 2030

With this scenario, the gross energy demand in 2030 is 270–300 Mtoe. The per capita indicators for energy and $CO_2$ emissions would be: e ≈ 1.2–1.3 kW/capita (+11 to +22 % relative to 2011) and $\alpha$ = 1.8–2.1 t $CO_2$ per capita (unchanged (var. *a*) to +19 % (var. *b*) relative to 2011).

The necessary change in the trend of the indicators to meet the 2-degree target is illustrated in Fig. 12.69 for variant *a*. A significant improvement of energy efficiency trends and a decided turnaround in the $CO_2$ intensity of energy are necessary.

**Fig. 12.69** Necessary trend change to achieve the 2-degrees target, variant *a*

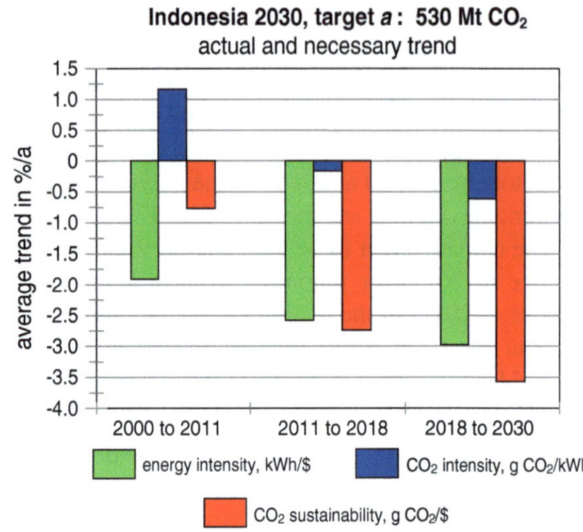


With variant *b* (Fig. 12.70), the improvement of energy intensity trends can occur a little later and the turnaround in the $CO_2$ intensity is less steep.

**Fig. 12.70** Necessary trend change to achieve the 2-degrees target, variant *b*

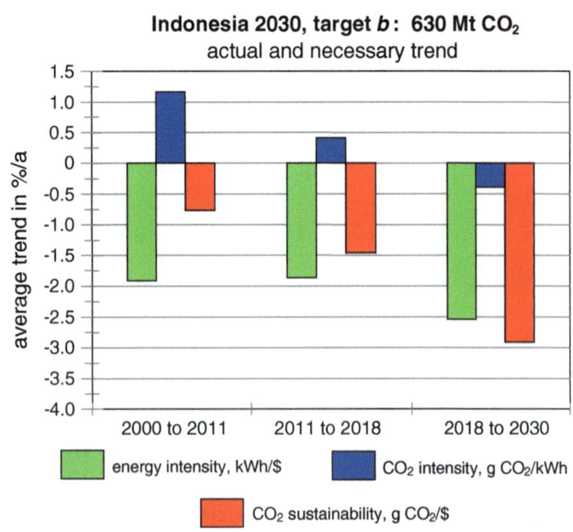

## 12.12.2 $CO_2$ Emissions until 2050

The actual $CO_2$ emissions in Indonesia from 1970 to 2011, and those allowed to 2050 for compliance with the 2 °C climate target (according to the indicators of Figs. 12.68, 12.69, 12.70) are shown in Fig. 12.71.

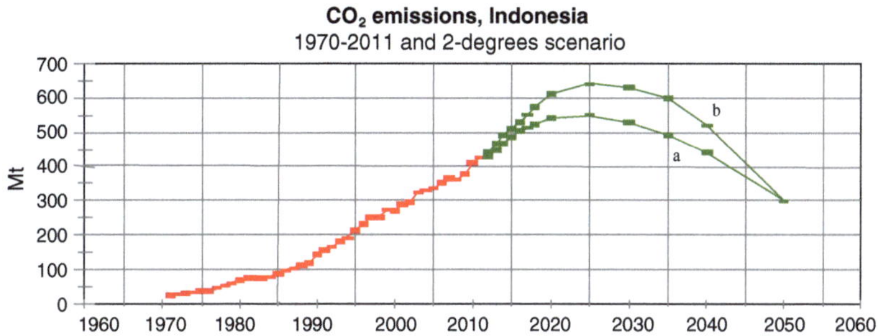


**Fig. 12.71** $CO_2$ emissions of Indonesia from 1970 to 2011 and climate protection scenario until 2050

## 12.12.3 Per Capita Indicators up to 2030

Figure 12.72 shows the per capita indicators of Indonesia from 1980 to 2011 as well as the profile up to 2030 resulting from the above considerations to meet the 2-degree target, for both variants *a* and *b*. The GDP (PPP) data up to 2018 correspond to the statistics and forecasts of the IMF.

**Fig. 12.72** Per capita indicators of Indonesia from 1980 to 2011 and 2-degrees scenario until 2030

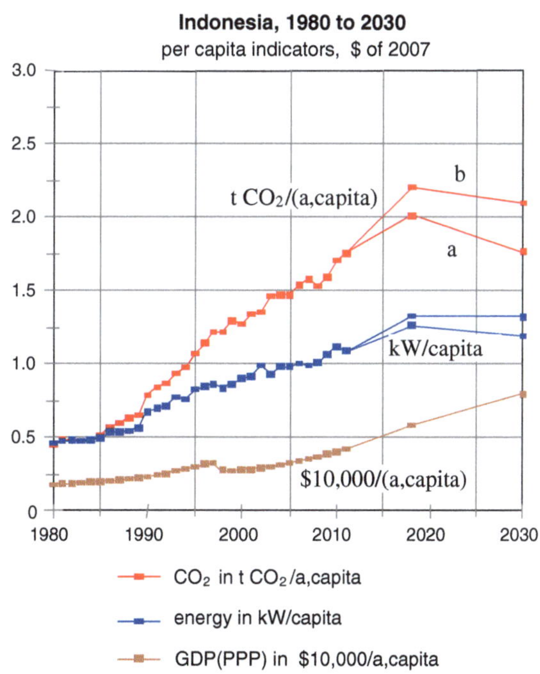

The following Fig. 12.73 shows, in more detail **per consumer sector**, the per capita emissions of the resident population in 2011, and those corresponding to the 2-degree scenario for 2030 (variant *a* and *b*).

The four consumer sectors are (total specific emissions of $CO_2$ in 2011: 1.8 t/capita):

- industry (0.5 t/capita)
- transport (0.5 t/capita)
- residential, services, agriculture, etc. (0.3 t/capita)
- energy-sector losses (0.5 t/capita).

The energy and emissions data for 2011 correspond to the Indonesia-Annex A5.

Final-energy carriers emit $CO_2$ directly or indirectly (via electricity and district heating). The assumed distribution of these energy carriers within the consumer sectors, for both variants in 2030, is ***one way among others*** to meet the 2-degree condition.

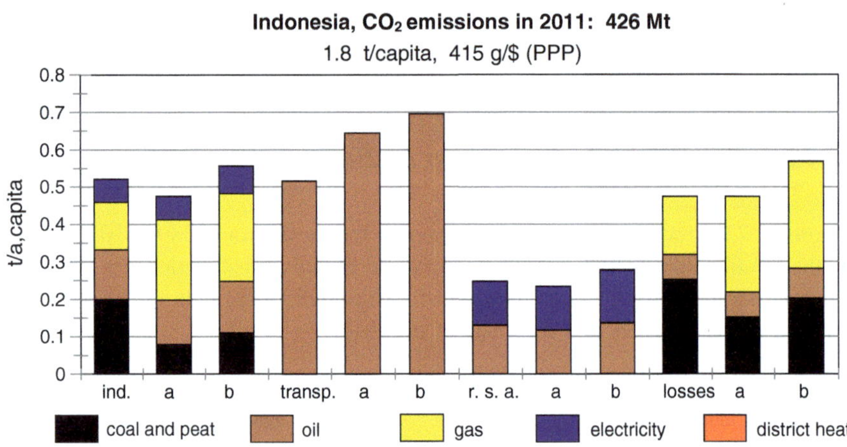


**Fig. 12.73** $CO_2$ emissions per capita of the 4 sectors: industry, transport, residential + services + agriculture, energy-sector losses (definitions see Appendix), each sector represented by three bars: *first bar* 2011 (data in the text and the diagram title, according to the Indonesia-Annex A5), *second bar* possible 2 °C scenario **variant *a*** for **2030**, 530 Mt, 1.8 t/capita, 221 g/$ (PPP), *third bar* possible 2 °C scenario **variant *b*** for **2030**, 630 Mt, 2.1 t/capita, 263 g/$ (PPP)

The target value for 2030 can, for example, be achieved if the per capita emissions:

- in the **electricity and district heat production sector, including energy-sector losses**, remain unchanged (variant *a*) or increase by not more than 20 % (variant *b*) from 0.65 t/capita in 2011 to 0.65–0.79 t/capita (efficiency improvements, replacement of coal by gas, CCS, renewable energy and nuclear power)
- in the **heating sector** (industry, residential, services etc.) through efficiency improvements, reduction of coal consumption, gas instead of coal, and the use of renewable energy sources (heat pump, waste management, solar energy, geothermal energy), are reduced by 10 % (*a*) or increase by not more than 5 % (*b*) (from 0.59 t/capita to 0.53–0.62 t/capita)
- in the **transport sector** increase by 23 % (*a*) but not more than 35 % (*b*), from 0.52 in 2011 to 0.64–0.70 t/capita, through efficiency improvements, gas fuels, bio-fuels and, as a more distant goal, electric mobility.

## 12.13 Brazil

### *12.13.1 Actual Indicators (Energy Intensity, $CO_2$ Intensity and $CO_2$ Sustainability) from 1980 to 2011 and Required up to 2030 to meet the 2 °C Target*

Brazil in 2011, with a figure of 195 g $CO_2$/\$, has the second-best $CO_2$ sustainability indicator of the G-20 countries, second only to France. This is due to the highly water-power based electricity production (see Appendix). The trend of the indicators from 1980 to 2011 and the necessary one for the 2-degree target by 2030, for variants *a* and *b* are shown in Fig. 12.74.

The climate target to be aimed at for 2030 is 147–171 g $CO_2$/\$ (−25 to −12 % relative to 2011) for a share of 500–580 Mt (2011: 408 Mt) of the total G-20 $CO_2$ emissions. The values for 2018 take into account the GDP (PPP) forecast by the International Monetary Fund (IMF) of around \$2,530 billion (\$ of 2007). The GDP for 2030 is estimated at \$3,400 billion.

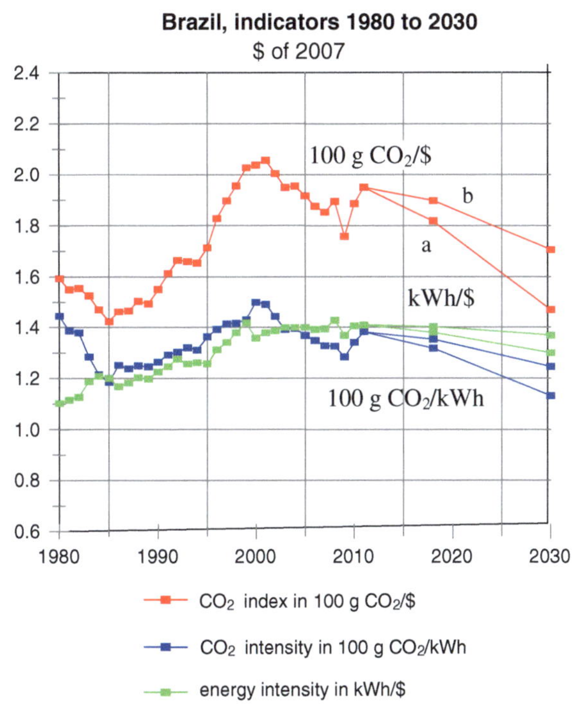


**Fig. 12.74** Brazil indicators 1980 to 2011 and 2-degrees scenario up to 2030

With this scenario, the gross energy demand in 2030 is 380–400 Mtoe (only energy use). The per capita indicators for energy and $CO_2$ emissions would be: $e \approx 2.1$–$2.2$ kW/capita (increase of 23 to 29 % relative to 2011) and $\alpha = 2.1$–$2.4$ t $CO_2$ per capita (unchanged (var. *a*) to increase of 17 % (var. *b*) relative to 2011).

The necessary change in the trend of the indicators, relative to the period 2000 to 2011, to meet the 2-degree target is illustrated in Fig. 12.75 for variant *a*. A turnaround in the energy efficiency trend, and a correction of the trend of the last 3 years and improvement of the trend of $CO_2$ intensity of energy from 2018 (−1.2 %/a) are necessary.

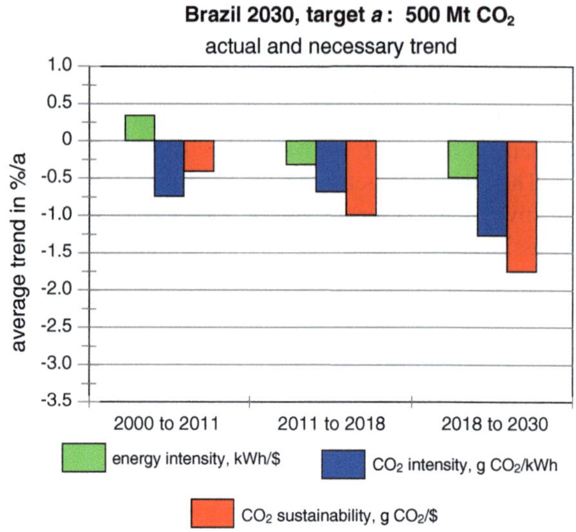


**Fig. 12.75** Necessary trend change to achieve the 2-degrees target, variant *a*

With variant *b* (Fig. 12.76) the improvement in the trend of energy intensity is gentler and the average trend of $CO_2$ intensity from 2000 to 2011 should be complied with if at all possible, particularly after 2018.

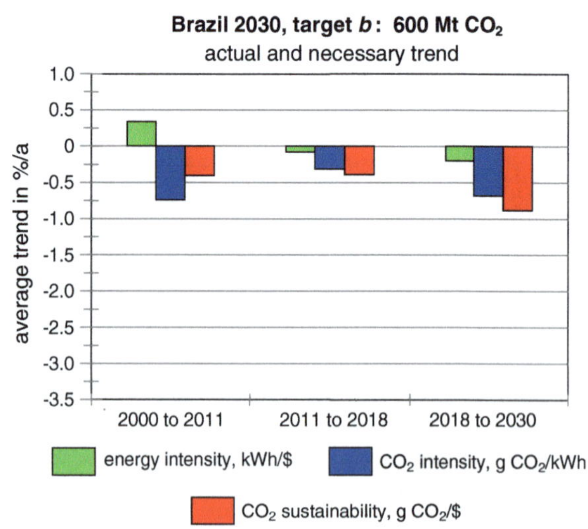


**Fig. 12.76** Necessary trend change to achieve the 2-degrees target, variant *b*

## 12.13.2 $CO_2$ Emissions until 2050

The actual $CO_2$ emissions in Brazil from 1970 to 2011, and those allowed to 2050 for compliance with the 2 °C climate target (according to the indicators of Figs. 12.74, 12.75, 12.76) are shown in Fig. 12.77.

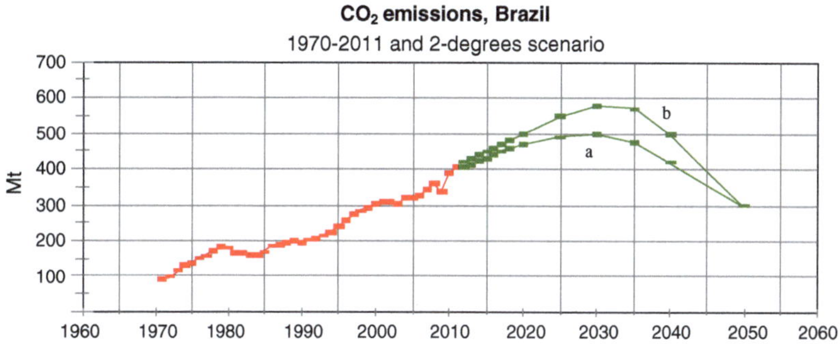


**Fig. 12.77** $CO_2$ emissions of Brazil from 1970 to 2011 and climate protection scenario until 2050

## 12.13.3 Per Capita Indicators up to 2030

Figure 12.78 shows the per capita indicators of Brazil from 1980 to 2011 as well as the profile up to 2030 resulting from the above considerations to meet the 2-degree target, for both variants *a* and *b*. The GDP (PPP) data up to 2018 correspond to the statistics and forecasts of the IMF.

**Fig. 12.78** Per capita indicators of Brazil from 1980 to 2011 and 2-degrees scenario until 2030

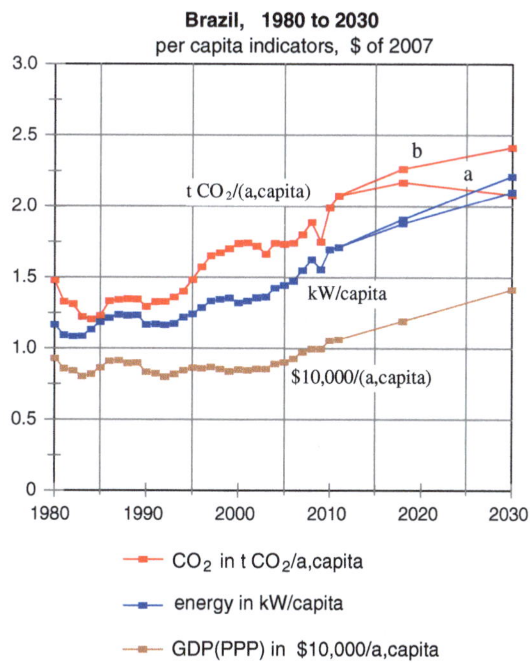

The following Fig. 12.79 shows, in more detail **per consumer sector**, the per capita emissions of the resident population in 2011, and those corresponding to the 2-degree scenario for 2030 (variant *a* and *b*).

The four consumer sectors are (total specific emissions of $CO_2$ in 2011: 2.1 t/capita):

- industry (0.5 t/capita)
- transport (1.0 t/capita)
- residential, services, agriculture, etc. (0.2 t/capita)
- energy-sector losses (0.4 t/capita).

The energy and emissions data for 2011 correspond to the Brazil-Annex A5.

Final-energy carriers emit $CO_2$ directly or indirectly (via electricity and district heating). The assumed distribution of these energy carriers within the consumer sectors, for both variants in 2030, is ***one way among others*** to meet the 2-degree condition.

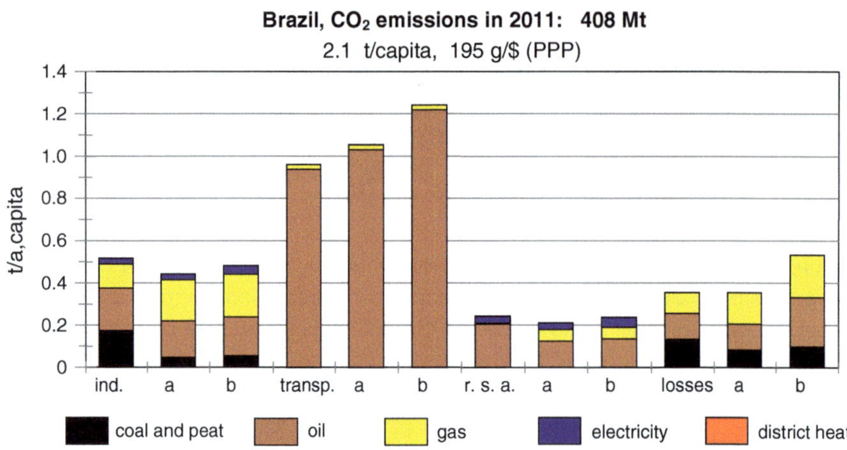


**Fig. 12.79** $CO_2$ emissions per capita of the 4 sectors: industry, transport, residential + services + agriculture, energy-sector losses (definitions see Appendix), each sector represented by three bars: *first bar* 2011 (data in the text and the diagram title, according to the Brazil-Annex A5), *second bar* possible 2 °C scenario **variant *a*** for **2030**, 500 Mt, 2.1 t/capita, 147 g/$ (PPP), *third bar* possible 2 °C scenario **variant *b*** for **2030**, 580 Mt, 2.4 t/capita, 171 g/$ (PPP)

The target value for 2030 can, for example, be achieved if the per capita emissions:

- in the **electricity and district heat production sector, including energy-sector losses**, remain unchanged (variant *a*) or increase by not more than 50 % (variant *b*) from 0.41 t/capita in 2011 to 0.41–0.62 t/capita (efficiency improvements, replacement of coal by gas, CCS, renewable energy and nuclear power)
- in the **heating sector** (industry, residential, services etc.) through efficiency improvements, reduction of coal consumption, gas instead of coal and oil, and the use of renewable energy sources (heat pump, waste management, solar energy, geothermal energy), are reduced by 16 % (*a*) or by 10 % (*b*) (from 0.7 t/capita to 0.59–0.63 t/capita)
- in the **transport sector** increase by 9 % (*a*) but not more than 29 % (*b*), from 0.96 in 2011 to 1.05–1.24 t/capita, through efficiency improvements, gas fuels, bio-fuels and electric mobility.

## 12.14 Australia

### *12.14.1 Actual Indicators (Energy Intensity, $CO_2$ Intensity and $CO_2$ Sustainability) from 1990 to 2011 and Required up to 2030 to meet the 2 °C Target*

For 2011, Australia with a figure of 472 g $CO_2$/\$, has one of the worst $CO_2$ sustainability indicators of the OECD group. The reason is the solely coal based production of electricity (see the appendix). The trend of the indicators from 1980 to 2011 and the necessary one for the 2-degree target by 2030, for variants *a* and *b* are shown in Fig. 12.80.

The climate target to be aimed at for 2030 is 214–243 g $CO_2$/\$ (−55 to −49 % relative to 2011) for a share of 300–340 Mt (−24 to −15 % relative to 2011) of the total G-20 $CO_2$ emissions. The values for 2018 take into account the GDP (PPP) forecast by the International Monetary Fund (IMF) of around \$1,030 billion (\$ of 2007). The GDP for 2030 is estimated at \$1,400 billion.

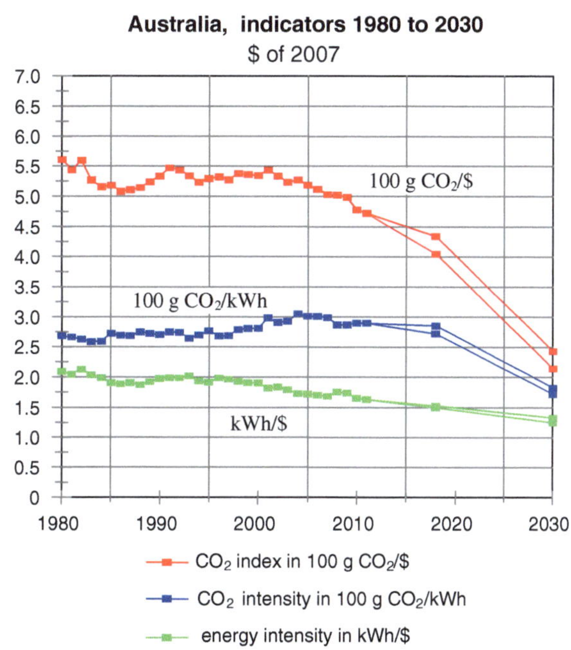


**Fig. 12.80** Australia indicators 1980 to 2011 and 2-degrees scenario up to 2030

With this scenario, the gross energy demand in 2030 is 150–160 Mtoe (only energy use). The per capita indicators for energy and $CO_2$ emissions would be: $e \approx 6.6$–7.1 kW/capita (−3 to +3 % relative to 2011) and $\alpha = 10.0$–11.3 t $CO_2$ per capita (−43 % (var. *a*) to −35 % (var. *b*) relative to 2011).

The necessary change in the trend of the indicators to meet the 2-degree target is illustrated in Fig. 12.81 for variant *a*. Maintaining the energy efficiency trend from 2000 to 2011 and a decided turnaround, especially after 2018, in the $CO_2$ intensity of energy (−3.5 %/a) are necessary.

**Fig. 12.81** Necessary trend change to meet the 2-degrees target, variant *a*

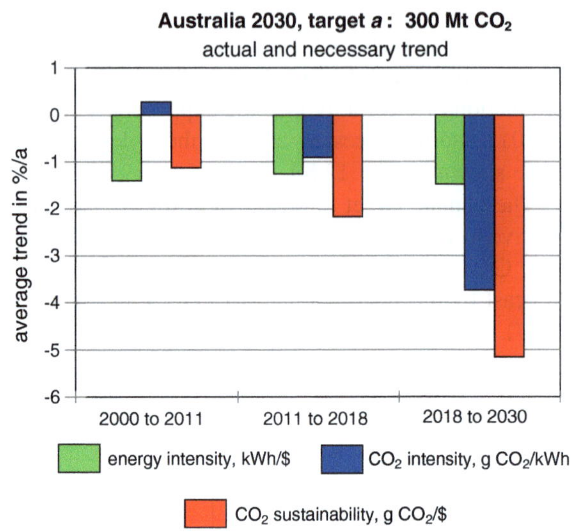


With variant *b* (Fig. 12.82) the trend change in $CO_2$ intensity starts a little later, but must be equally marked.

**Fig. 12.82** Necessary trend change to meet the 2-degrees target, variant *b*

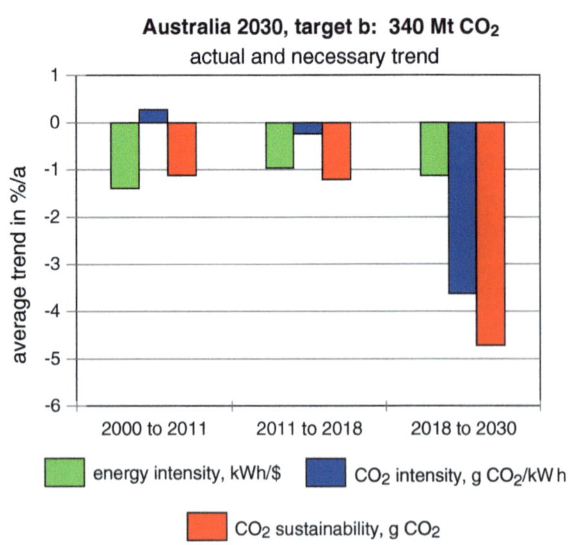

## 12.14.2 $CO_2$ Emissions until 2050

The actual $CO_2$ emissions in Australia from 1970 to 2011, and those allowed to 2050 for compliance with the 2 °C climate target (according to the indicators of Figs. 12.80, 12.81, 12.82) are shown in Fig. 12.83.

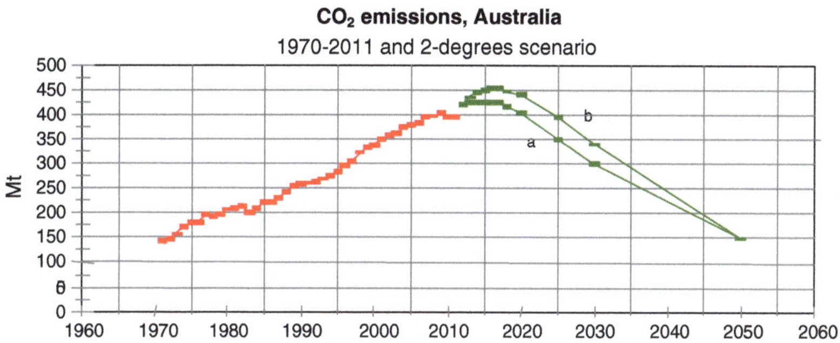


**Fig. 12.83** $CO_2$ emissions of Australia from 1970 to 2011 and climate protection scenario until 2050

## 12.14.3 Per Capita Indicators up to 2030

Figure 12.84 shows the per capita indicators of Australia from 1980 to 2011 as well as the profile up to 2030 resulting from the above considerations to meet the 2-degree target, for both variants *a* and *b*. The GDP (PPP) data up to 2018 correspond to the statistics and forecasts of the IMF.

**Fig. 12.84** Per capita indicators of Australia from 1980 to 2011 and 2-degrees scenario until 2030

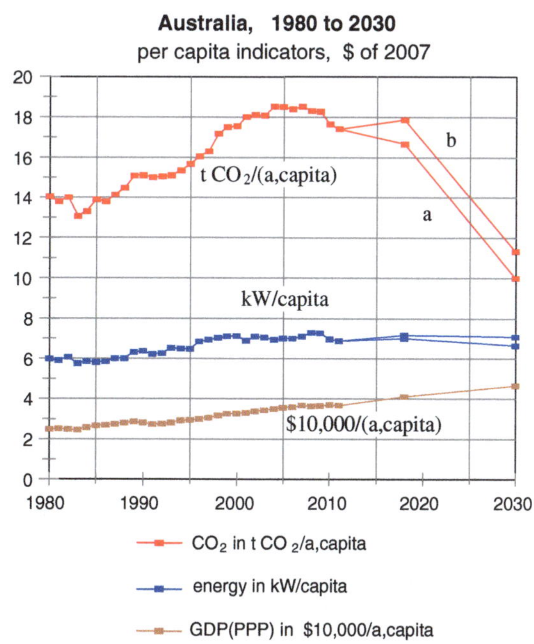

The following Fig. 12.85 shows, in more detail **per consumer sector**, the per capita emissions of the resident population in 2011, and those corresponding to the 2-degree scenario for 2030 (variant *a* and *b*).

The four consumer sectors are (total specific emissions of $CO_2$ in 2011: 17.4 t/capita):

- industry (3.1 t/capita)
- transport (4.2 t/capita)
- residential, services, agriculture, etc. (2.6 t/capita)
- energy-sector losses (7.6 t/capita).

The energy and emissions data for 2011 correspond to the Australia-Annex A5.

Final-energy carriers emit $CO_2$ directly or indirectly (via electricity and district heating). The assumed distribution of these energy carriers within the consumer sectors, for both variants in 2030, is ***one way among others*** to meet the 2-degree condition.

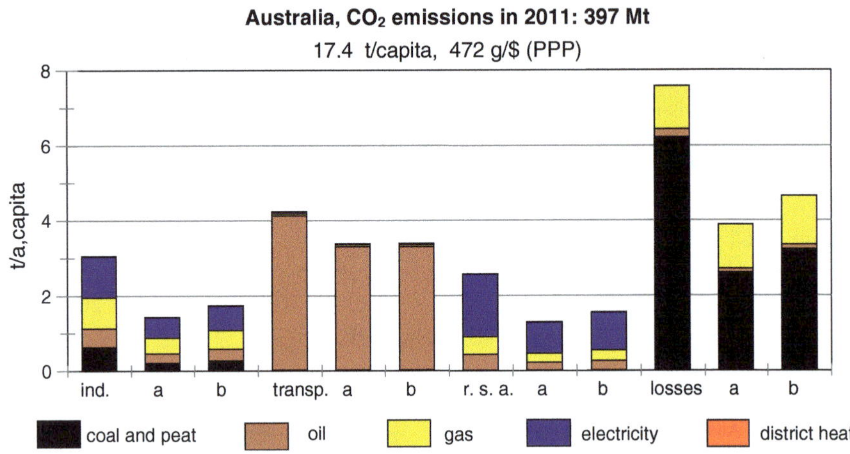


**Fig. 12.85** $CO_2$ emissions per capita of the 4 sectors: industry, transport, residential + services + agriculture, energy-sector losses (definitions see Appendix), each sector represented by three bars: *first bar*: 2011 (data in the text and the diagram title, according to the Australia-Annex A5), *second bar* possible 2 °C scenario **variant *a*** for **2030**, 300 Mt, 10.0 t/capita, 214 g/$ (PPP), *third bar* possible 2 °C scenario **variant *b*** for **2030**, 340 Mt, 11.3 t/capita, 243 g/$ (PPP)

The target value for 2030 can, for example, be achieved if the per capita emissions:

- in the **electricity and district heat production sector, including energy-sector losses**, are reduced by 49 % (variant *a*) or 39 % (variant *b*) from 10.4 t/capita in 2011 to 5.3–6.3 t/capita (efficiency improvements, replacement of coal by gas, CCS, renewable energy and eventually nuclear power)
- in the **heating sector** (industry, residential, services etc.) through efficiency improvements, reduction of coal consumption, gas instead of coal, and the use of renewable energy sources (heat pump, waste management, solar energy, geothermal energy), are reduced by 53 % (*a*) or 43 % (*b*) (from 2.9 t/capita to 1.3–1.6 t/capita)
- in the **transport sector** are reduced by 20 % (*a* and *b*) from 4.2 in 2011 to 3.4 t/capita, through efficiency improvements, gas fuels, bio-fuels and electric mobility.

# 12.15 Italy

## 12.15.1 Actual Indicators (Energy Intensity, $CO_2$ Intensity and $CO_2$ Sustainability) from 1980 to 2011 and Required up to 2030 to meet the 2 °C Target

Italy in 2011, with a $CO_2$ sustainability indicator of 234 g $CO_2$/\$, has the fourth-best position of the G-20 group and thus is also slightly below the EU-27 average. This is mainly due to gas-based electricity production but also to hydropower and other renewable energies (see also the appendix). The trend of the indicators from 1980 to 2011 and the necessary one for the 2-degree target by 2030, for variants *a* and *b* are shown in Fig. 12.86.

The climate target to be aimed at for 2030 is 174–189 g $CO_2$/\$ (−26 to −19 % relative to 2011) for a share of 313–340 Mt (−20 to −13 % relative to 2011) of the total G-20 $CO_2$ emissions. The values for 2018 take into account the GDP (PPP) forecast by the International Monetary Fund (IMF) of around \$1,710 billion (\$ of 2007). The GDP for 2030 is estimated at \$1,800 billion.

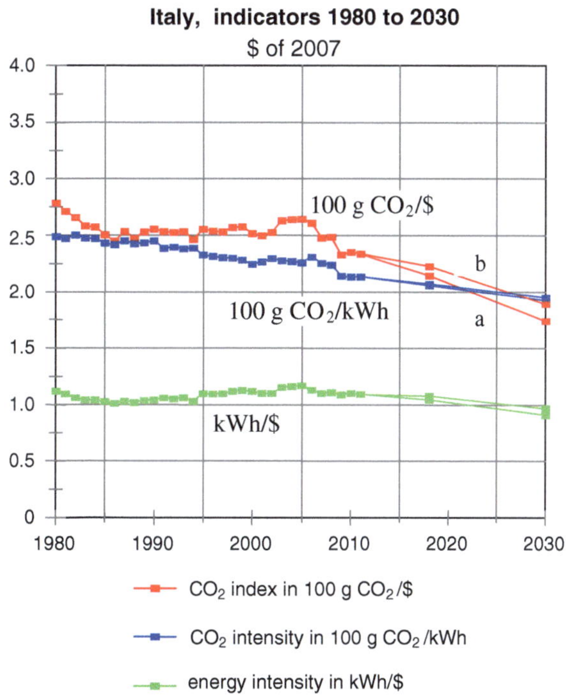


**Fig. 12.86** Italy indicators 1980 to 2011 and 2-degrees scenario up to 2030

With this scenario, the gross energy demand in 2030 is 140–150 Mtoe (only energy use). The per capita indicators for energy and $CO_2$ emissions would be: e ≈ 3.0–3.2 kW/capita (−13 to −7 % relative to 2011) and $\alpha$ = 5.1–5.5 t $CO_2$ per capita (−22 % (var. *a*) to −15 % (var. *b*) relative to 2011).

The change in the trend of the indicators to meet the 2-degree target is illustrated in Fig. 12.87 for variant *a*. A significant improvement of energy efficiency trend and maintaining and slightly increasing the trend in the $CO_2$ intensity of energy (to −0.6 %/a) are required.

**Fig. 12.87** Necessary trend change to achieve the 2-degrees target, variant *a*

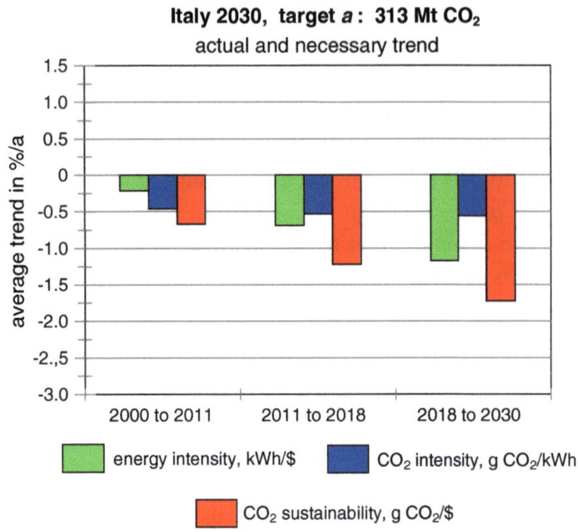


With variant *b* (Fig. 12.88) the improvement in the energy intensity trend starts a little later.

**Fig. 12.88** Necessary trend change to achieve the 2-degrees target, variant *b*

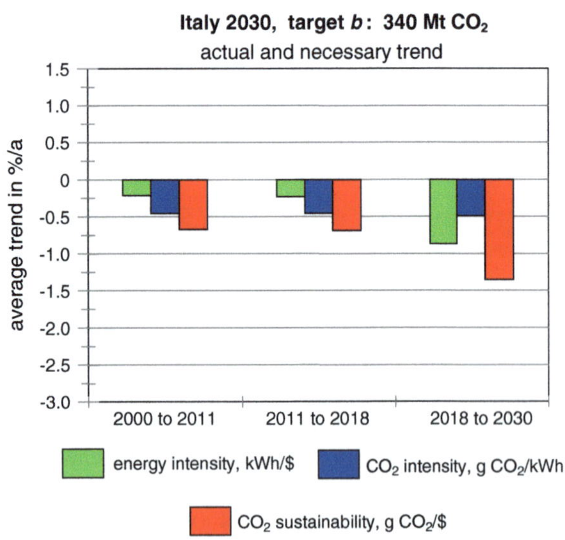

### 12.15.2 $CO_2$ Emissions until 2050

The actual $CO_2$ emissions in Italy from 1970 to 2011, and those allowed to 2050 for compliance with the 2 °C climate target (according to the indicators of Figs. 12.86, 12.87, 12.88) are shown in Fig. 12.89.

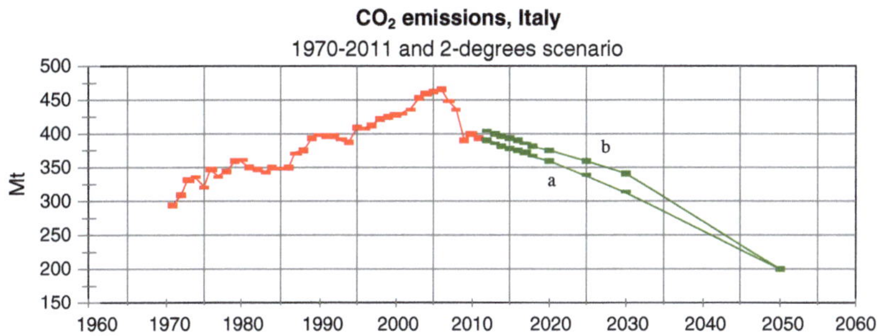


**Fig. 12.89** $CO_2$ emissions of Italy from 1970 to 2011 and climate protection scenario until 2050

### 12.15.3 Per Capita Indicators up to 2030

Figure 12.90 shows the per capita indicators of Italy from 1980 to 2011 as well as the profile up to 2030 resulting from the above considerations to meet the 2-degree target, for both variants *a* and *b*. The GDP (PPP) data up to 2018 correspond to the statistics and forecasts of the IMF.

**Fig. 12.90** Per capita indicators of Italy from 1980 to 2011 and 2-degrees scenario until 2030

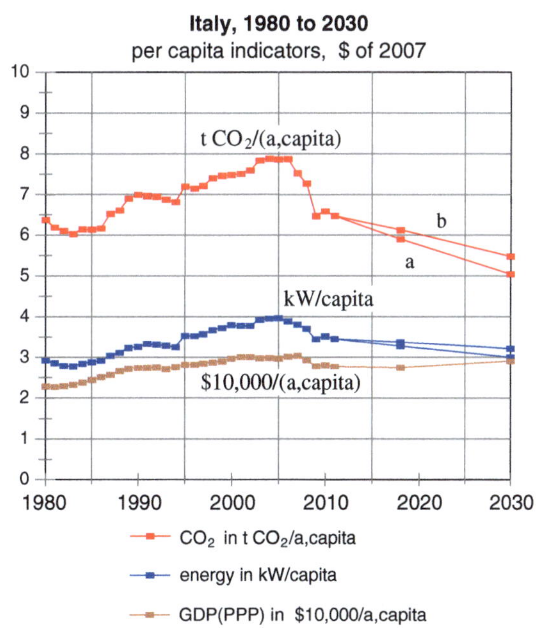

The following Fig. 12.91 shows, in more detail **per consumer sector**, the per capita emissions of the resident population in 2011, and those corresponding to the 2-degree scenario for 2030 (variant *a* and *b*).

The four consumer sectors are (total specific emissions of $CO_2$ in 2011: 6.5 t/capita):

- industry (1.2 t/capita)
- transport (1.9 t/capita)
- residential, services, agriculture, etc. (1.9 t/capita)
- energy-sector losses (1.5 t/capita).

The energy and emissions data for 2011 correspond to the Italy-Annex A5.

Final-energy carriers emit $CO_2$ directly or indirectly (via electricity and district heating). The assumed distribution of these energy carriers within the consumer sectors, for both variants in 2030, is ***one way among others*** to meet the 2-degree condition.

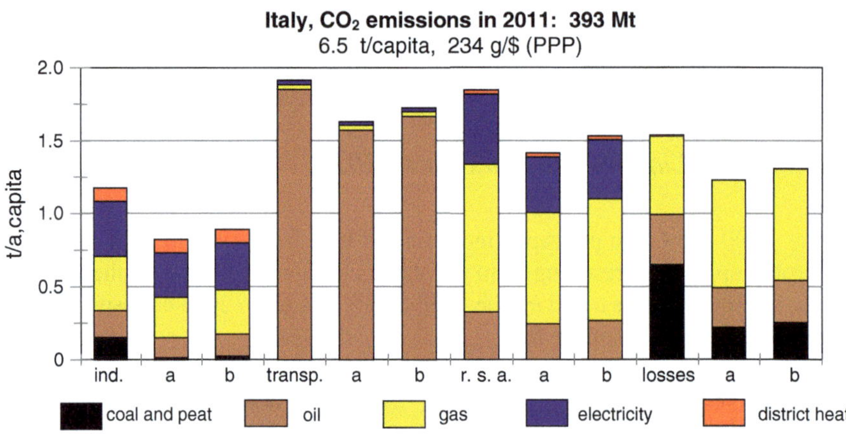


**Fig. 12.91** $CO_2$ emissions per capita of the 4 sectors: industry, transport, residential + services + agriculture, energy-sector losses (definitions see Appendix), each sector represented by three bars: *first bar* 2011 (data in the text and the diagram title, according to the Italy-Annex A5), *second bar* possible 2 °C scenario **variant *a*** for **2030**, 313 Mt, 5.1 t/capita, 174 g/$ (PPP), *third bar* possible 2 °C scenario **variant *b*** for **2030**, 340 Mt, 5.5 t/capita, 189 g/$ (PPP)

The target value for 2030 can, for example, be achieved if the per capita emissions:

- in the **electricity and district heat production sector, including energy-sector losses**, are reduced by 19 % (variant *a*) or 14 % (variant *b*) from 2.5 t/capita in 2011 to 2.1–2.2 t/capita (efficiency improvements, replacement of coal by gas, CCS and renewable energy)
- in the **heating sector** (industry, residential, services etc.) through efficiency improvements (better insulation), reduction of coal consumption, gas instead of oil, and the use of renewable energy sources (heat pump, waste management, solar energy, geothermal energy), are reduced by 30 % (*a*) or 10 % (*b*) (from 2.1 t/capita to 1.4–1.6 t/capita)
- in the **transport sector** are reduced by 15 % (*a*) or 10 % (*b*), from 1.9 in 2011 to 1.6–1.7 t/capita, through efficiency improvements, gas fuels, bio-fuels and electric mobility.

## 12.16 South Africa

### 12.16.1 Actual Indicators (Energy Intensity, $CO_2$ Intensity and $CO_2$ Sustainability) from 1980 to 2011 and Required up to 2030 to meet the 2 °C Target

In 2011, South Africa had a very bad $CO_2$ sustainability indicator of 722 g $CO_2$/\$, putting it in the third-last position of the G-20 group. The reasons are the mainly coal-based electricity production and the poor energy efficiency (see also the appendix). The trend of the indicators from 1980 to 2011 and the necessary one for the 2-degree target by 2030, for variants *a* and *b* are shown in Fig. 12.92.

The climate target to be aimed at for 2030 is 353–412 g $CO_2$/\$ (−51 % to −43 % relative to 2011) for a share of 300–350 Mt (−18 to −5 % relative to 2011) of the total G-20 $CO_2$ emissions. The values for 2018 take into account the GDP (PPP) forecast by the International Monetary Fund (IMF) of around \$630 billion (\$ of 2007). The GDP for 2030 is estimated at \$850 billion.

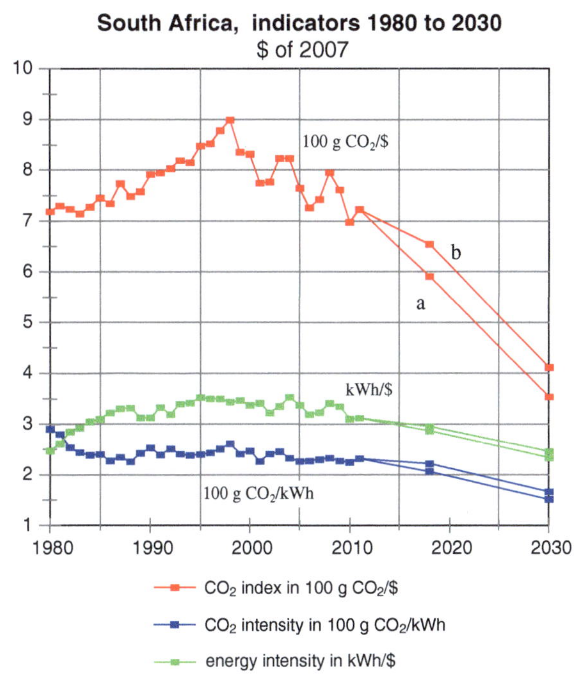


**Fig. 12.92** South Africa indicators 1980 to 2011 and 2-degrees scenario up to 2030

With this scenario, the gross energy demand in 2030 is 170–180 Mtoe (only energy use). The per capita indicators for energy and $CO_2$ emissions would be: e ≈ 3.5–3.7 kW/capita (unchanged (var. *a*) to an increase of 5 % (var. *b*) relative to 2011) and $\alpha = 4.7$–5.5 t $CO_2$ per capita (−35 % (var. *a*) to −25 % (var. *b*) relative to 2011).

The change in the trend of the indicators required to meet the 2-degree target is illustrated in Fig. 12.93 for variant *a*. A significant improvement both of the trends in energy efficiency and in $CO_2$ intensity of energy is necessary even before 2018.

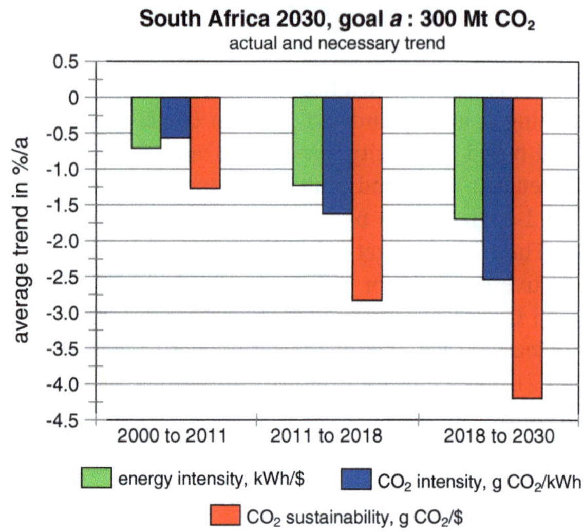


**Fig. 12.93** Necessary trend change to meet the 2-degrees target, variant *a*

With variant *b* (Fig. 12.94), the strong trend change occurs only after 2018.

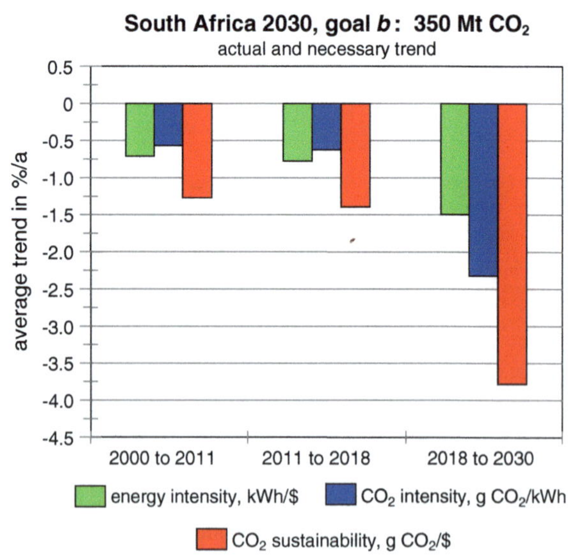


**Fig. 12.94** Necessary trend change to meet the 2-degrees target, variant *b*

## 12.16.2 $CO_2$ Emissions until 2050

The actual $CO_2$ emissions in South Africa from 1970 to 2011, and those allowed to 2050 for compliance with the 2 °C climate target (according to the indicators of Figs. 12.92, 12.93, 12.94) are shown in Fig. 12.95.

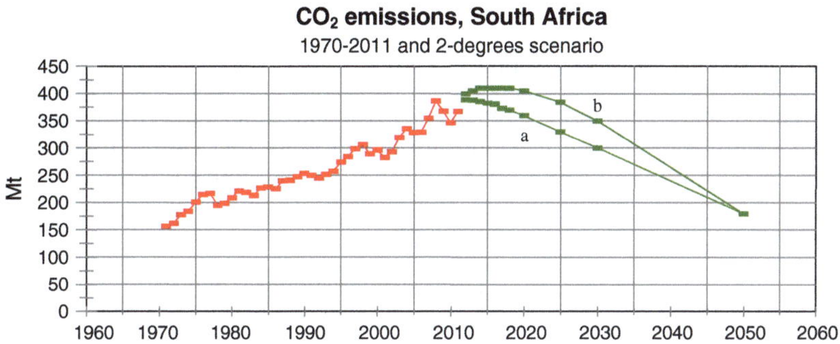


**Fig. 12.95** $CO_2$ emissions of South Africa from 1970 to 2011 and climate protection scenario until 2050

## 12.16.3 Per Capita Indicators up to 2030

Figure 12.96 shows the per capita indicators of South Africa from 1980 to 2011 as well as the profile up to 2030 resulting from the above considerations to meet the 2-degree target, for both variants *a* and *b*. The GDP (PPP) data up to 2018 correspond to the statistics and forecasts of the IMF.

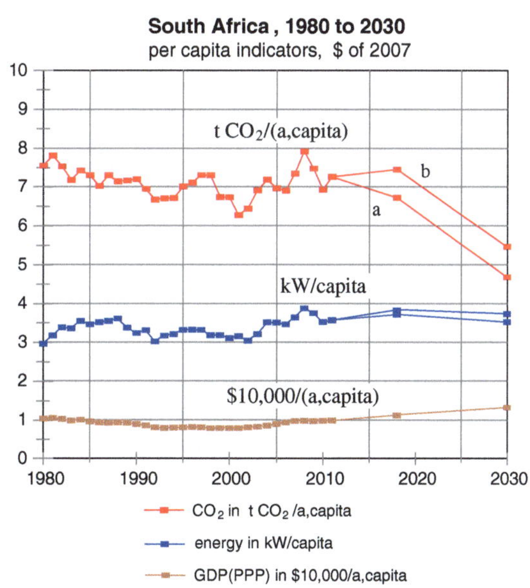


**Fig. 12.96** Per capita indicators of South Africa from 1980 to 2011 and 2-degrees scenario until 2030

The following Fig. 12.97 shows, in more detail **per consumer sector**, the per capita emissions of the resident population in 2011, and those corresponding to the 2-degree scenario for 2030 (variant *a* and *b*).

The four consumer sectors are (total specific emissions of $CO_2$ in 2011: 7.3 t/capita):

- industry (1.4 t/capita)
- transport (0.9 t/capita)
- residential, services, agriculture, etc. (0.9 t/capita)
- energy-sector losses (4.1 t/capita).

The energy and emissions data for 2011 correspond to the South Africa-Annex A5.

Final-energy carriers emit $CO_2$ directly or indirectly (via electricity and district heating). The assumed distribution of these energy carriers within the consumer sectors, for both variants in 2030, is ***one way among others*** to meet the 2-degree condition.

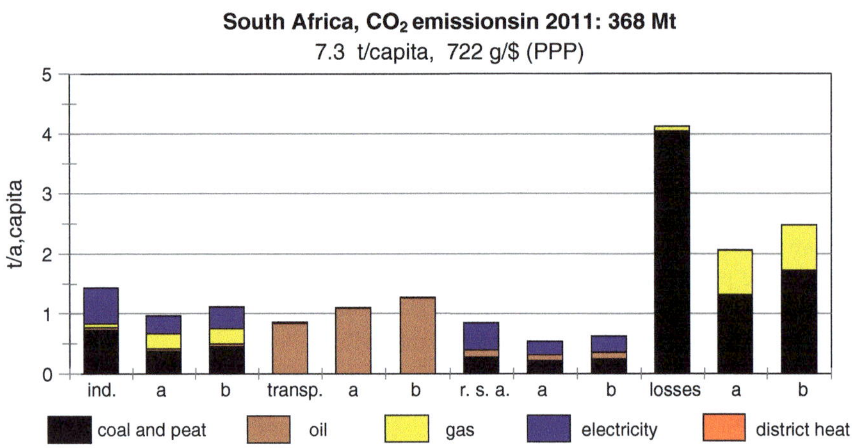


**Fig. 12.97** $CO_2$ emissions per capita of the 4 sectors: industry, transport, residential + services + agriculture, energy-sector losses (definitions see Appendix), each sector represented by three bars: *first bar* 2011 (data in the text and the diagram title, according to the South Africa-Annex A5), *second bar* possible 2 °C scenario **variant *a*** for **2030**, 300 Mt, 4.7 t/capita, 353 g/$ (PPP), *third bar* possible 2 °C scenario **variant *b*** for **2030**, 350 Mt, 5.5 t/capita, 412 g/$ (PPP)

The target value for 2030 can, for example, be achieved if the per capita emissions:

- in the **electricity and district heat production sector, including energy-sector losses**, are reduced by 50 % (variant *a*) or 40 % (variant *b*) from 5.2 t/capita in 2011 to 2.6–3.1 t/capita (efficiency improvements, replacement of coal by gas, CCS, renewable energy and nuclear power)
- in the **heating sector** (industry, residential, services etc.) through efficiency improvements, reduction of coal consumption, gas instead of coal, and the use of renewable energy sources (heat pump, waste management, solar energy, geothermal energy), are reduced by 20 % (*a*) or 11 % (*b*) (from 1.2 t/capita to 1.0–1.1 t/capita)
- in the **transport sector** increase by 16 % (*a*) but not more than 48 % (*b*), from 0.9 in 2011 to 1.1–1.3 t/capita, through efficiency improvements, gas fuels, biofuels and electric mobility.

## 12.17 France

### *12.17.1 Actual Indicators (Energy Intensity, $CO_2$ Intensity and $CO_2$ Sustainability) from 1990 to 2011 and Required up to 2030 to meet the 2 °C Target*

France is, in 2011, with a $CO_2$ indicator of sustainability of 162 g $CO_2$/\$, in the first place of the G20 countries, thanks to the low-$CO_2$ emissions of the electricity industry (see Appendix). The actual graph of the indicators from 1980 to 2011 and the values necessary for meeting the 2-degree target by 2030, for the variants *a* and *b*, are shown in Fig. 12.98.

The climate target to be aimed at for 2030 is 116–122 g $CO_2$/\$ (−28 to −25 % relative to 2011) for a share of 209 to 305 Mt (−12 to −7 % relative to 2011) of the total G-20 $CO_2$ emissions. The values for 2018 take into account the GDP (PPP) forecast by the International Monetary Fund (IMF) of around \$2,200 billion (\$ of 2007). The GDP for 2030 is estimated at \$2,500 billion.

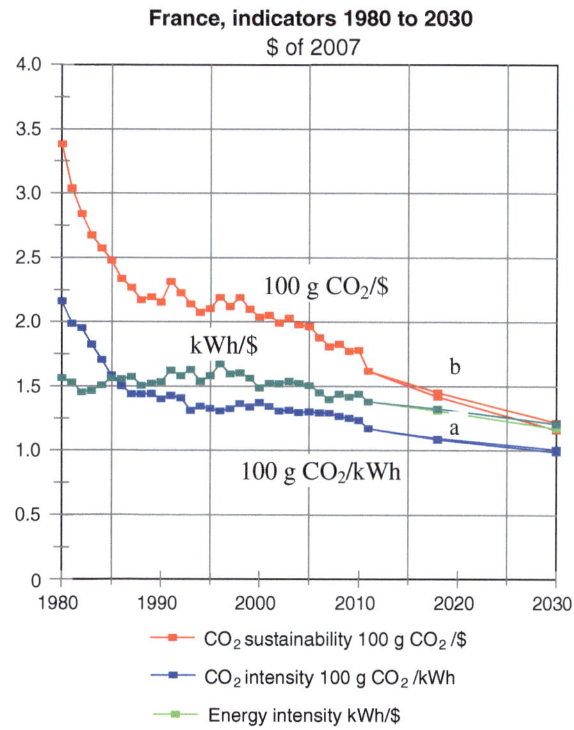


**Fig. 12.98** Indicators of France from 1980 to 2011 and 2-degrees scenario up to 2030

With this scenario, the gross energy demand in 2030 is 252–260 Mtoe (only energy use). The per capita indicators for energy and $CO_2$ emissions would be:

e ≈ 4.5–4.6 kW/capita (−9 to −6 % relative to 2011) and $\alpha$ = 3.9–4.1 t $CO_2$ per capita (−23 % (var. *a*) to −19 % (var. *b*) relative to 2011), see also Sect. 12.17.3.

The change in the trend of the indicators required to meet the 2-degree target is illustrated in Fig. 12.99 for variant *a*: maintaining the overall trend ensures compliance with the 2-degree target.

This is even more the case, for the very similar variant *b* (Fig. 12.100).

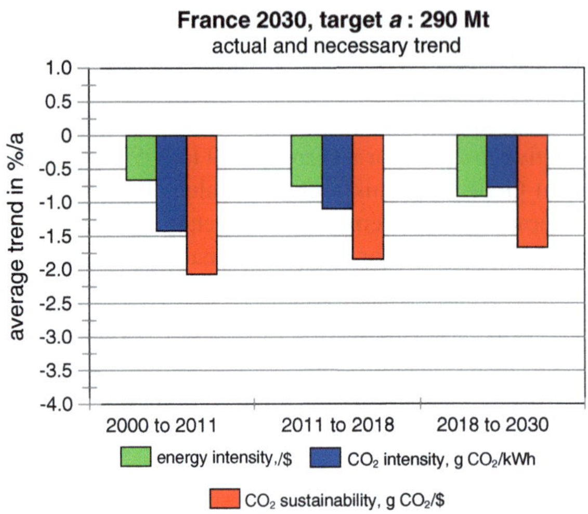


**Fig. 12.99** Necessary trend change to meet the 2-degrees target, variant *a*

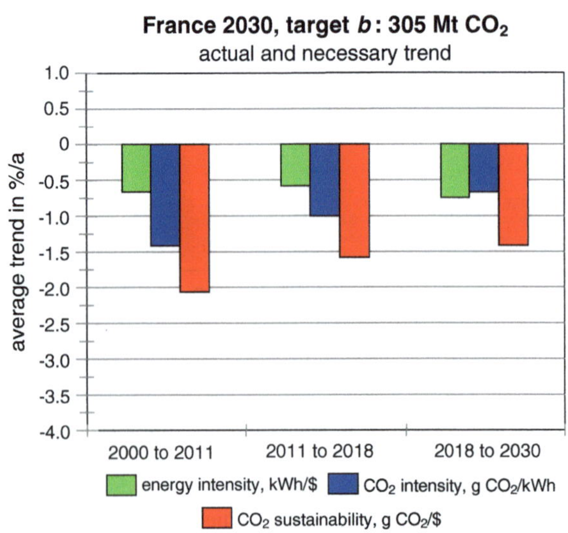


**Fig. 12.100** Necessary trend change to meet the 2-degrees target, variant *b*

## 12.17.2 $CO_2$ Emissions until 2050

The actual $CO_2$ emissions in France from 1970 to 2011, and those allowed to 2050 for compliance with the 2 °C climate target (according to the indicators of Figs. 12.98, 12.99, 12.100) are shown in Fig. 12.101.

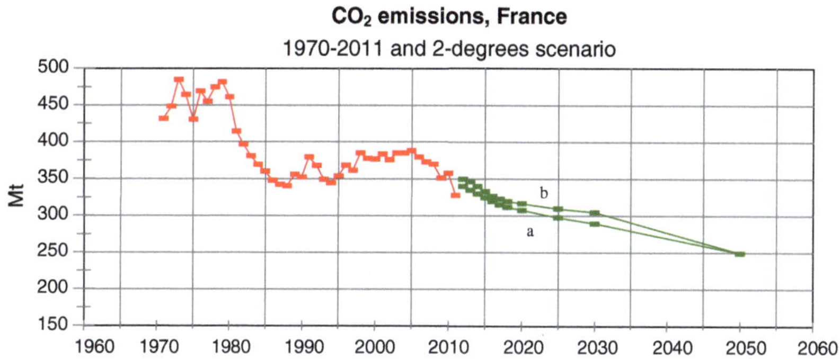


**Fig. 12.101** $CO_2$ emissions of France from 1970 to 2011 and climate protection scenario until 2050

## 12.17.3 Per Capita Indicators up to 2030

Figure 12.102 shows the per capita indicators of France from 1980 to 2011 as well as the profile up to 2030 resulting from the above considerations to meet the 2-degree target, for both variants *a* and *b*. The GDP (PPP) data up to 2018 correspond to the statistics and forecasts of the IMF.

**Fig. 12.102** Per capita indicators of France from 1980 to 2011 and 2-degrees scenario until 2030

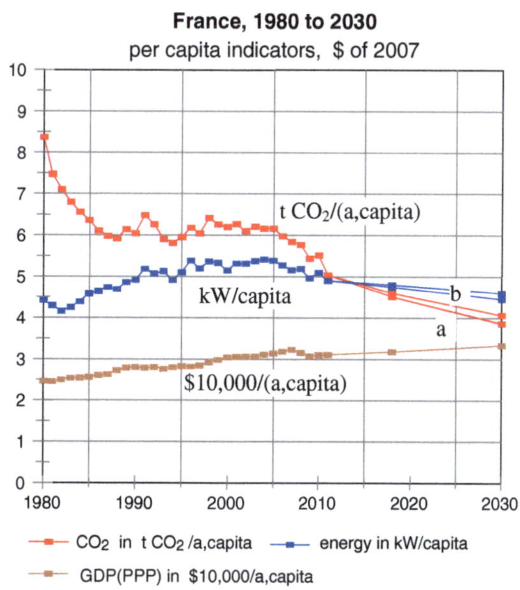

The following Fig. 12.103 shows, in more detail **per consumer sector**, the per capita emissions of the resident population in 2011, and those corresponding to the 2-degree scenario for 2030 (variant *a* and *b*).

The four consumer sectors are (total specific emissions of $CO_2$ in 2011: 5.0 t/capita):

- industry (0.8 t/capita)
- transport (1.9 t/capita)
- residential, services, agriculture, etc. (1.5 t/capita)
- energy-sector losses (0.8 t/capita).

The energy and emissions data for 2011 correspond to the France-Annex A5.

Final-energy carriers emit $CO_2$ directly or indirectly (via electricity and district heating). The assumed distribution of these energy carriers within the consumer sectors, for both variants in 2030, is ***one way among others*** to meet the 2-degree condition.

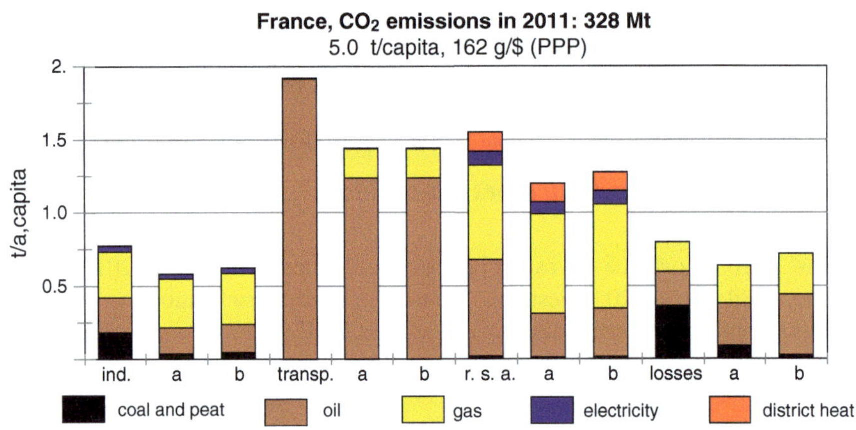


**Fig. 12.103** $CO_2$ emissions per capita of the 4 sectors: industry, transport, residential + services + agriculture, energy-sector losses (definitions see Appendix), each sector represented by three bars: *first bar* 2011 (data in the text and the diagram title, according to the France-Annex A5), *second bar* possible 2 °C scenario **variant *a*** for **2030**, 290 Mt, 3.9 t/capita, 116 g/$ (PPP), *third bar* possible 2 °C scenario **variant *b*** for **2030**, 305 Mt, 4.1 t/capita, 122 g/$ (PPP)

The target value for 2030 can, for example, be achieved if the per capita emissions:

- in the **electricity and district heat production sector, including energy-sector losses**, are reduced by 40 % (variant *a*) or 8 % (variant *b*) from 1.07 t/capita in 2011 to 0.64–0.98 t/capita (efficiency improvements, replacement of coal by gas, CCS, renewable energy and nuclear power)
- in the **heating sector** (industry, residential, services etc.) through efficiency improvements, reduction of coal consumption, gas instead of coal, and the use of renewable energy sources (heat pump, waste management, solar energy, geothermal energy), are reduced by 25 % (*a* and *b*) (from 2.1 t/capita in 2011 to 1.54 t/capita)
- in the **transport sector** are reduced 25 % (*a* and *b*) from 1.9 in 2011 to 1.4 t/capita, through efficiency improvements, gas fuels, bio-fuels and electric mobility.

## 12.18 Turkey

### *12.18.1 Actual Indicators (Energy Intensity, $CO_2$ Intensity and $CO_2$ Sustainability) from 1980 to 2011 and Required up to 2030 to meet the 2 °C Target*

For 2011, Turkey has an indicator of $CO_2$ sustainability of 291 g $CO_2$/\$, which is slightly below the world average. Electricity and heat production depend heavily on fossil fuels (see Appendix). The actual graph of the indicators from 1980 to 2011 and the necessary one for the 2-degree target by 2030, for the variants *a* and *b*, are shown in Fig. 12.104.

The climate target to be aimed at for 2030 is 174–195 g $CO_2$/\$ (−40 to −33 % relative to 2011) for a share of 330 to 370 Mt (+15 to +29 % relative to 2011) of the total G-20 $CO_2$ emissions. The values for 2018 take into account the GDP (PPP) forecast by the International Monetary Fund (IMF) of around \$1,280 billion (\$ of 2007). The GDP for 2030 is estimated at \$1,900 billion.

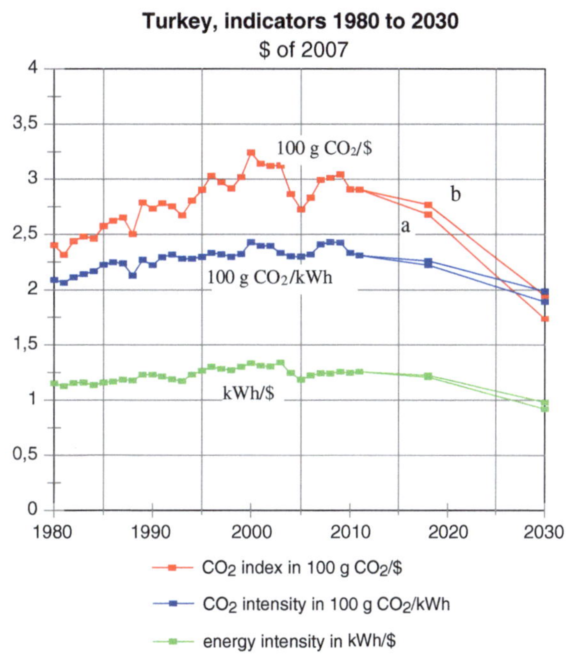


**Fig. 12.104** Turkey indicators 1980 to 2011 and 2-degrees scenario up to 2030

With this scenario, the gross energy demand in 2030 is 150–160 Mtoe (only energy use). The per capita indicators for energy and $CO_2$ emissions would be: e ≈ 2.3–2.7 kW/capita (increase of 22–39 % relative to 2011) and $\alpha = 4.1$–$4.6$ t $CO_2$ per capita (increase of 7 % (var. *a*) to 20 % (var. *b*) relative to 2011); see also Sect. 12.18.3.

The necessary change in the trend of the indicators to meet the 2-degree target is illustrated in Fig. 12.105 for variant *a*. A significant improvement of the trend after 2018, both in energy efficiency and $CO_2$ intensity of energy is required.

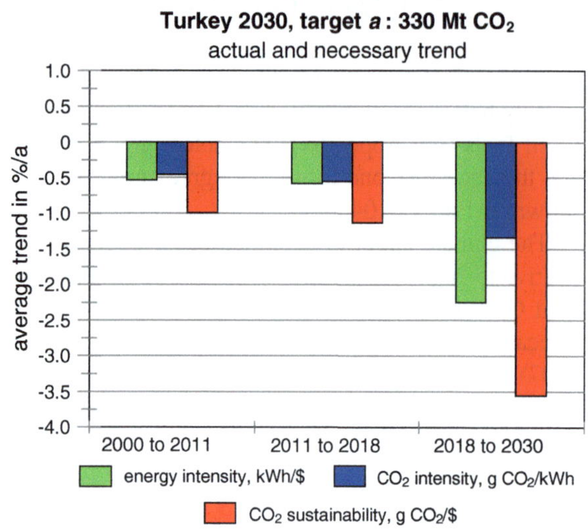


**Fig. 12.105** Necessary trend change to meet the 2-degrees target, variant *a*

With variant *b* (Fig. 12.106) the necessary strengthening of the trend occurs delayed and slightly weakened.

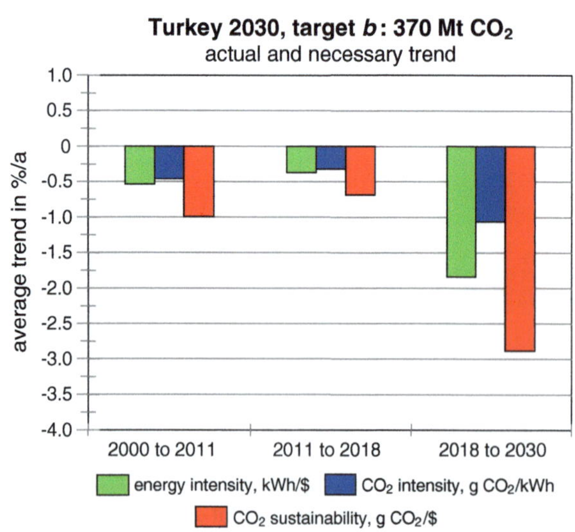


**Fig. 12.106** Necessary trend change to meet the 2-degrees target, variant *b*

## 12.18.2 $CO_2$ Emissions until 2050

The actual $CO_2$ emissions in Turkey from 1970 to 2011, and those allowed to 2050 for compliance with the 2 °C climate target (according to the indicators of Figs. 12.104, 12.105, 12.106) are shown in Fig. 12.107.

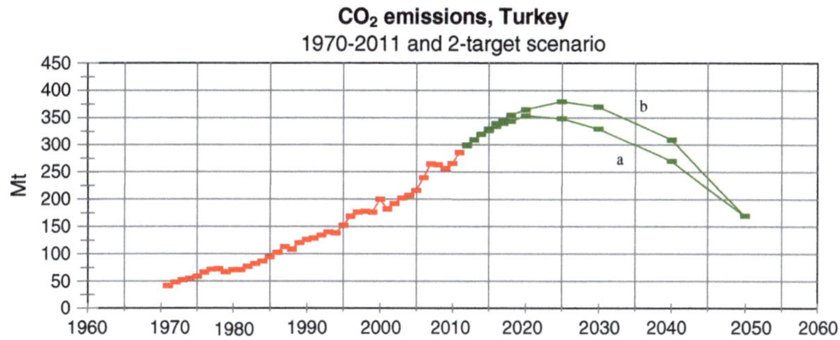


**Fig. 12.107** $CO_2$ emissions of Turkey from 1970 to 2011 and climate protection scenario until 2050

## 12.18.3 Per Capita Indicators up to 2030

Figure 12.108 shows the per capita indicators of Turkey from 1980 to 2011 as well as the profile up to 2030 resulting from the above considerations to meet the 2-degree target, for both variants *a* and *b*. The GDP (PPP) data up to 2018 correspond to the statistics and forecasts of the IMF.

**Fig. 12.108** Per capita indicators of Turkey from 1980 to 2011 and 2-degrees scenario until 2030

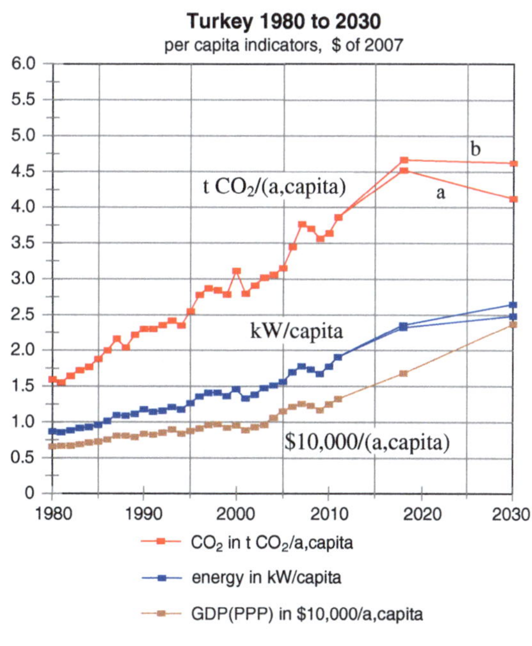

The following Fig. 12.109 shows, in more detail **per consumer sector**, the per capita emissions of the resident population in 2011, and those corresponding to the 2-degree scenario for 2030 (variant *a* and *b*).

The four consumer sectors are (total specific emissions of $CO_2$ in 2011: 5.9 t/capita):

- industry (2.0 t/capita)
- transport (0.4 t/capita)
- residential, services, agriculture, etc. (0.7 t/capita)
- energy-sector losses (2.8 t/capita).

The energy and emissions data for 2011 correspond to the Turkey-Annex A5.

Final-energy carriers emit $CO_2$ directly or indirectly (via electricity and district heating). The assumed distribution of these energy carriers within the consumer sectors, for both variants in 2030, is ***one way among others*** to meet the 2-degree condition.

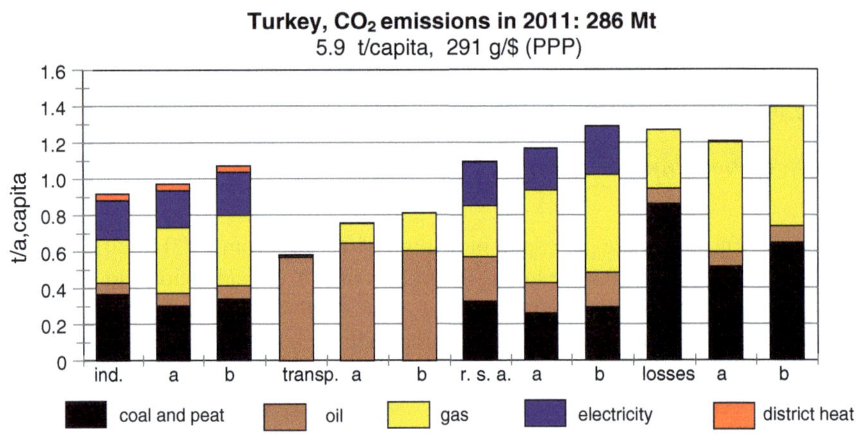


**Fig. 12.109** $CO_2$ emissions per capita of the 4 sectors: industry, transport, residential + services + agriculture, energy-sector losses (definitions see Appendix), each sector represented by three bars: *first bar* 2011 (data in the text and the diagram title, according to the Turkey-Annex A5), *second bar* possible 2 °C scenario **variant *a*** for **2030**, 330 Mt, 4.1 t/capita, 174 g/$ (PPP), *third bar* possible 2 °C scenario **variant *b*** for **2030**, 370 Mt, 4.6 t/capita, 195 g/$ (PPP)

The target value for 2030 can, for example, be achieved if the per capita emissions:

- in the **electricity and district heat production sector, including energy-sector losses**, are reduced by 5 % (variant *a*) or increase by not more than 10 % (variant *b*) from 1.77 t/capita in 2011 to 1.68–1.94 t/capita (efficiency improvements, replacement of coal by gas, CCS, renewable energy and nuclear power)
- in the **heating sector** (industry, residential, services etc.) through efficiency improvements (better insulation), reduction of coal consumption, gas instead of coal, and the use of renewable energy sources (heat pump, waste management, solar energy, geothermal energy), increase by not more than 21 % (*a* and *b*) from 1.5 t/capita in 2011 to 1.8 t/capita
- in the **transport sector** increase by 31 % (*a*) but not more than 40 % (*b*), from 0.58 in 2011 to 0.76–0.81 t/capita, through efficiency improvements, gas fuels, bio-fuels and electric mobility.

# 12.19 Argentina

## 12.19.1 Actual Indicators (Energy Intensity, $CO_2$ Intensity and $CO_2$ Sustainability) from 1980 to 2011 and Required up to 2030 to meet the 2 °C Target

In 2011, Argentina, with an indicator of $CO_2$ sustainability of 281 g $CO_2$/\$, was well below the world average, thanks to hydropower and gas (see Appendix). The trend of the indicators from 1980 to 2011 and the necessary one for the 2-degree target by 2030, for the variants *a* and *b* are shown in Fig. 12.110.

The climate target to be aimed at for 2030 is 168–182 g $CO_2$/\$ (−40 to −35 % relative to 2011) for a share of 185–200 Mt (unchanged to +9 % relative to 2011) of the total G-20 $CO_2$ emissions. The values for 2018 take into account the GDP (PPP) forecast by the International Monetary Fund (IMF) of around \$790 billion (\$ of 2007). The GDP for 2030 is estimated at \$1,100 billion.

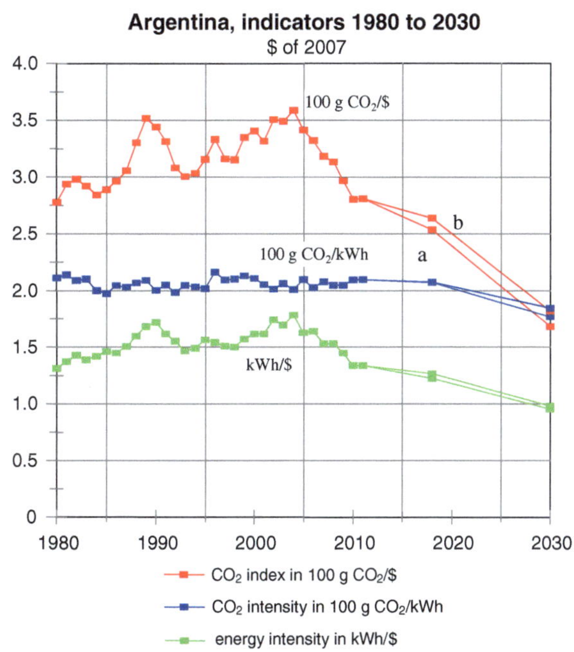


**Fig. 12.110** Argentina indicators 1980 to 2011 and 2-degrees scenario up to 2030

With this scenario, the gross energy demand in 2030 is 90–93 Mtoe (only energy use). The per capita indicators for energy and $CO_2$ emissions would be: e ≈ 2.4–2.5 kW/capita (decrease of 3 % to unchanged, relative to 2011) and $\alpha$ = 3.7–4.0 t $CO_2$ per capita (−18 % (var. *a*) to −12 % (var. *b*) relative to 2011), see also Sect. 12.19.3

The necessary change in the trend of the indicators to meet the 2-degree target is illustrated in Fig. 12.111 for variant *a*. The good trend in energy intensity must be maintained, and overcoming the trend stagnation in $CO_2$ intensity of energy after 2018 is required.

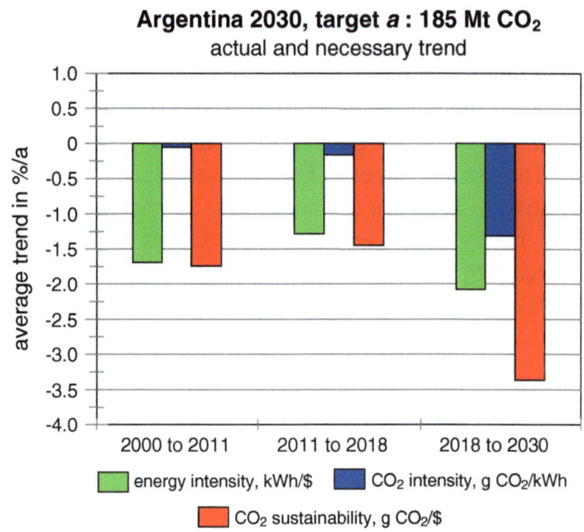


**Fig. 12.111** Necessary trend change to meet the 2-degrees target, variant *a*

Variant *b* (Fig. 12.112) allows a regression of the energy intensity trend until 2018 and the turnaround in the trend of $CO_2$ intensity of energy is less severe after 2018.

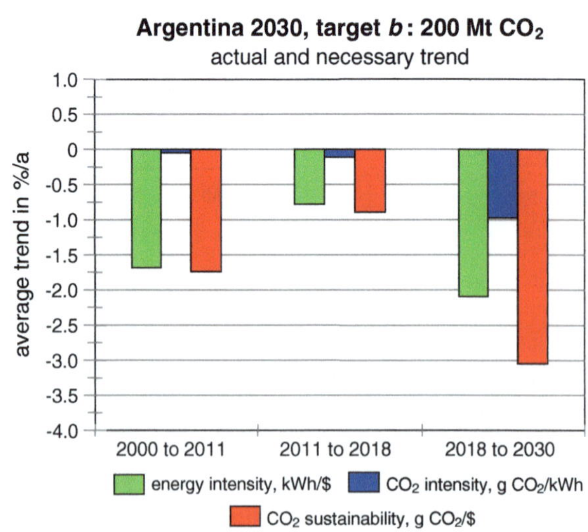


**Fig. 12.112** Necessary trend change to meet the 2-degrees target, variant *b*

## 12.19.2 $CO_2$ Emissions until 2050

The actual $CO_2$ emissions in Argentina from 1970 to 2011, and those allowed to 2050 for compliance with the 2 °C climate target (according to the indicators of Figs. 12.110, 12.111, 12.112) are shown in Fig. 12.113.

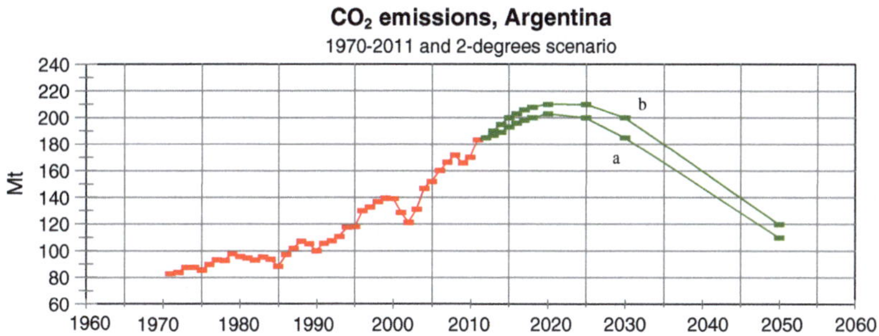


**Fig. 12.113** $CO_2$ emissions of Argentina from 1970 to 2011 and climate protection scenario until 2050

## 12.19.3 Per Capita Indicators up to 2030

Figure 12.114 shows the per capita indicators of Argentina from 1980 to 2011 as well as the profile up to 2030 resulting from the above considerations to meet the 2-degree target, for both variants *a* and *b*. The GDP (PPP) data up to 2018 correspond to the statistics and forecasts of the IMF.

**Fig. 12.114** Per capita indicators of Argentina from 1980 to 2011 and 2-degrees scenario until 2030

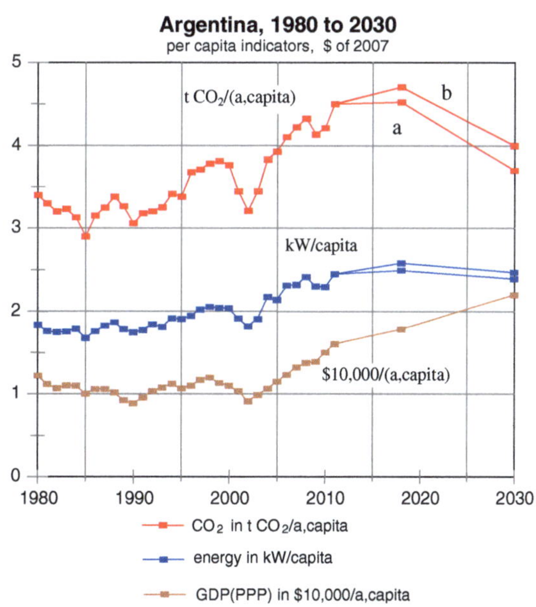

The following Fig. 12.115 shows, in more detail **per consumer sector**, the per capita emissions of the resident population in 2011, and those corresponding to the 2-degree scenario for 2030 (variant *a* and *b*).

The four consumer sectors are (total specific emissions of $CO_2$ in 2011: 4.5 t/capita):

- industry (1.0 t/capita)
- transport (1.2 t/capita)
- residential, services, agriculture, etc. (1.1 t/capita)
- energy-sector losses (1.2 t/capita).

The energy and emissions data for 2011 correspond to the Argentina-Annex A5.

Final-energy carriers emit $CO_2$ directly or indirectly (via electricity and district heating). The assumed distribution of these energy carriers within the consumer sectors, for both variants in 2030, is ***one way among others*** to meet the 2-degree condition.

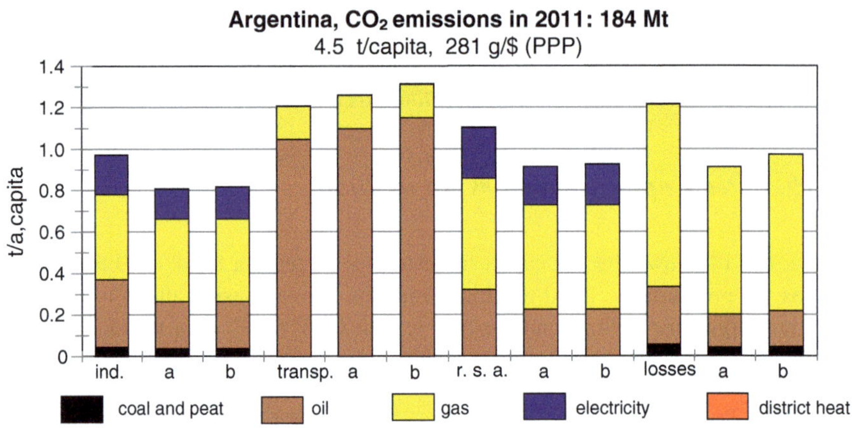


**Fig. 12.115** $CO_2$ emissions per capita of the 4 sectors: industry, transport, residential + services + agriculture, energy-sector losses (definitions see Appendix), each sector represented by three bars: *first bar* 2011 (data in the text and the diagram title, according to the Argentina-Annex A5), *second bar* possible 2 °C scenario **variant *a*** for **2030**, 185 Mt, 3.7 t/capita, 168 g/$ (PPP), *third bar* possible 2 °C scenario **variant *b*** for **2030**, 200 Mt, 4.0 t/capita, 182 g/$ (PPP)

The target value for 2030 can, for example, be achieved if the per capita emissions:

- in the **electricity and district heat production sector, including energy-sector losses**, are reduced by 25 % (variant *a*) or 20 % (variant *b*) from 1.66 t/capita in 2011 to 1.24–1.33 t/capita (efficiency improvements, replacement of oil by gas, CCS, renewable energy and nuclear power)
- in the **heating sector** (industry, residential, services etc.) through efficiency improvements (better insulation), gas instead of oil, and the use of renewable energy sources (heat pump, waste management, solar energy, geothermal energy), are reduced by 20 % (*a*) or 15 % (*b*) (from 1.64 t/capita in 2011 to 1.3–1.4 t/capita)
- in the **transport sector** increase by 4 % (*a*) but not more than 8 % (*b*), from 1.21 t/capita in 2011 to 1.26–1.31 t/capita, through efficiency improvements, gas fuels, bio-fuels and electric mobility.

# Chapter 13
# Population, GDP, Energy and $CO_2$ Emissions of World, OECD, Non-OECD, G-20

See Figs. 13.1, 13.2, 13.3, 13.4, 13.5 and 13.6.

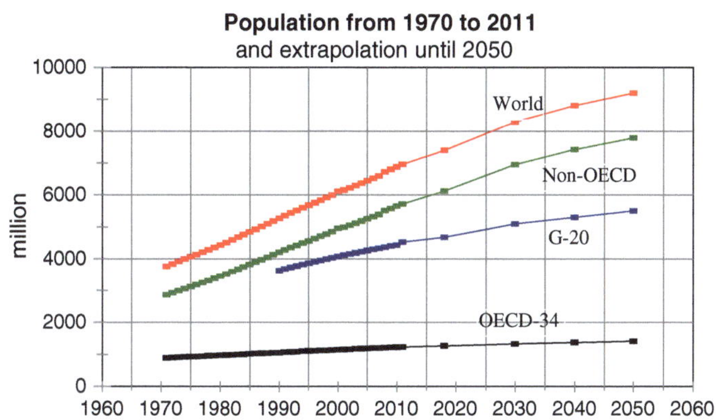


**Fig. 13.1** World wide population, 1970–2011 and extrapolation until 2050



V. Crastan, *Global Energy Demand and 2-degree Target, Report 2014*,
DOI 10.1007/978-3-319-12310-3_13

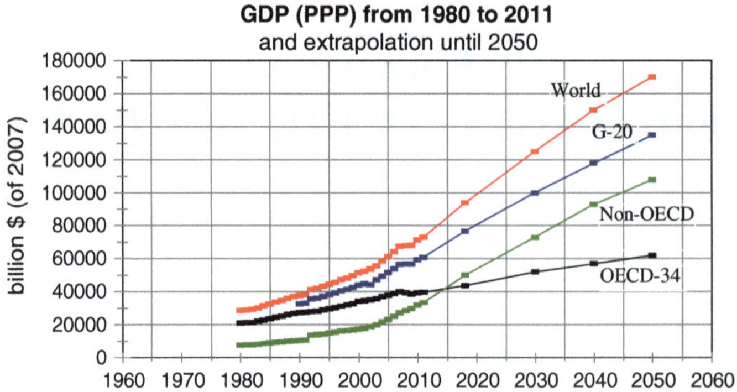


**Fig. 13.2** World wide gross domestic product, with purchasing power parity, in Dollar of 2007, from 1980 to 2011 and extrapolation until 2050. The values for 2018 correspond to the IMF forecast

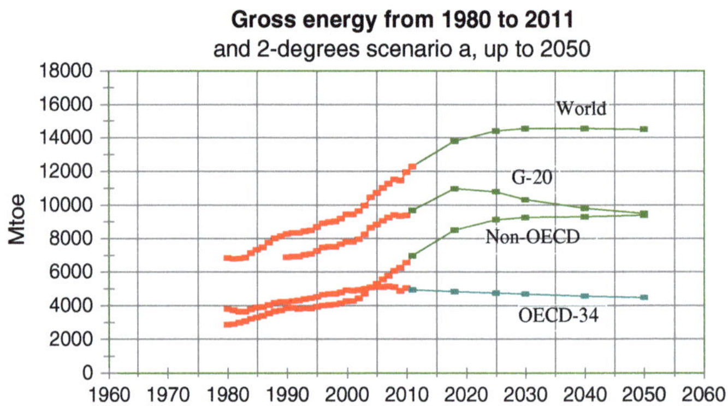


**Fig. 13.3** Actual gross energy demand (energy use) in Mtoe from 1980 to 2011 and 2-degree scenario up to 2050, variant *a*

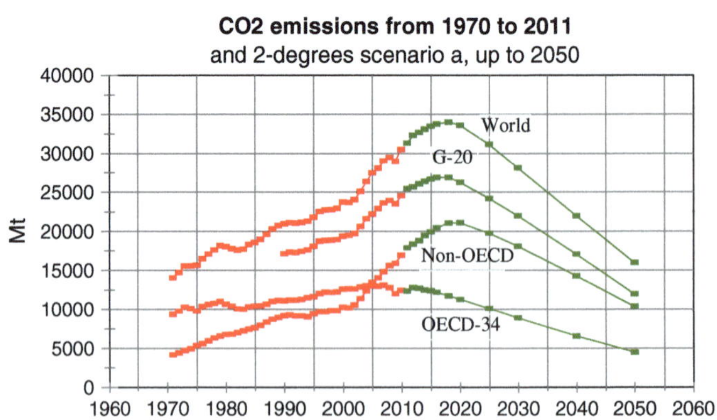


**Fig. 13.4** Actual $CO_2$ emissions from energy from 1970 to 2011 and 2-degrees scenario up to 2050, variant *a* (World including marine and aviation bunkers)

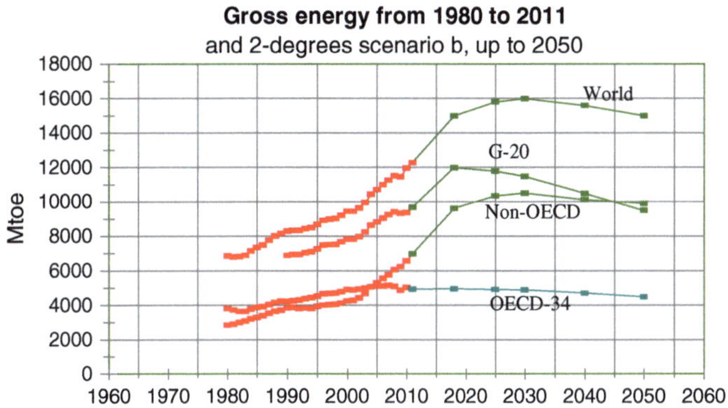


**Fig. 13.5** Actual gross energy demand (energy use) in Mtoe from 1980 to 2011 and 2-degree scenario up to 2050, variant *b*

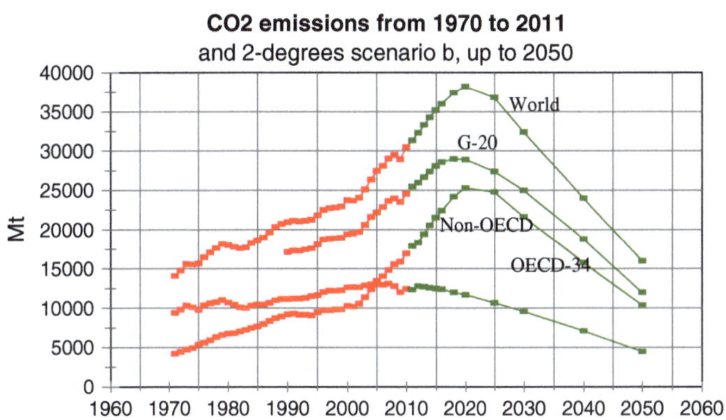


**Fig. 13.6** Actual $CO_2$ emissions from energy from 1970 to 2011 and 2-degrees scenario up to 2050, variant *b* (World including marine and aviation bunkers)

# Appendix
# Energy Flow, Energy Demand and $CO_2$ Emissions of the World's-Regions and of the G-20 Countries in 2011

**Valentin Crastan**

## Description of Diagrams

The energy consumption of the world, for the individual regions of the world and for all G-20 countries is illustrated by 6 diagrams A1 ⋯ A6, described below (data based on: Energy Statistics from the IEA, International Energy Agency [5]).

## Figure A1: Energy Flow in the Energy Sector

Figure A1 describes the flow of energy in the energy sector from primary energy to gross energy (or gross inland consumption) and to final energy. Primary energy and gross energy are illustrated by the **energy carriers** used. Energy is expressed in Mtoe.

The **primary energy** is the sum of local production and, for regions, net imports minus net exports of energy carriers (for countries, effective imports/exports instead of net imports/net exports per energy carrier).

The **gross energy** follows from the primary energy after subtracting the non-energy use demand (e.g. for the chemical industry) and any stock changes. For the regions of the world, the amounts of energy for the international marine and aviation bunkers are also subtracted, since the corresponding $CO_2$ emissions are recorded only worldwide.

It is the task of the **energy sector** to provide consumers with energy in the form of **final energy**. In this diagram we distinguish 4 forms of final energy: **electricity**, **district heating**, **motor fuels** and "**heat**". The latter consists mainly of non-electric heating and process heat (from fossil and renewable energies) and without district heating. Stationary work, of non-electrical origin, can also be included (e.g. stationary gas, petrol or diesel engines and pumps); at least in developed countries, this share is minimal. With the conversion of gross energy into final energy, losses occur, which we refer to as **losses of the energy sector**.

These losses consist of **thermal losses** in power stations (thermodynamic reasons) as well as in combined heat and power plants and heating plants, and



V. Crastan, *Global Energy Demand and 2-degree Target, Report 2014*,
DOI 10.1007/978-3-319-12310-3

additionally **electrical losses** in the transmission and distribution network, including the energy sector's own energy requirement and, finally, of the **remaining losses** in the energy sector (refining, liquefaction and gasification plants, heat transfer losses, in-house use of heat, etc.).

The diagram also shows the **$CO_2$ emissions in Mt** associated with the final energy consumption and the losses of the energy sector (together corresponding to the gross inland consumption). Most of the losses of the energy sector are normally associated with the electricity and district heat production, which is why the $CO_2$ emissions of these three factors are combined. Separation can be done using Figure A4.

## Figure A2: Shares of Energy Carriers

Figure A2 shows the shares of energy carriers, in percentage of total final energy demand, required for obtaining the 4 final energy types, and for covering the losses of the energy sector.

This diagram thus complements diagram A1, with respect to the energy carrier distribution in the final energy area. The energy-carriers-colors are shown in the legend, and also apply to the energy flow diagram A1.

## Figure A3: Losses of the Energy Sector

The losses of the energy sector represent a significant part of gross energy demand. Worldwide, in 2011, they were about 54 % and in some countries they even exceed 100 % of final energy demand. For the $CO_2$ emissions they are therefore of primary importance.

Figure A3 shows the percentage composition of these losses and the resulting $CO_2$ emissions. A characteristic parameter and index of the $CO_2$-efficiency of the energy sector in the world region or country is the ratio Mt $CO_2$ relative to Mtoe losses.

## Figure A4: Energy Flow of Final Energy to the Final Users

The chart shows the breakdown of the 4 types of final energy over the three end-user categories. Similarly, the $CO_2$ emissions are allocated to these consumer groups.

The end users are (according to IEA statistics)

- industry,
- residential, services, agriculture, etc.,
- transport.

To obtain the total emissions, the $CO_2$ emissions of the energy sector (i.e. of the corresponding losses) have to be added to this.

## Figure A5: $CO_2$ Emissions and Originating End-Energy Carriers

The figure gives the total $CO_2$ emissions in Mt, the emissions per capita and its distribution over the consumer groups, detailed per end-energy carrier.

In addition, the indicator of $CO_2$ sustainability of the country or region is stated.

The causes of the $CO_2$ emissions of the end energy carriers electricity and district heat can be taken from the diagrams A2 and A6.

## Figure A6: Electricity Production and Consumption

The large, often decisive importance of electricity consumption and of its generation for the amount of $CO_2$ emissions is clear from the flow charts A1 and A4, and also from A5. The share of the losses of the energy sector (mostly coupled with the electricity production) is also to be taken into account.

Diagram A6 shows in detail the percentage composition of the energy sources used for the production of electricity.

Similarly, imports and exports are given in percentage of final consumption.

The final consumption follows after deduction of losses (network losses and the energy sector's own use).

## Regions of the World

### World

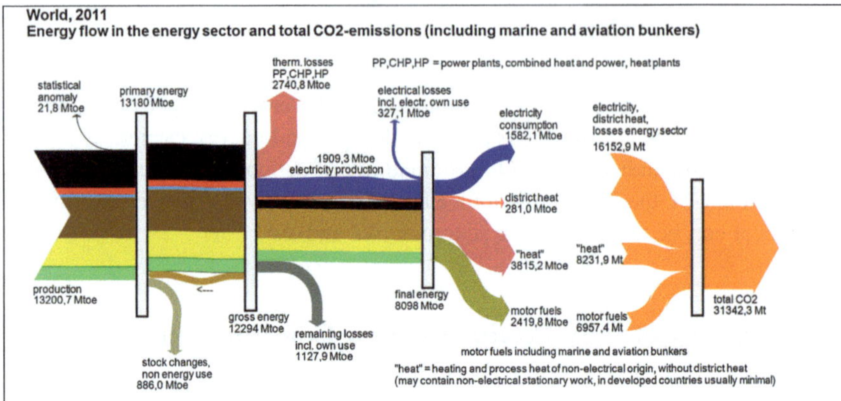


A1. World: Energy flow in the energy sector from primary to final energy and total $CO_2$ emissions. Color of the energy carriers as in A2 and A5 (petroleum dark brown, oil products light brown)

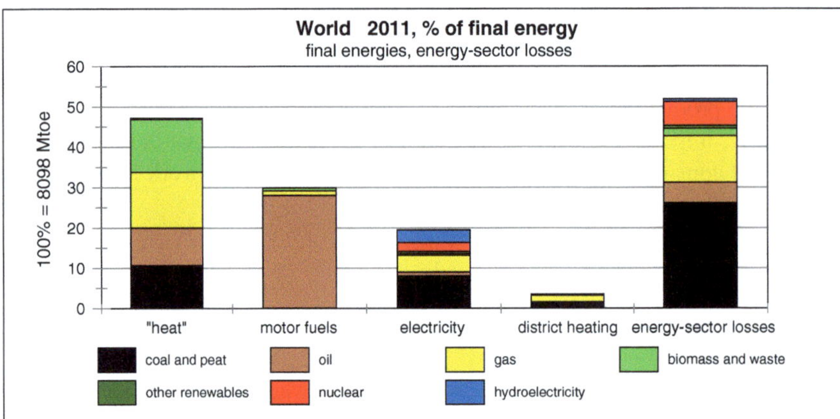


A2. World: : Shares of energy carriers for obtaining the final energies ("heat", motor fuels, electricity, district heating) and to cover the losses of the energy sector

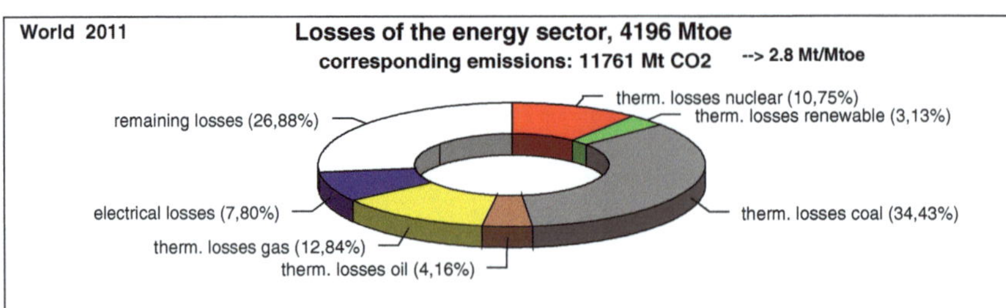


A3. World: Percentage distribution of losses in the energy sector; contributors to $CO_2$ emissions include thermal losses of fossil-fuel plants, the electrical losses and the residual losses

## World

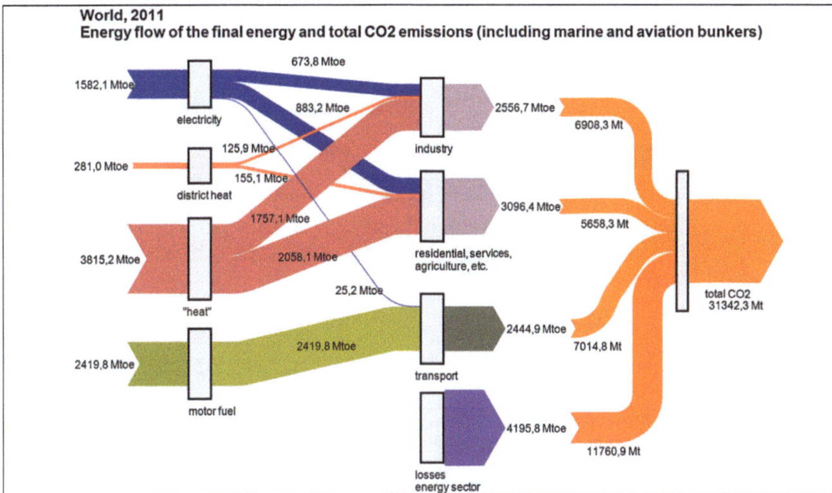


A4. World: Flow of final energy to the final users and $CO_2$ emissions of the economy sectors

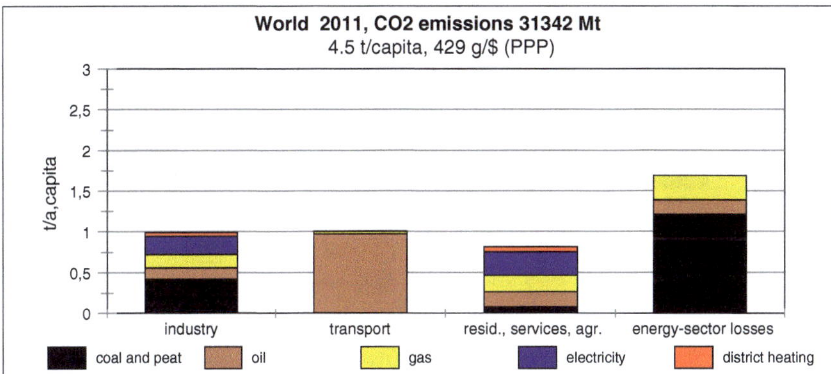


A5. World: Energy carriers responsible for the $CO_2$ emissions of sectors of the economy ;
(for electricity see also A2 or A6 and for district heating also A2)

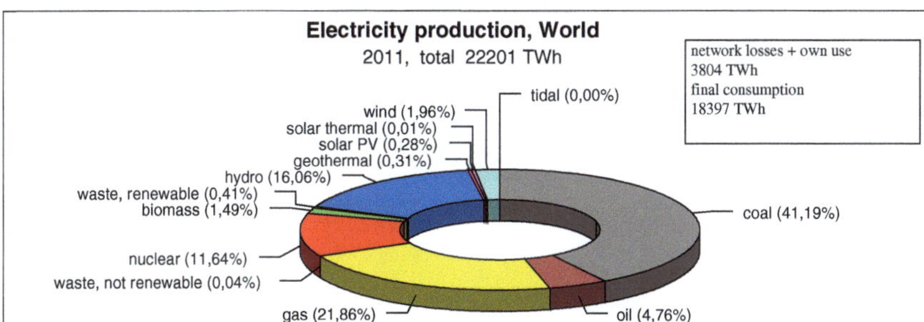


A6. World: Generation of electrical energy, final consumption = production - losses

## OECD-34

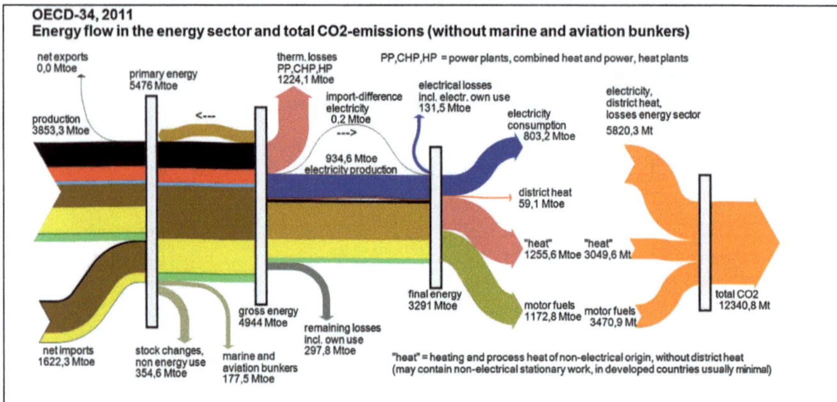


A1. OECD-34: Energy flow in the energy sector from primary to final energy and total $CO_2$ emissions. Color of the energy carriers as in A2 and A5 (petroleum dark brown, oil products light brown)

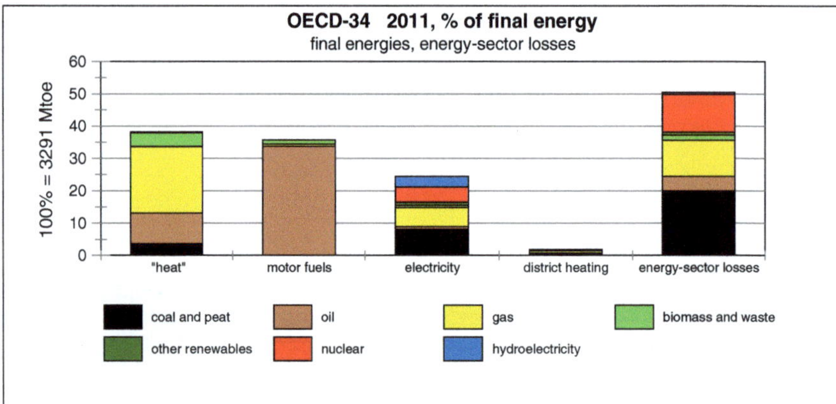


A2. OECD-34: Shares of energy carriers for obtaining the final energies ("heat", motor fuels, electricity, district heating) and to cover the losses of the energy sector

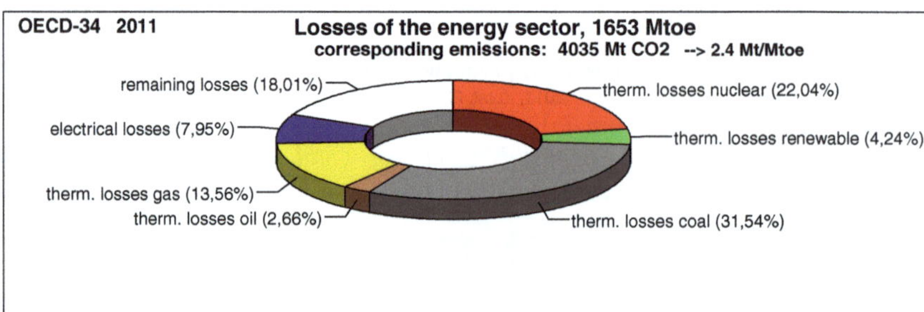


A3. OECD-34: Percentage distribution of losses in the energy sector; contributors to $CO_2$ emissions include thermal losses of fossil-fuel plants, the electrical losses and the residual losses

## OECD-34

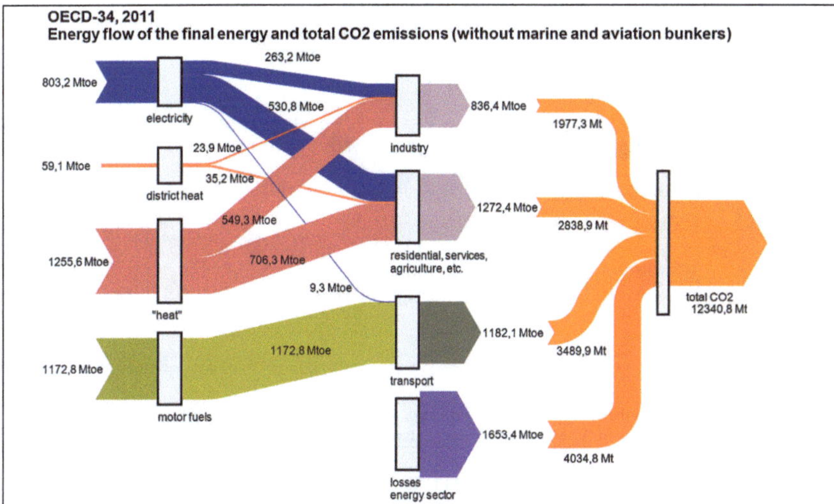


A4. OECD-34: Flow of final energy to the final users and $CO_2$ emissions of the economy sectors

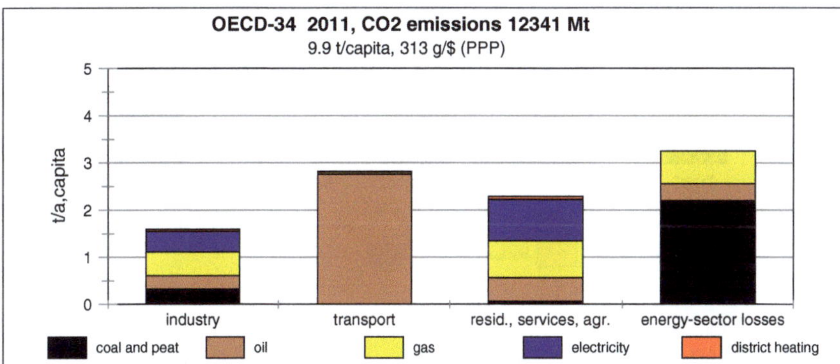


A5. OECD-34: Energy carriers responsible for the $CO_2$ emissions of sectors of the economy ; (for electricity see also A2 or A6 and for district heating also A2)

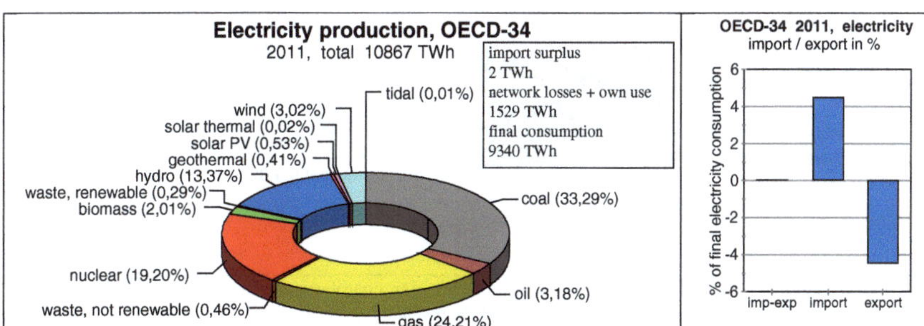


A6. OECD-34: Generation of electrical energy, final consumption = production + imports - exports - losses

## EU-27

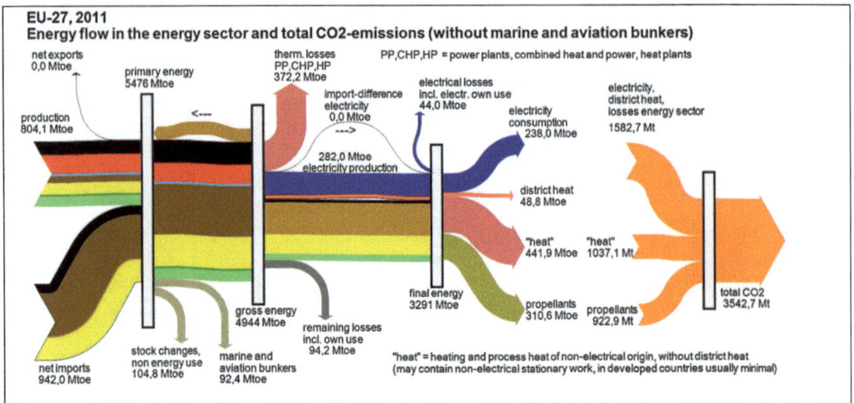


A1. EU-27: Energy flow in the energy sector from primary to final energy and total $CO_2$ emissions. Color of the energy carriers as in A2 and A5 (petroleum dark brown, oil products light brown)

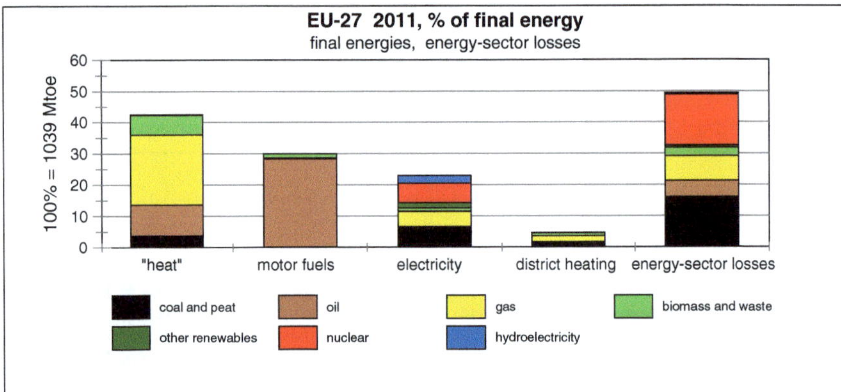


A2. EU-27: Shares of energy carriers for obtaining the final energies ("heat", motor fuels, electricity, district heating) and to cover the losses of the energy sector

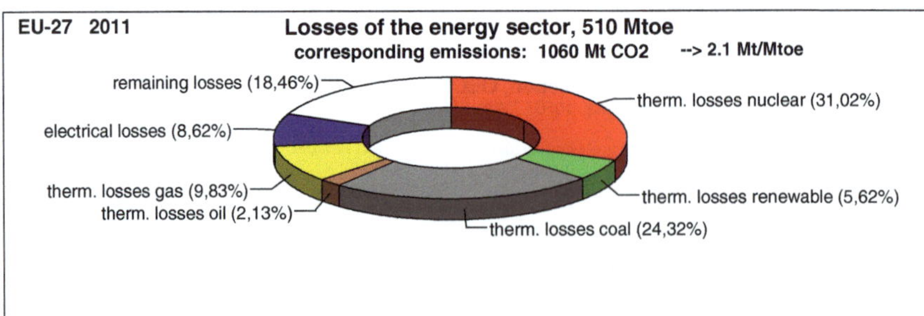


A3. EU-27: Percentage distribution of losses in the energy sector; contributors to $CO_2$ emissions include thermal losses of fossil-fuel plants, the electrical losses and the residual losses

## EU-27

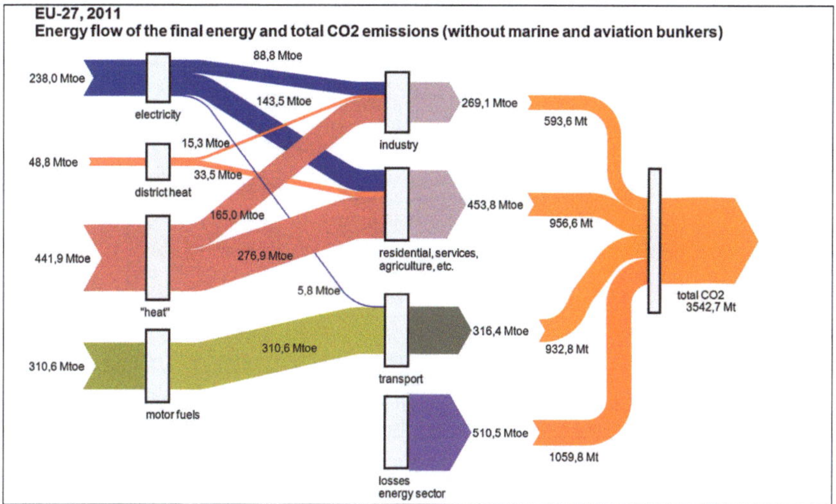


A4. EU-27: Flow of final energy to the final users and $CO_2$ emissions of the economy sectors

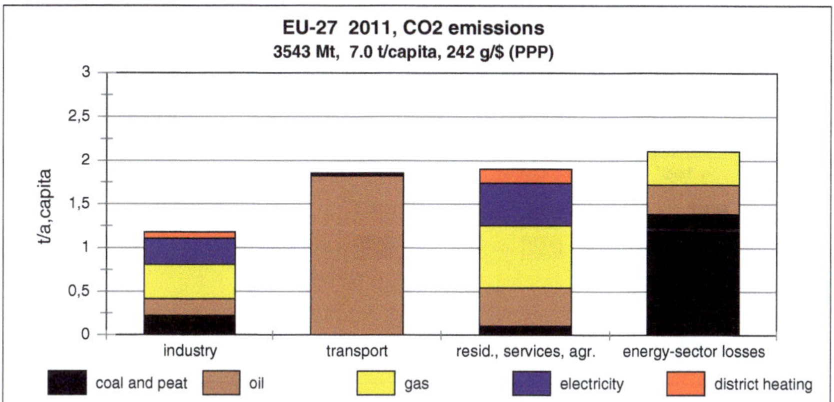


A5. EU-27: Energy carriers responsible for the $CO_2$ emissions of sectors of the economy ; (for electricity see also A2 or A6 and for district heating also A2)

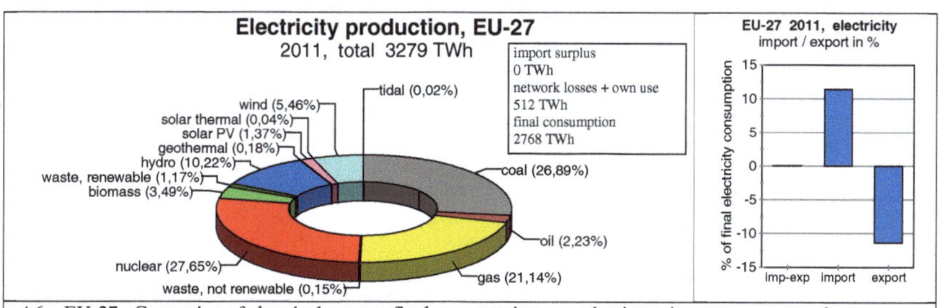


A6. EU-27: Generation of electrical energy, final consumption = production + imports - exports - losses

## Middle East

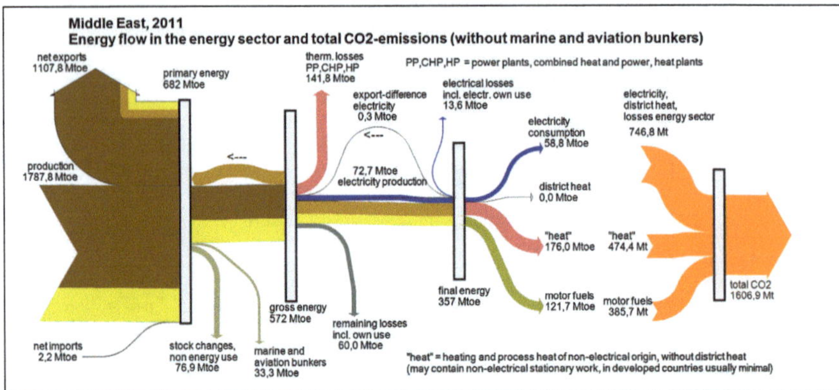


A1. Middle East: Energy flow in the energy sector from primary to final energy and total $CO_2$ emissions. Color of the energy carriers as in A2 and A5 (petroleum dark brown, oil products light brown)

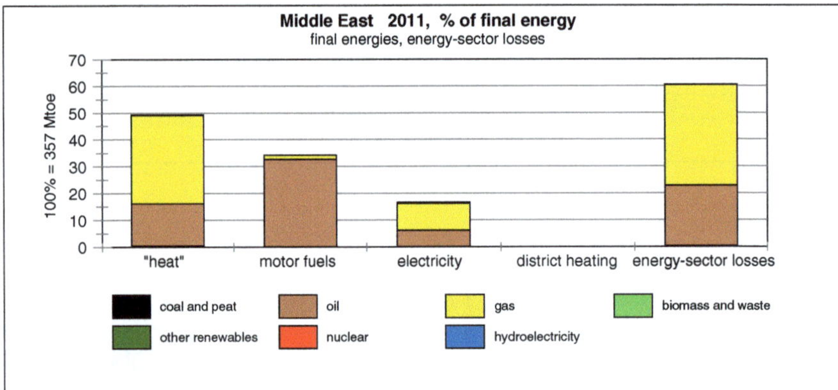


A2. Middle East: Shares of energy carriers for obtaining the final energies ("heat", motor fuels, electricity, district heating) and to cover the losses of the energy sector

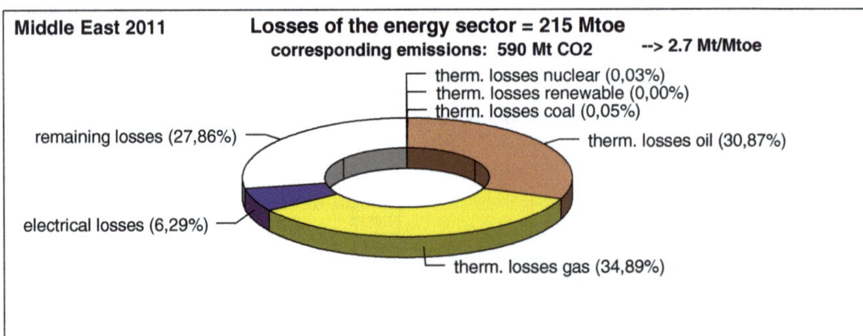


A3. Middle East: Percentage distribution of losses in the energy sector; contributors to $CO_2$ emissions include thermal losses of fossil-fuel plants, the electrical losses and the residual losses

## Middle East

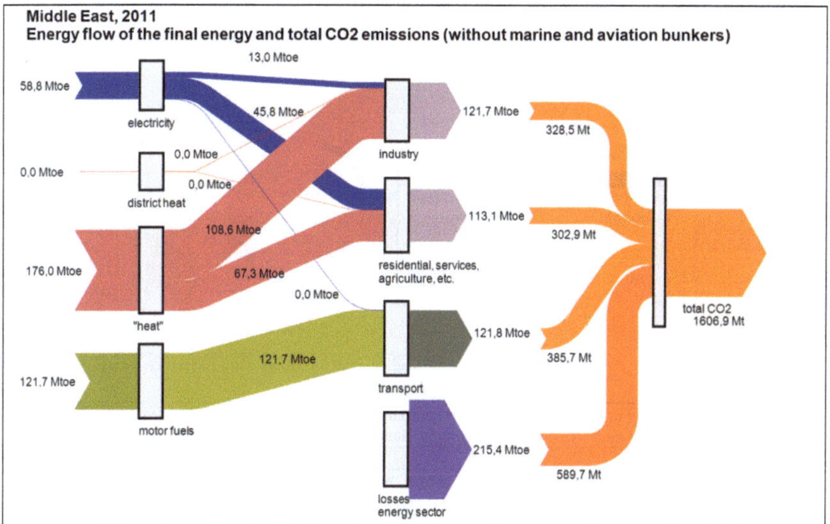


A4. Middle East: Flow of final energy to the final users and $CO_2$ emissions of the economy sectors

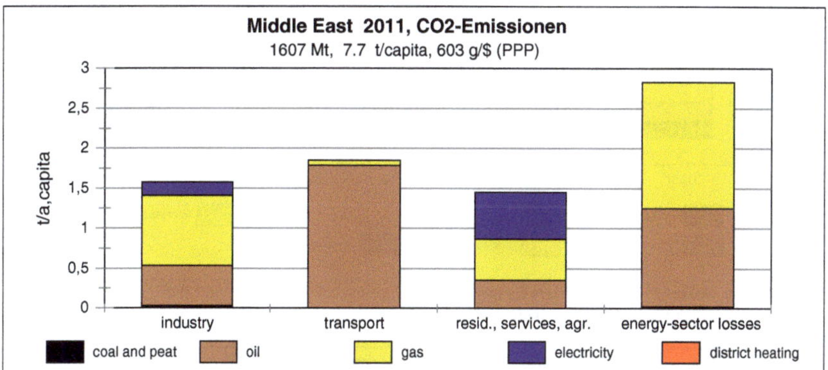


A5. Middle East: Energy carriers responsible for the $CO_2$ emissions of sectors of the economy ; (for electricity see also A2 or A6 and for district heating also A2)

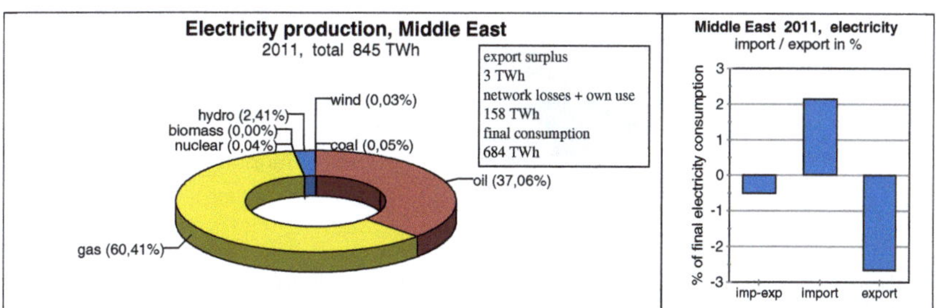


A6. Middle East: Generation of electrical energy, final consumption = production + imports - exports - losses

## Eurasia+

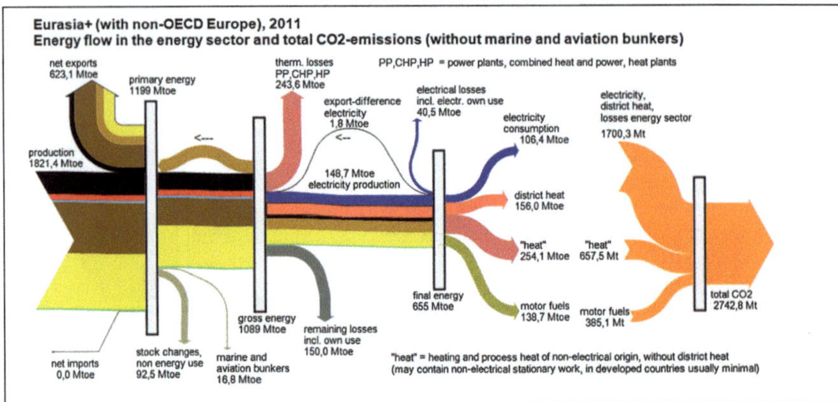


A1. Eurasia+: Energy flow in the energy sector from primary to final energy and total $CO_2$ emissions. Color of the energy carriers as in A2 and A5 (petroleum dark brown, oil products light brown)

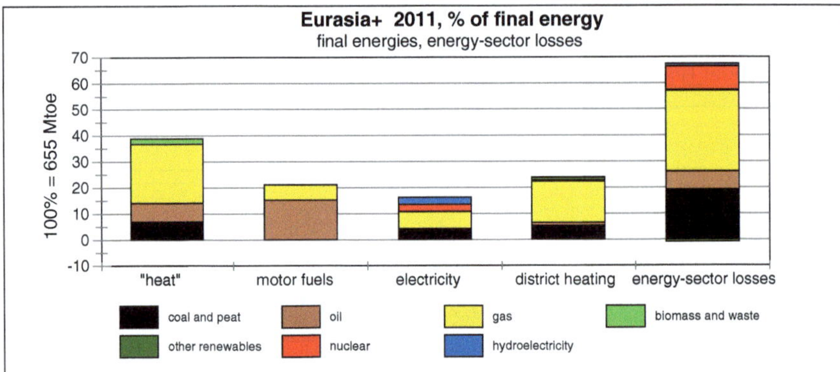


A2. Eurasia+: Shares of energy carriers for obtaining the final energies ("heat", motor fuels, electricity, district heating) and to cover the losses of the energy sector

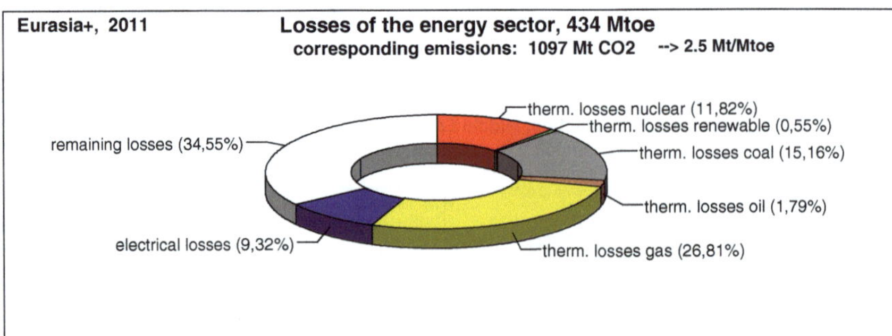


A3. Eurasia+: Percentage distribution of losses in the energy sector; contributors to $CO_2$ emissions include thermal losses of fossil-fuel plants, the electrical losses and the residual losses

## Eurasia+

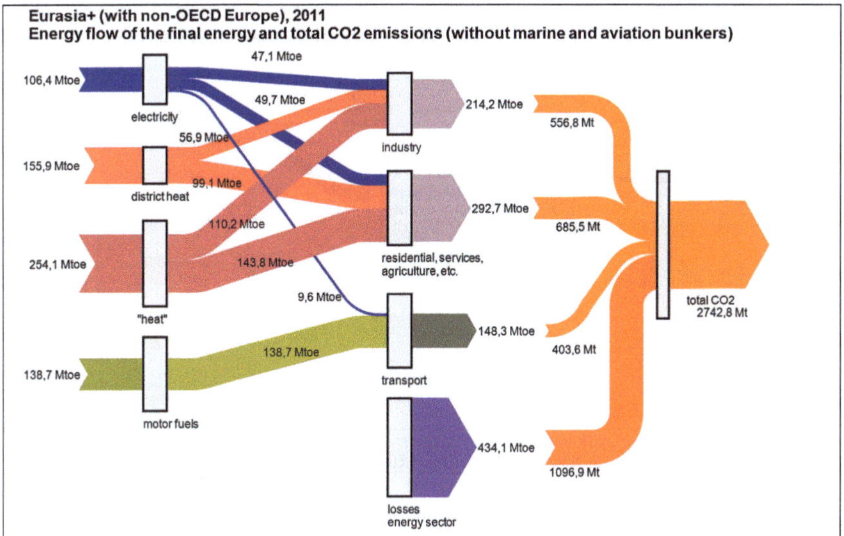


A4. Eurasia+: Flow of final energy to the final users and $CO_2$ emissions of the economy sectors

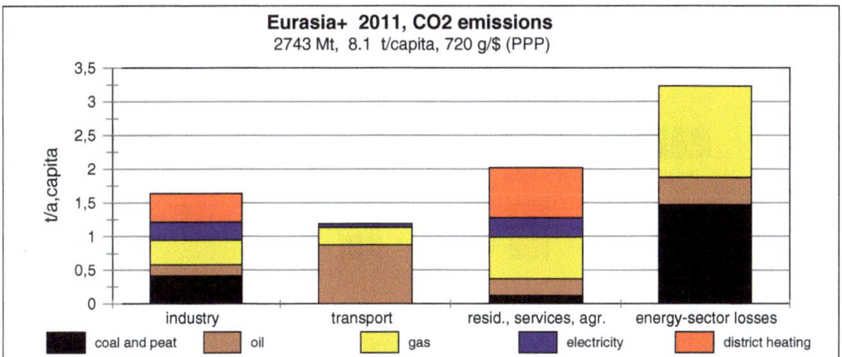


A5. Eurasia+: Energy carriers responsible for the $CO_2$ emissions of sectors of the economy ; (for electricity see also A2 or A6 and for district heating also A2)

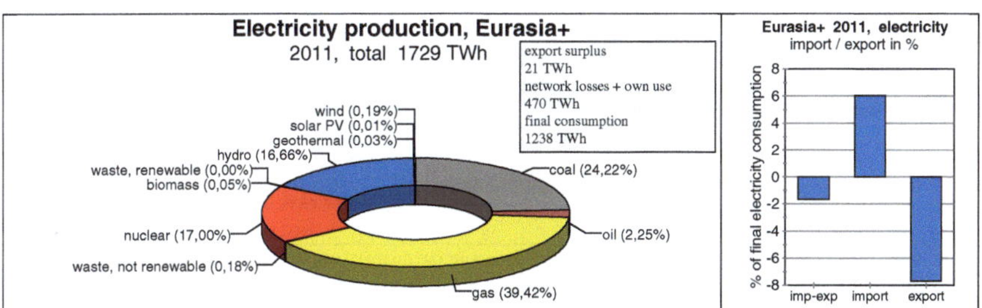


A6. Eurasia+: Generation of electrical energy, final consumption = production + imports - exports - losses

## Rest-Asia/Oceania

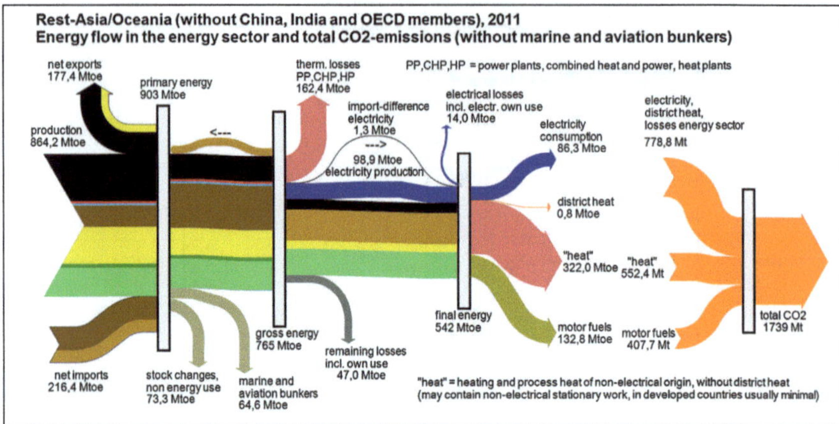


A1. Rest-Asia/Oceania: Energy flow in the energy sector from primary to final energy and total $CO_2$ emissions. Color of the energy carriers as in A2 and A5 (petroleum dark brown, oil products light brown)

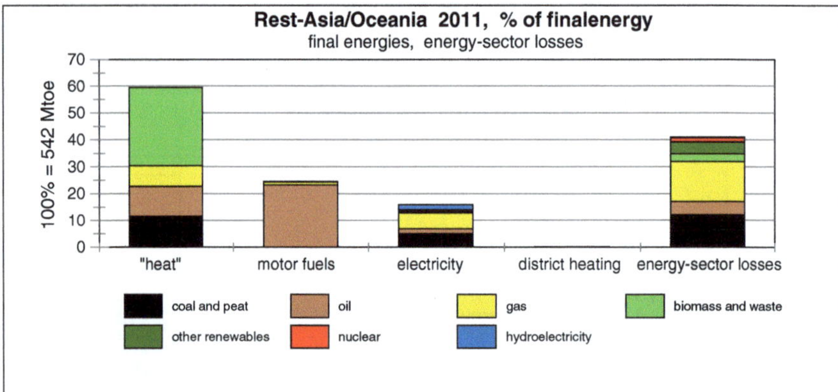


A2. Rest-Asia/Oceania: Shares of energy carriers for obtaining the final energies ("heat", motor fuels, electricity, district heating) and to cover the losses of the energy sector

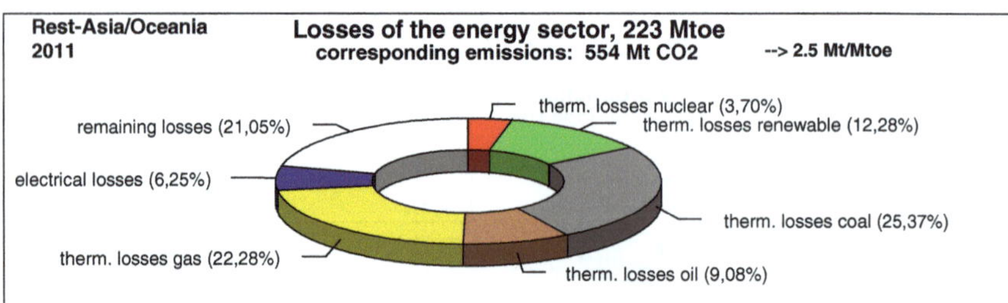


A3. Rest-Asia/Oceania: Percentage distribution of losses in the energy sector; contributors to $CO_2$ emissions include thermal losses of fossil-fuel plants, the electrical losses and the residual losses

## Rest-Asia/Oceania

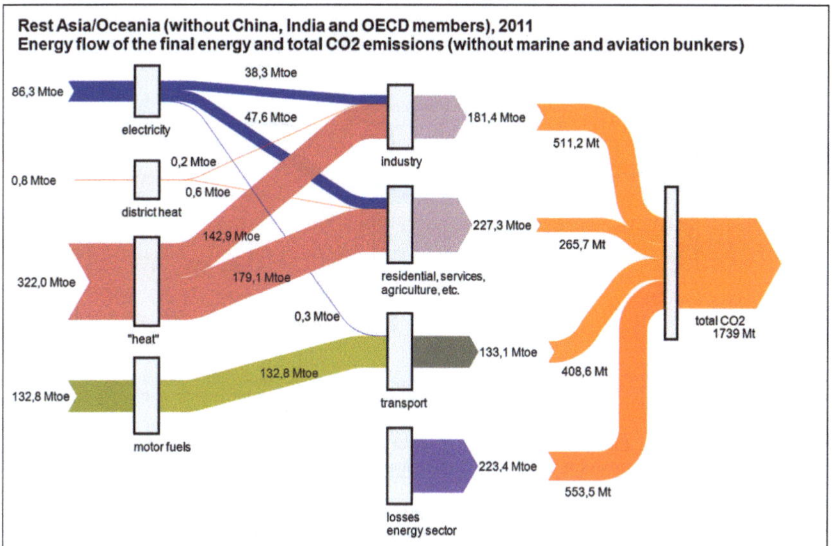


A4. Rest-Asia/Oceania: Flow of final energy to the final users and $CO_2$ emissions of the economy sectors

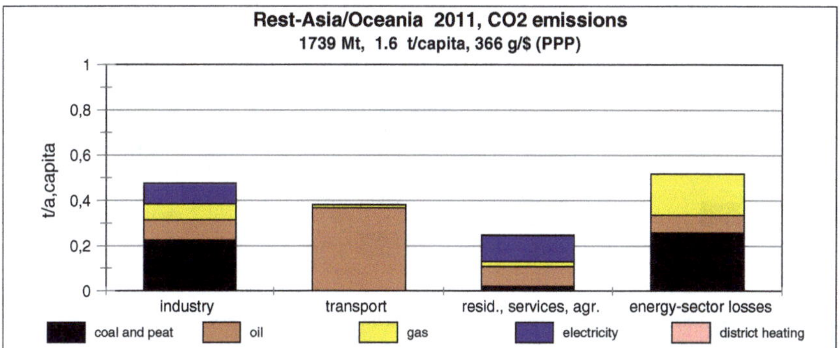


A5. Rest-Asia/Oceania: Energy carriers responsible for the $CO_2$ emissions of sectors of the economy ; (for electricity see also A2 or A6 and for district heating also A2)

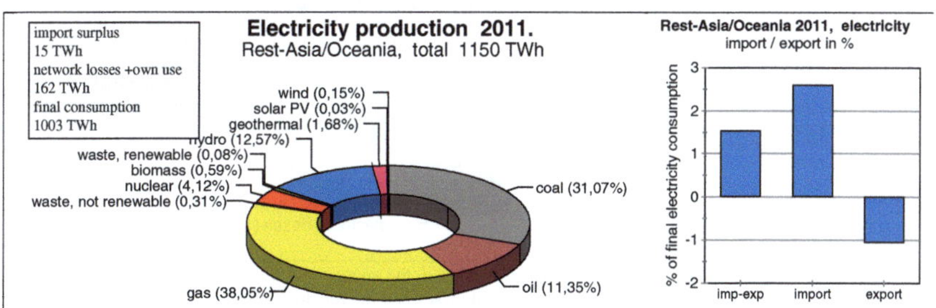


A6. Rest-Asia/Oceania: Generation of electrical energy, final consumption = production + imp. – exp. - losses

## Non-OECD America

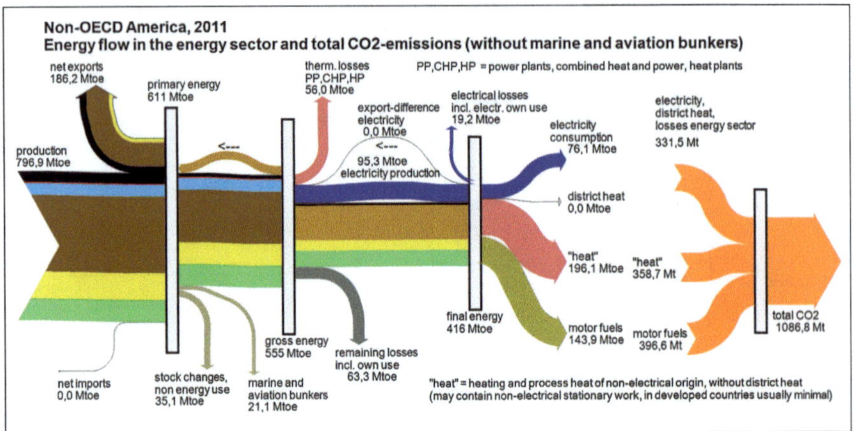


A1. Non-OECD America: Energy flow in the energy sector from primary to final energy and total $CO_2$ emissions. Color of the energy carriers as in A2 and A5 (petroleum dark brown, oil products light brown)

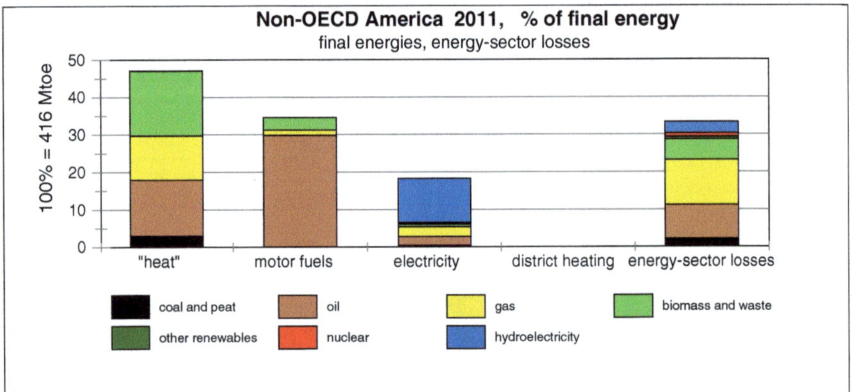


A2. Non-OECD America: Shares of energy carriers for obtaining the final energies ("heat", motor fuels, electricity, district heating) and to cover the losses of the energy sector

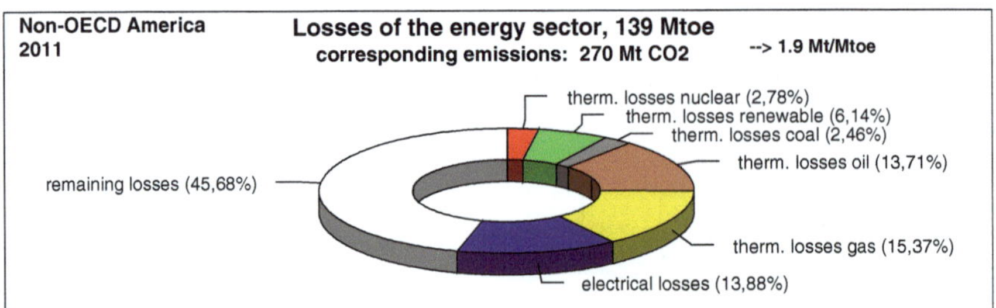


A3. Non-OECD America: Percentage distribution of losses in the energy sector; contributors to $CO_2$ emissions include thermal losses of fossil-fuel plants, the electrical losses and the residual losses

## Non-OECD America

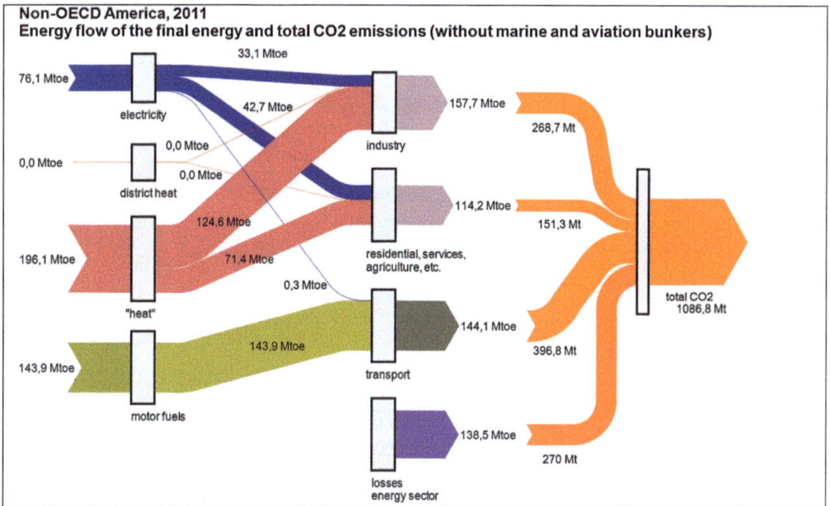


A4. Non-OECD America: Flow of final energy to the final users and $CO_2$ emissions of the economy sectors

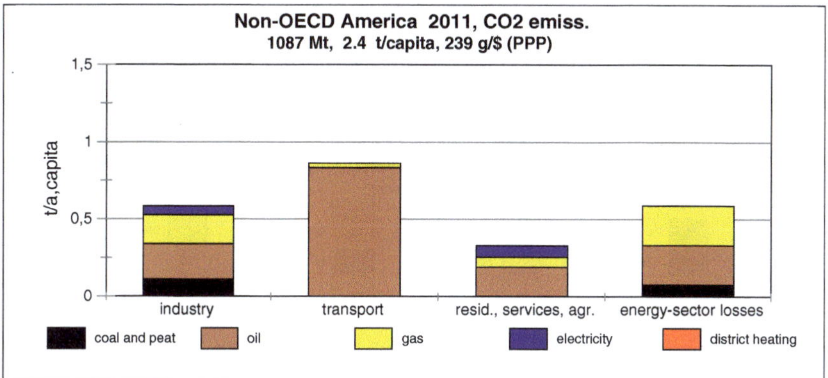


A5. Non-OECD America: Energy carriers responsible for the $CO_2$ emissions of sectors of the economy ; (for electricity see also A2 or A6 and for district heating also A2)

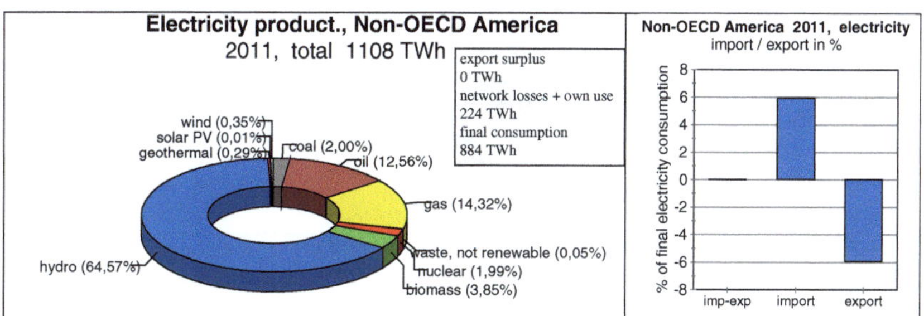


A6. Non-OECD America: Generation of electrical energy, final consumption = production + imp. – exp. - losses

## Africa

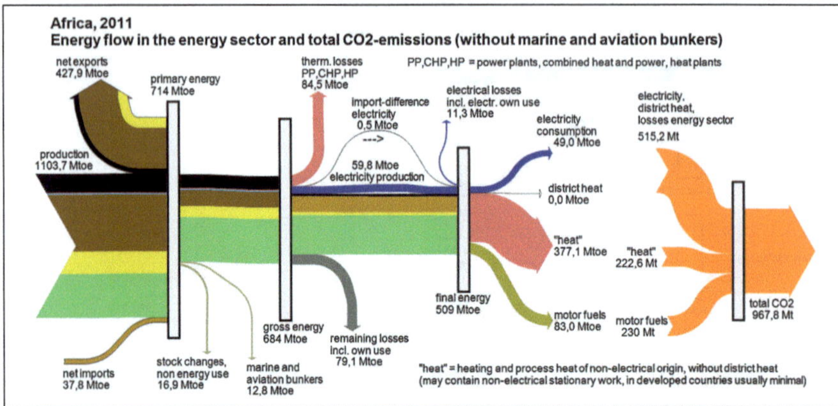


A1. Africa: Energy flow in the energy sector from primary to final energy and total $CO_2$ emissions. Color of the energy carriers as in A2 and A5 (petroleum dark brown, oil products light brown)

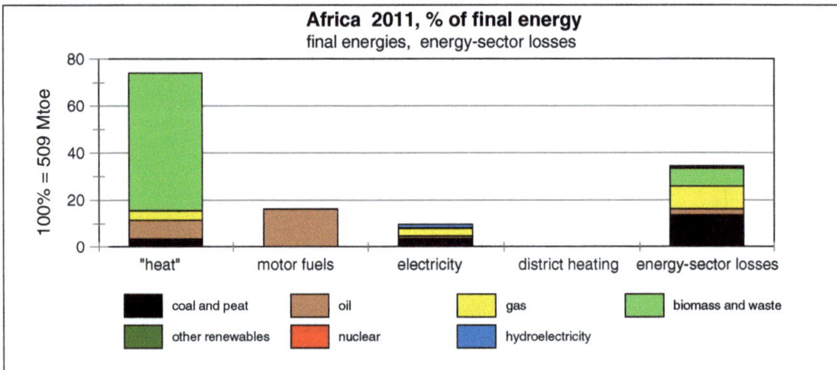


A2. Africa: Shares of energy carriers for obtaining the final energies ("heat", motor fuels, electricity, district heating) and to cover the losses of the energy sector

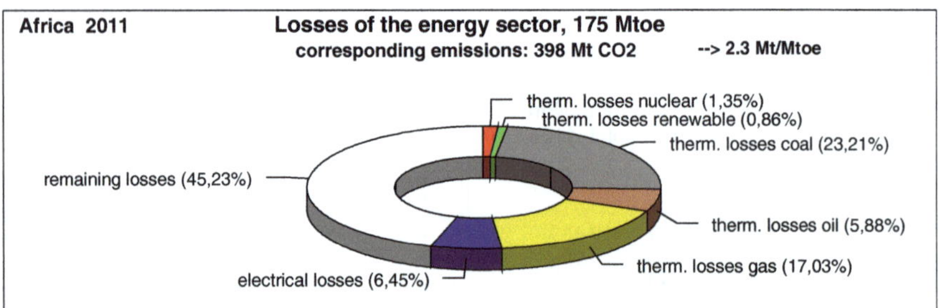


A3. Africa: Percentage distribution of losses in the energy sector; contributors to $CO_2$ emissions include thermal losses of fossil-fuel plants, the electrical losses and the residual losses

## Africa

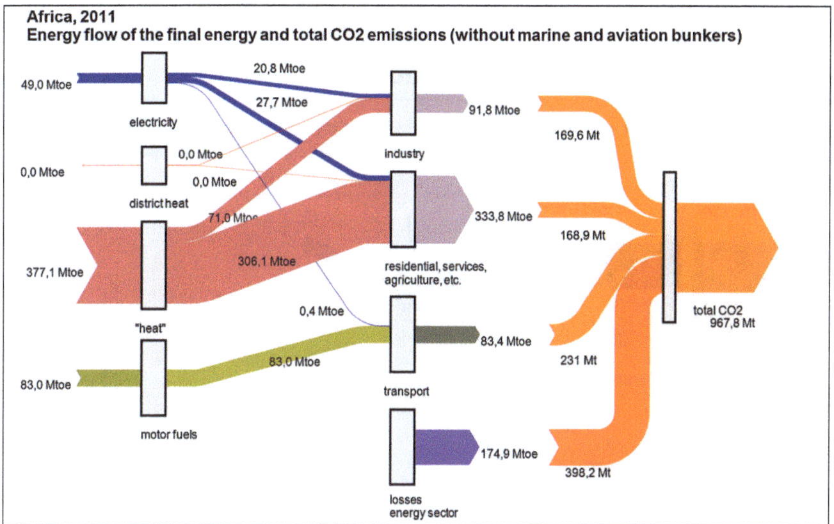


A4. Africa: Flow of final energy to the final users and $CO_2$ emissions of the economy sectors

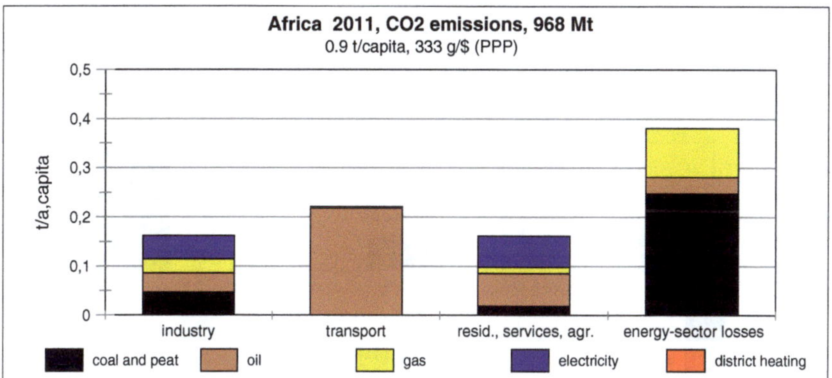


A5. Africa: Energy carriers responsible for the $CO_2$ emissions of sectors of the economy ;
(for electricity see also A2 or A6 and for district heating also A2)

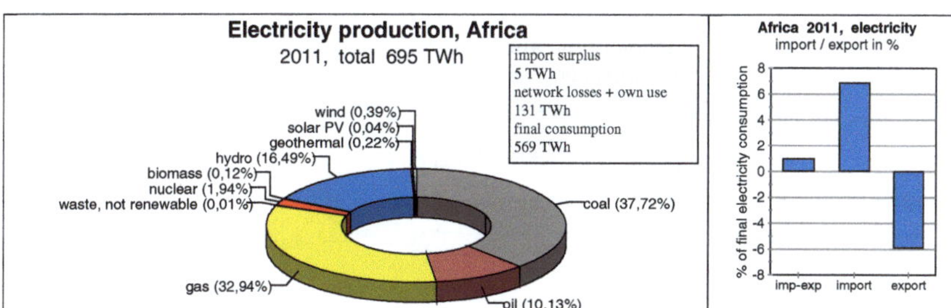


A6. Africa: Generation of electrical energy, final consumption = production + imports - exports - losses

## G-20 Countries

### G-20

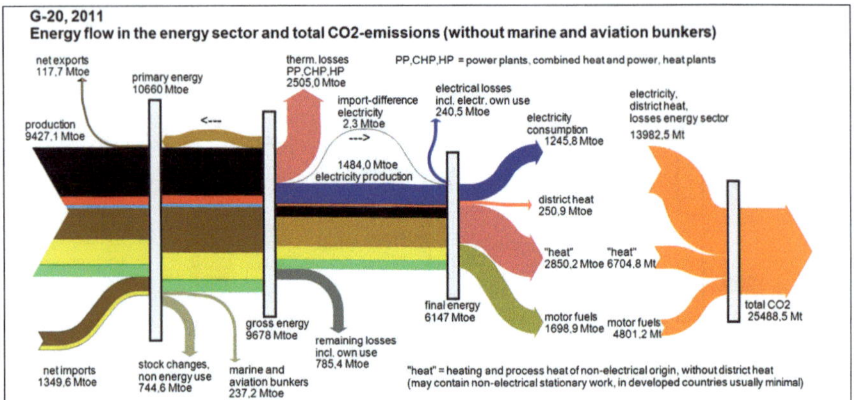


A1. G-20: Energy flow in the energy sector from primary to final energy and total $CO_2$ emissions. Color of the energy carriers as in A2 and A5 (petroleum dark brown, oil products light brown)

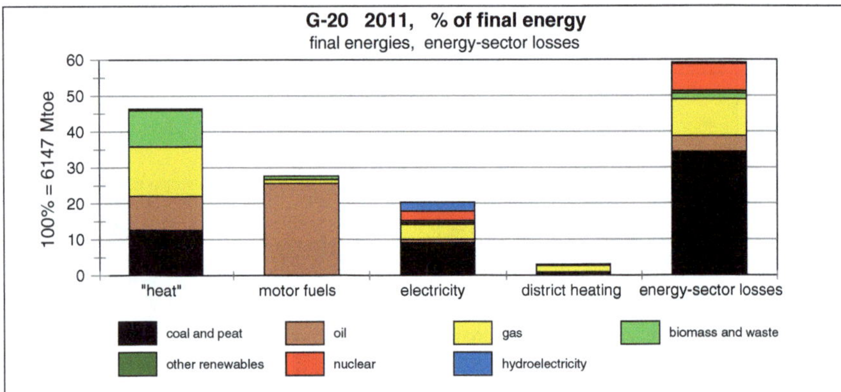


A2. G-20: Shares of energy carriers for obtaining the final energies ("heat", motor fuels, electricity, district heating) and to cover the losses of the energy sector

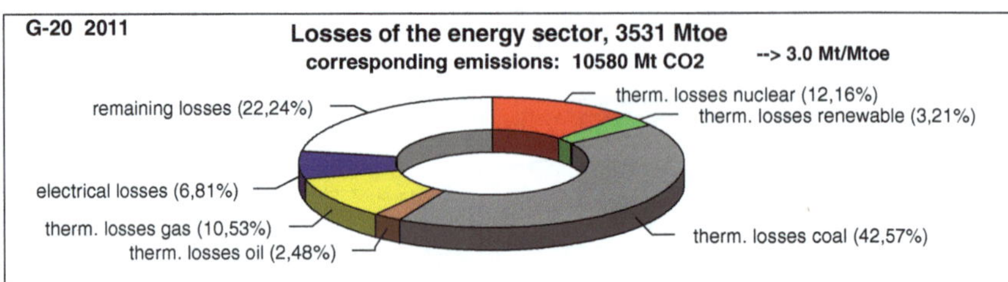


A3. G-20: Percentage distribution of losses in the energy sector; contributors to $CO_2$ emissions include thermal losses of fossil-fuel plants, the electrical losses and the residual losses

## G-20

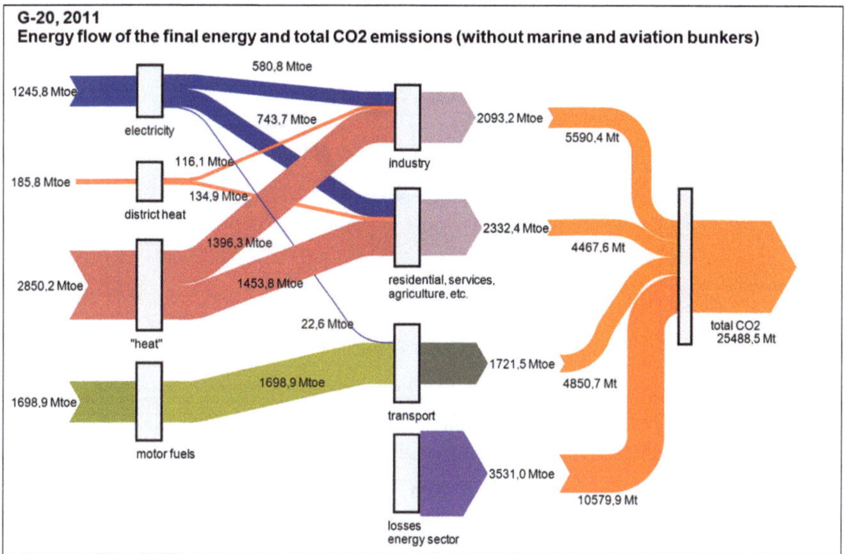


A4. G-20: Flow of final energy to the final users and $CO_2$ emissions of the economy sectors

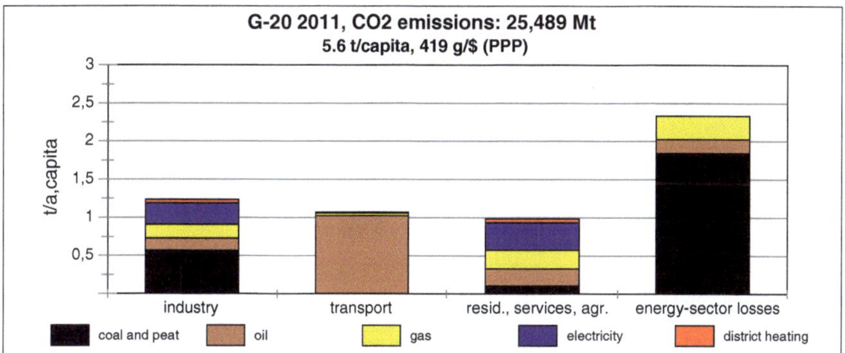


A5. G-20: Energy carriers responsible for the $CO_2$ emissions of sectors of the economy ; (for electricity see also A2 or A6 and for district heating also A2)

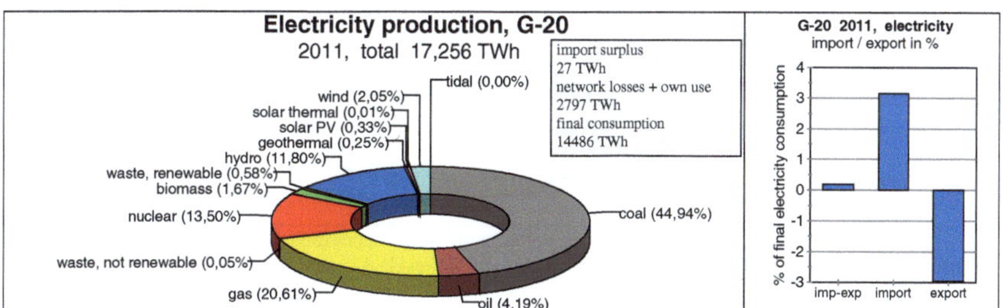


A6. G-20: Generation of electrical energy, final consumption = production + imports - exports - losses

## Argentina

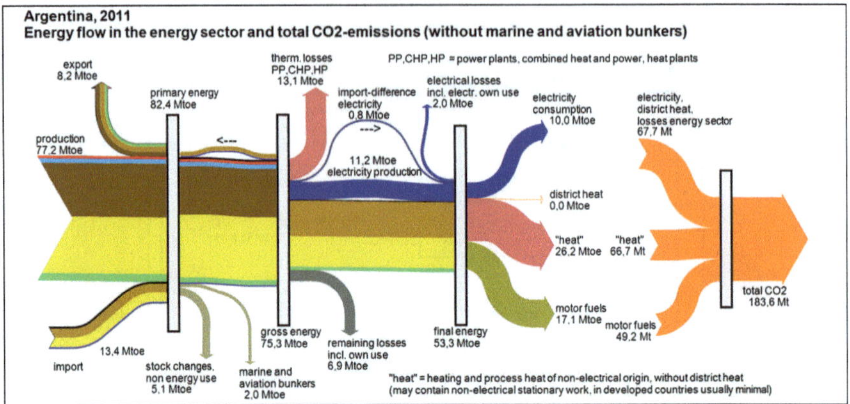


A1. Argentina: Energy flow in the energy sector from primary to final energy and total $CO_2$ emissions. Color of the energy carriers as in A2 and A5 (petroleum dark brown, oil products light brown)

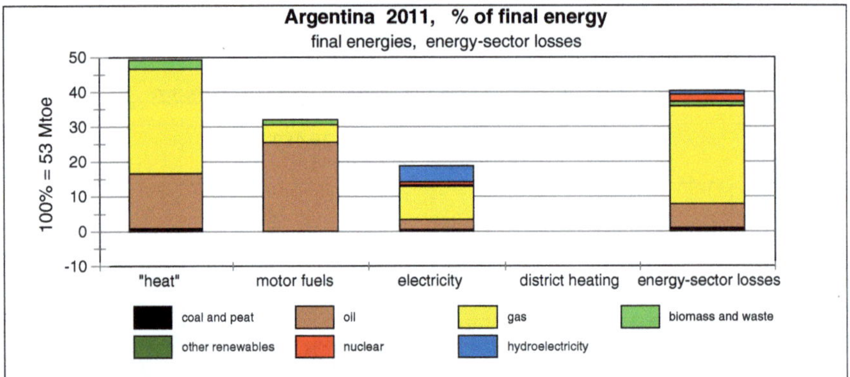


A2. Argentina: Shares of energy carriers for obtaining the final energies ("heat", motor fuels, electricity, district heating) and to cover the losses of the energy sector

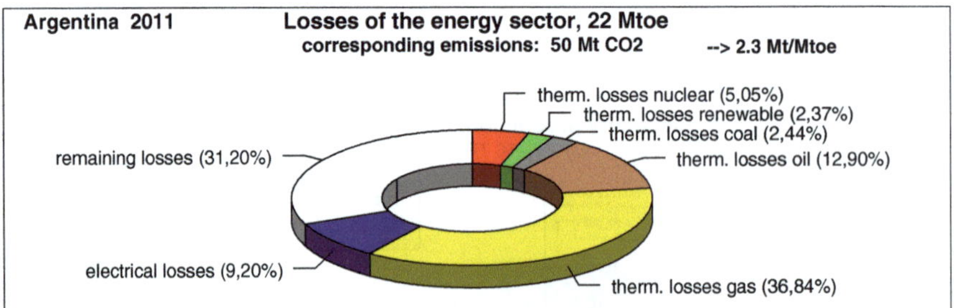


A3. Argentina: Percentage distribution of losses in the energy sector; contributors to $CO_2$ emissions include thermal losses of fossil-fuel plants, the electrical losses and the residual losses

## Argentina

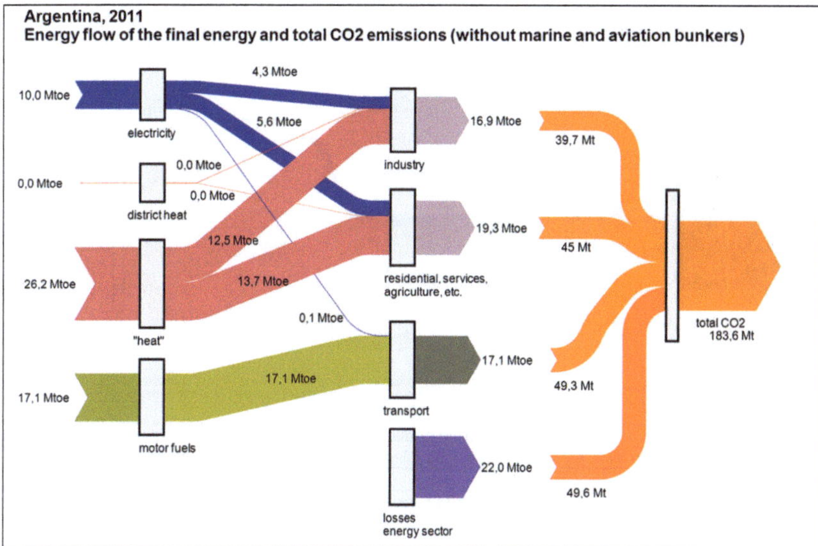


A4. Argentina: Flow of final energy to the final users and $CO_2$ emissions of the economy sectors

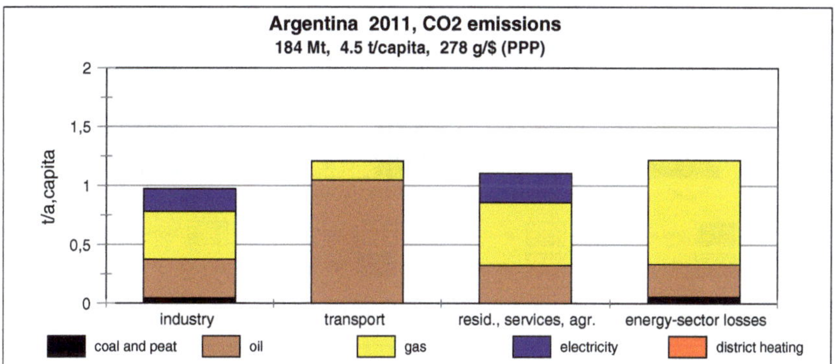


A5. Argentina: Energy carriers responsible for the $CO_2$ emissions of sectors of the economy ; (for electricity see also A2 or A6 and for district heating also A2)

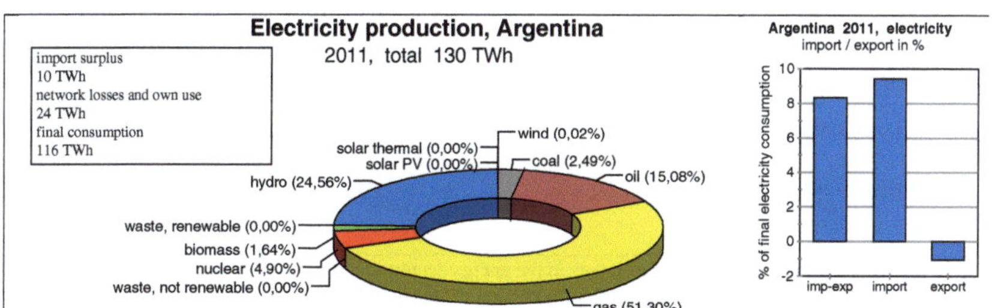


A6. Argentina: Generation of electrical energy, final consumption = production + imports - exports - losses

## Australia

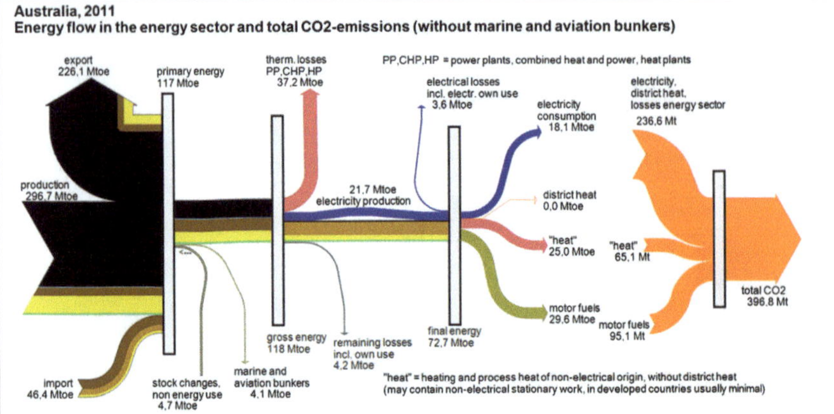


A1. Australia: Energy flow in the energy sector from primary to final energy and total $CO_2$ emissions. Color of the energy carriers as in A2 and A5 (petroleum dark brown, oil products light brown)

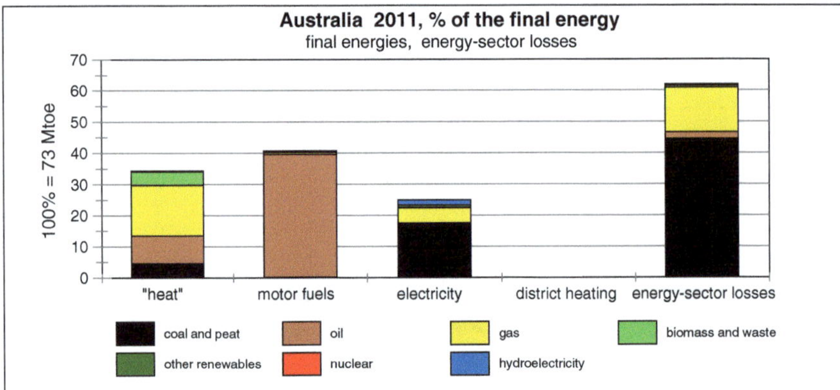


A2. Australia: Shares of energy carriers for obtaining the final energies ("heat", motor fuels, electricity, district heating) and to cover the losses of the energy sector

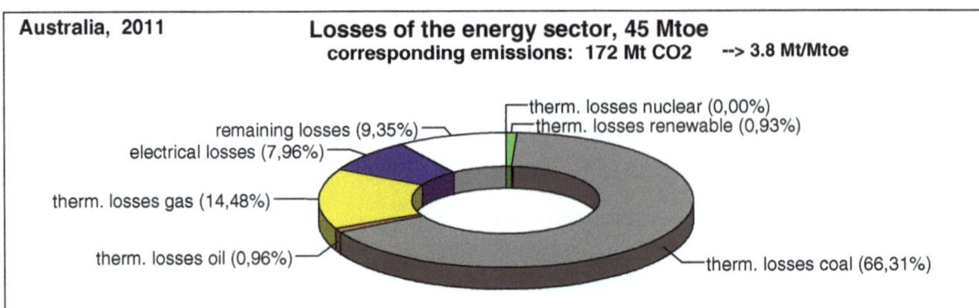


A3. Australia: Percentage distribution of losses in the energy sector; contributors to $CO_2$ emissions include thermal losses of fossil-fuel plants, the electrical losses and the residual losses

## Australia

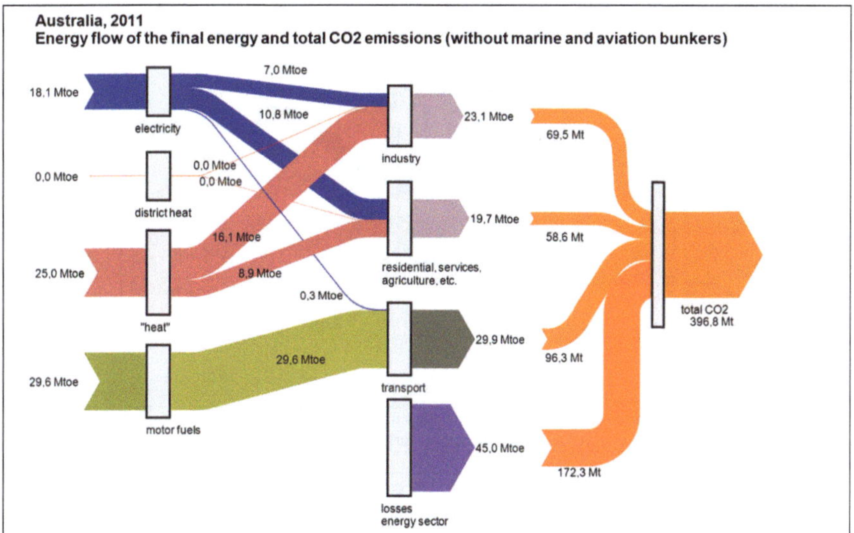


A4. Australia: Flow of final energy to the final users and $CO_2$ emissions of the economy sectors

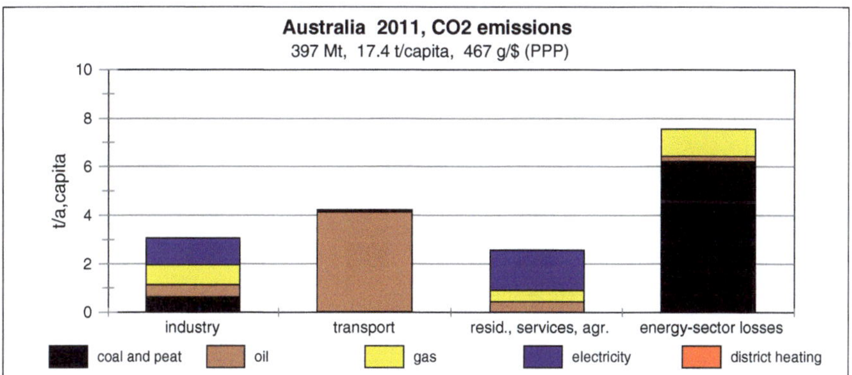


A5. Australia: Energy carriers responsible for the $CO_2$ emissions of sectors of the economy ; (for electricity see also A2 or A6 and for district heating also A2)

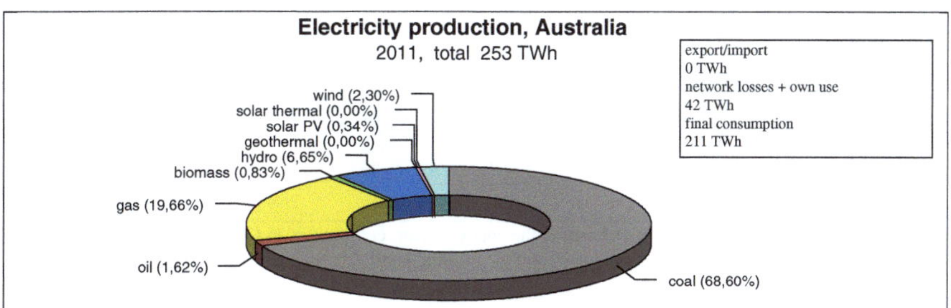


A6. Australia: Generation of electrical energy, final consumption = production + imports - exports - losses

## Brazil

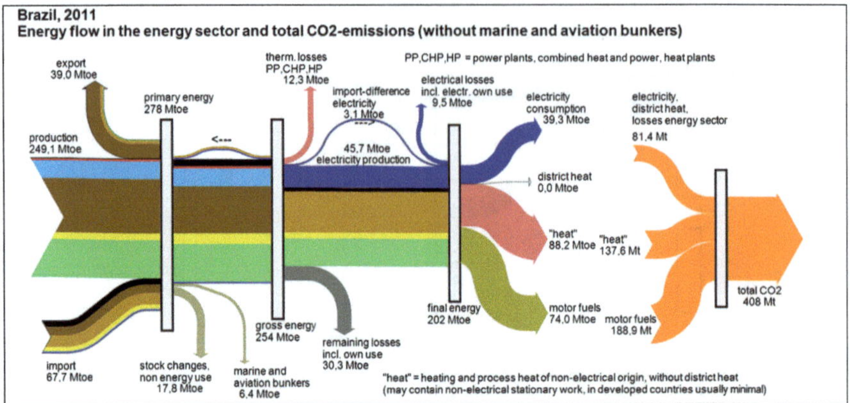


A1. Brazil: Energy flow in the energy sector from primary to final energy and total $CO_2$ emissions. Color of the energy carriers as in A2 and A5 (petroleum dark brown, oil products light brown)

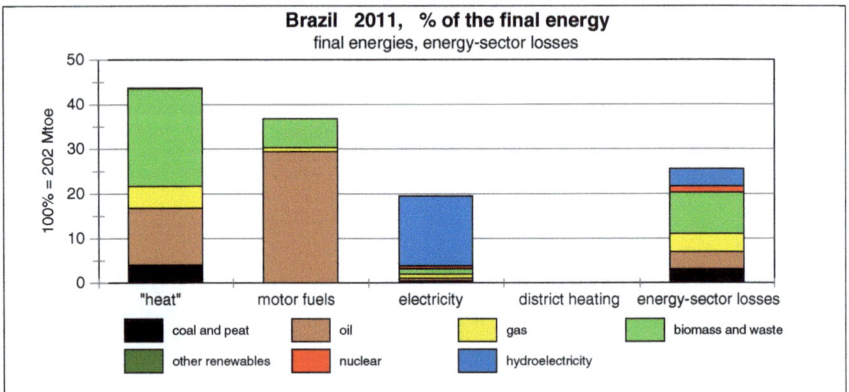


A2. Brazil: Shares of energy carriers for obtaining the final energies ("heat", motor fuels, electricity, district heating) and to cover the losses of the energy sector

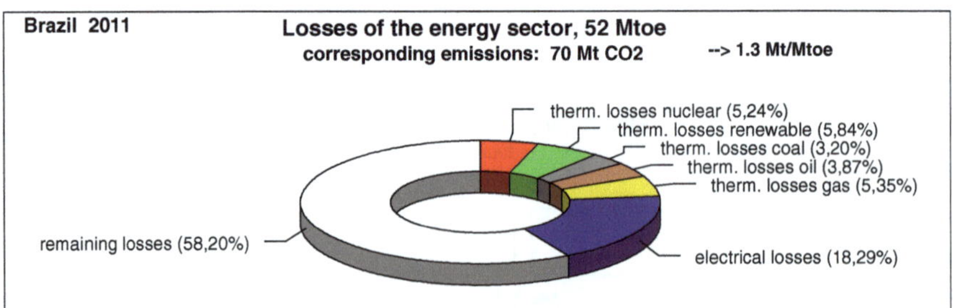


A3. Brazil: Percentage distribution of losses in the energy sector; contributors to $CO_2$ emissions include thermal losses of fossil-fuel plants, the electrical losses and the residual losses

## Brazil

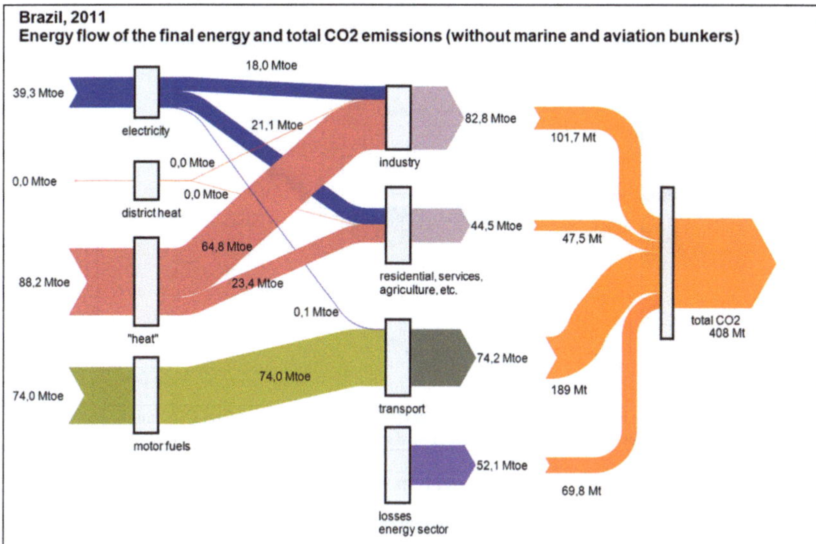


A4. Brazil: Flow of final energy to the final users and $CO_2$ emissions of the economy sectors

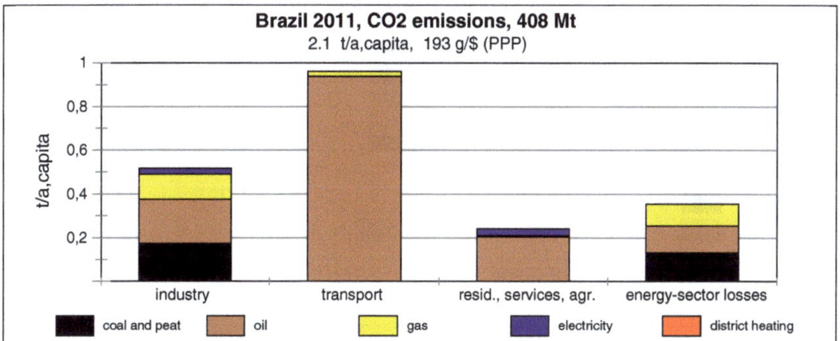


A5. Brazil: Energy carriers responsible for the $CO_2$ emissions of sectors of the economy ;
(for electricity see also A2 or A6 and for district heating also A2)

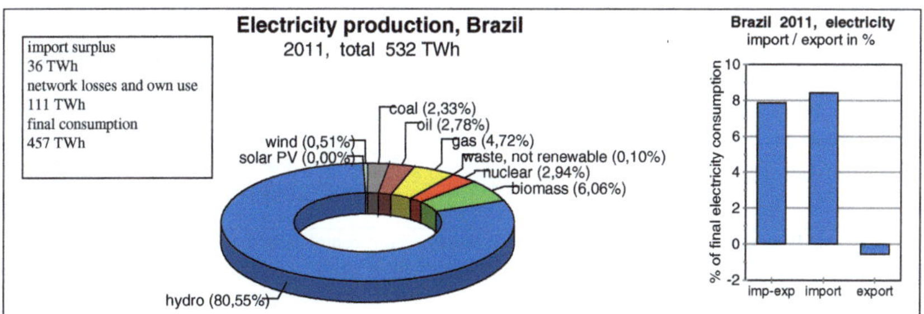


6. Brazil: Generation of electrical energy, final consumption = production + imports - exports - losses

## Canada

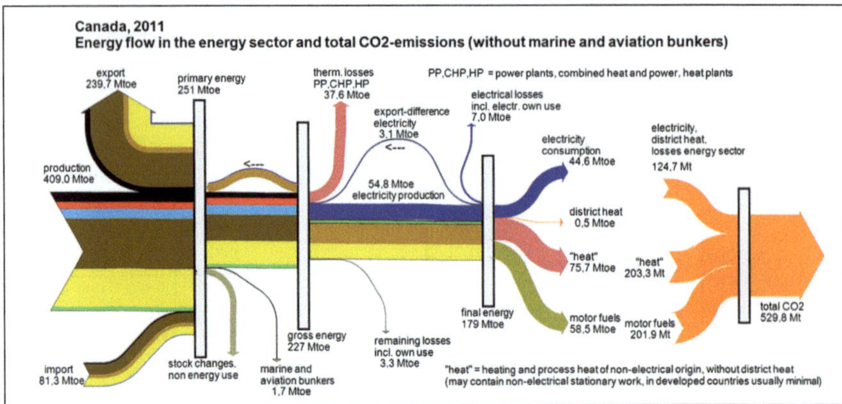


A1. Canada: Energy flow in the energy sector from primary to final energy and total $CO_2$ emissions. Color of the energy carriers as in A2 and A5 (petroleum dark brown, oil products light brown)

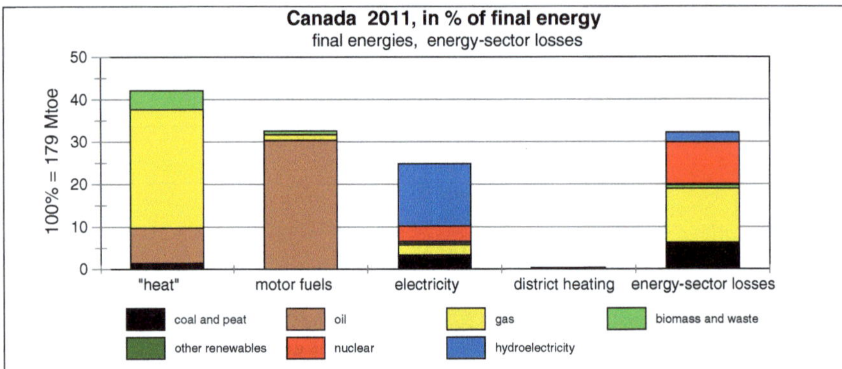


A2. Canada: Shares of energy carriers for obtaining the final energies ("heat", motor fuels, electricity, district heating) and to cover the losses of the energy sector

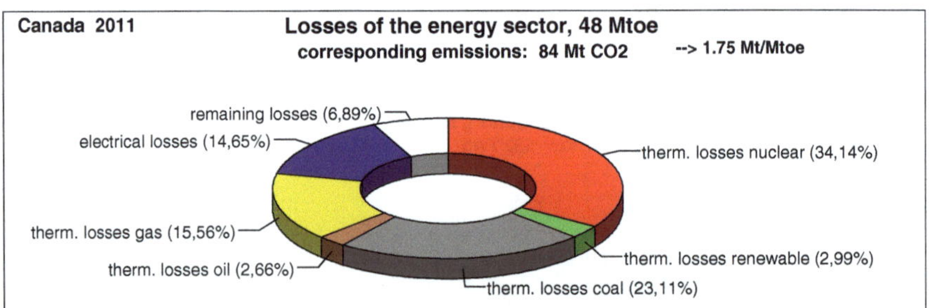


A3. Canada: Percentage distribution of losses in the energy sector; contributors to $CO_2$ emissions include thermal losses of fossil-fuel plants, the electrical losses and the residual losses

# Canada

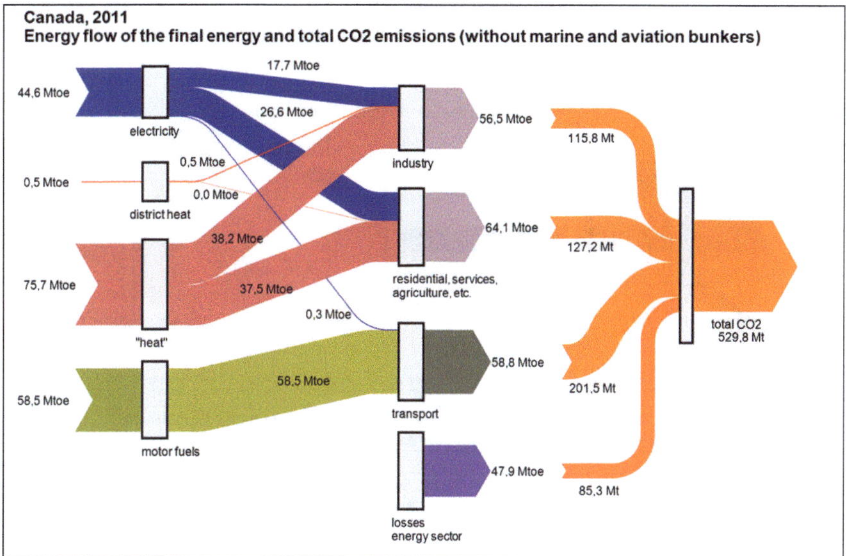


A4. Canada: Flow of final energy to the final users and $CO_2$ emissions of the economy sectors

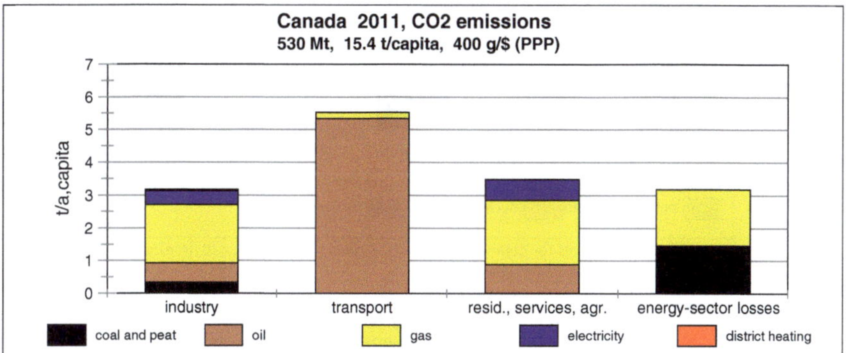


A5. Canada: Energy carriers responsible for the $CO_2$ emissions of sectors of the economy ;
(for electricity see also A2 or A6 and for district heating also A2)

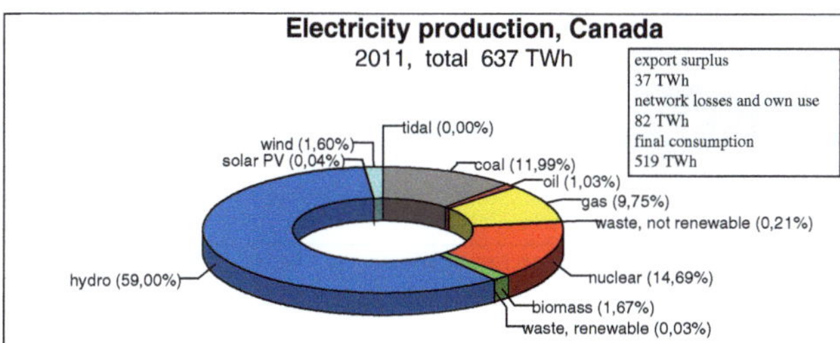


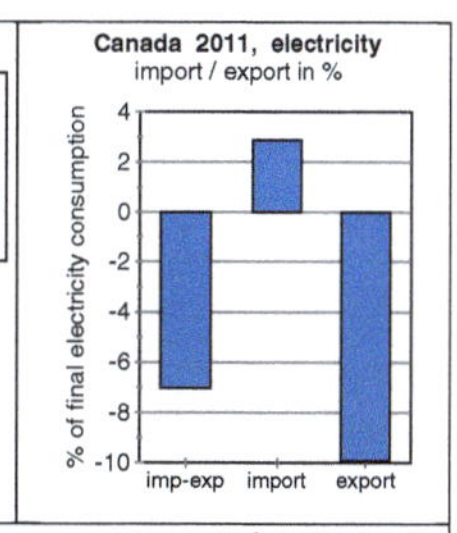


A6. Canada: Generation of electrical energy, final consumption = production + imports - exports - losses

## China (with Hongkong)

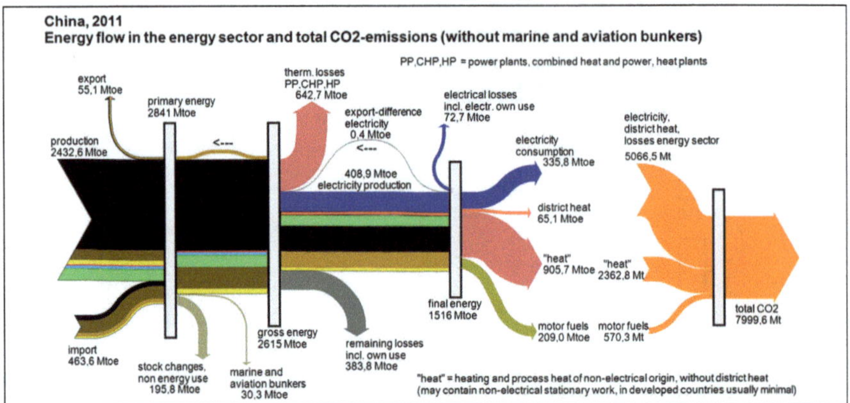


A1. China: Energy flow in the energy sector from primary to final energy and total $CO_2$ emissions. Color of the energy carriers as in A2 and A5 (petroleum dark brown, oil products light brown)

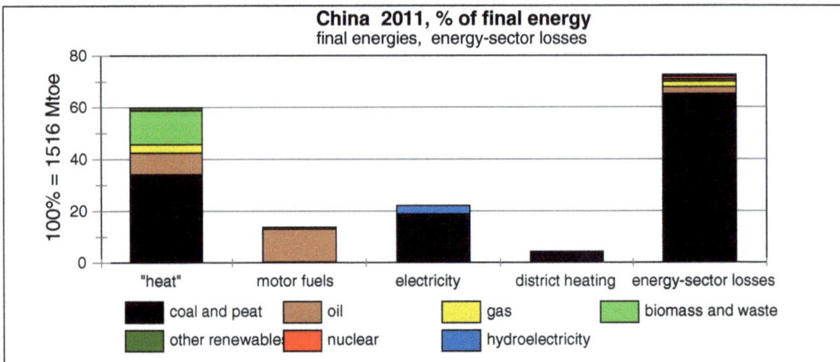


A2. China: Shares of energy carriers for obtaining the final energies ("heat", motor fuels, electricity, district heating) and to cover the losses of the energy sector

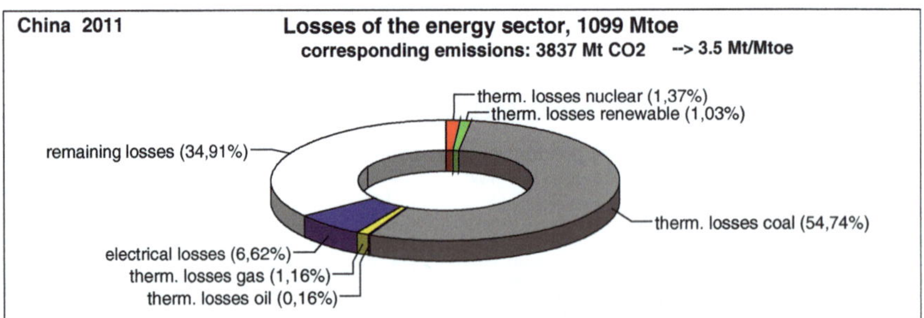


A3. China: Percentage distribution of losses in the energy sector; contributors to $CO_2$ emissions include thermal losses of fossil-fuel plants, the electrical losses and the residual losses

## China (with Hongkong)

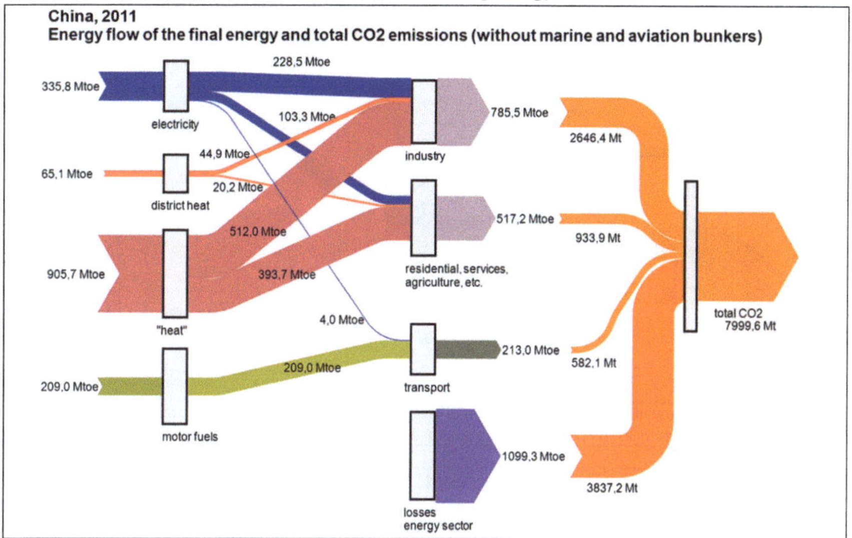


A4. China: Flow of final energy to the final users and $CO_2$ emissions of the economy sectors

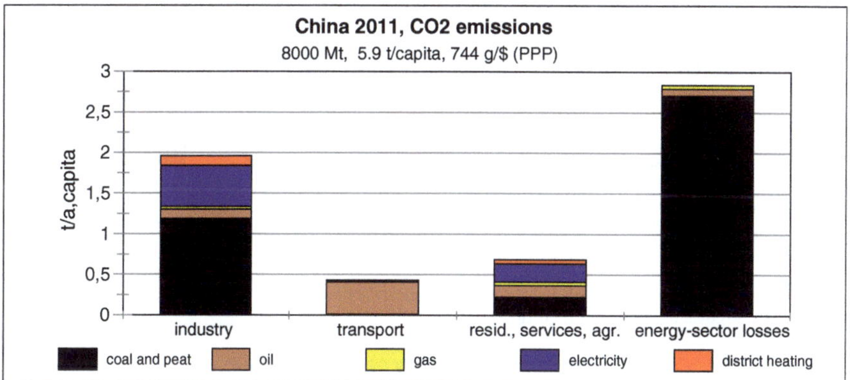


A5. China: Energy carriers responsible for the $CO_2$ emissions of sectors of the economy ;
(for electricity see also A2 or A6 and for district heating also A2)

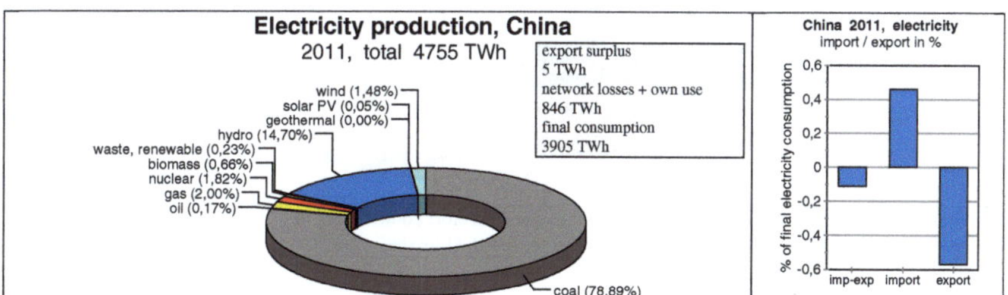


A6. China: Generation of electrical energy, final consumption = production + imports - exports - losses

## France

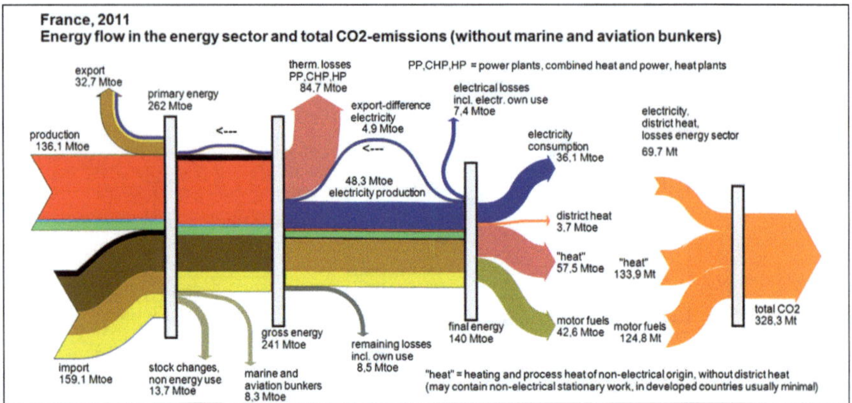


A1. France: Energy flow in the energy sector from primary to final energy and total $CO_2$ emissions. Color of the energy carriers as in A2 and A5 (petroleum dark brown, oil products light brown)

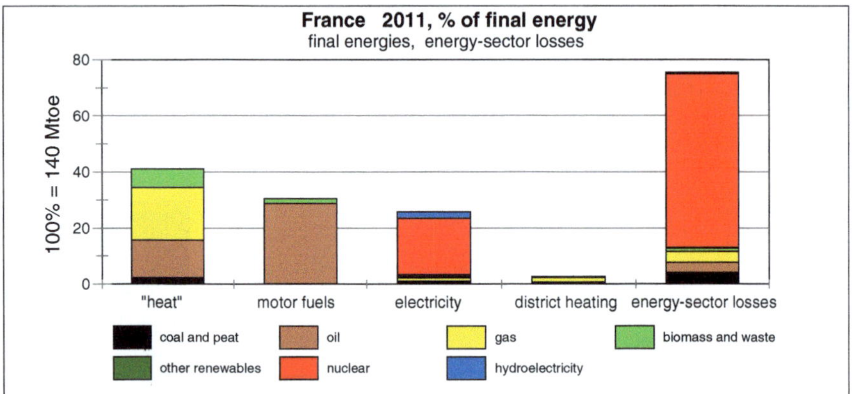


A2. France: Shares of energy carriers for obtaining the final energies ("heat", motor fuels, electricity, district heating) and to cover the losses of the energy sector

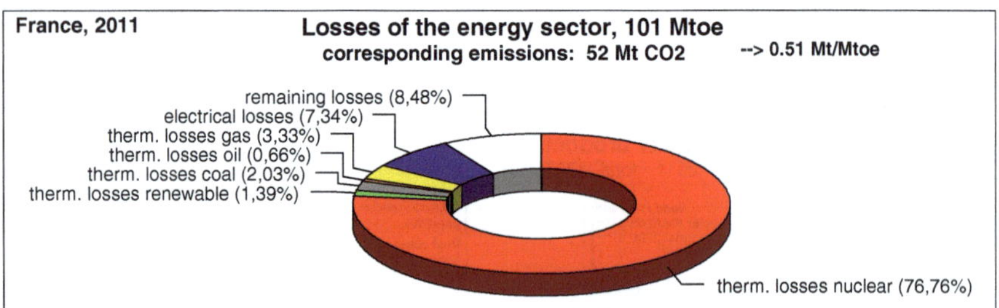


3. France: Percentage distribution of losses in the energy sector; contributors to $CO_2$ emissions include thermal losses of fossil-fuel plants, the electrical losses and the residual losses

# France

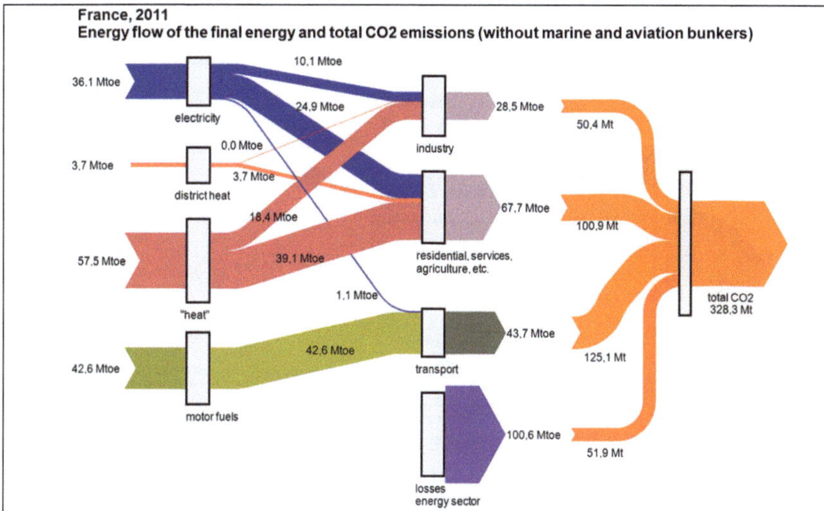


A4. France: Flow of final energy to the final users and $CO_2$ emissions of the economy sectors

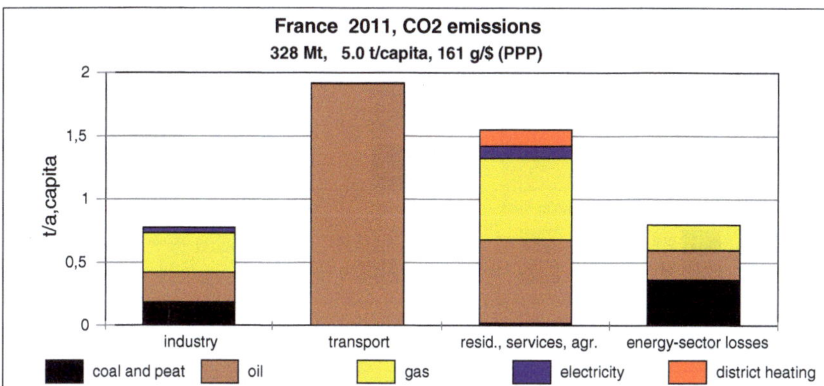


A5. France: Energy carriers responsible for the $CO_2$ emissions of sectors of the economy ; (for electricity see also A2 or A6 and for district heating also A2)

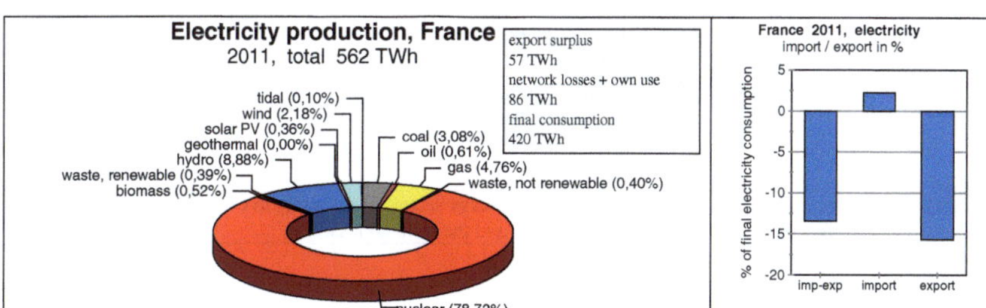


A6. France: Generation of electrical energy, final consumption = production + imports - exports - losses

## Germany

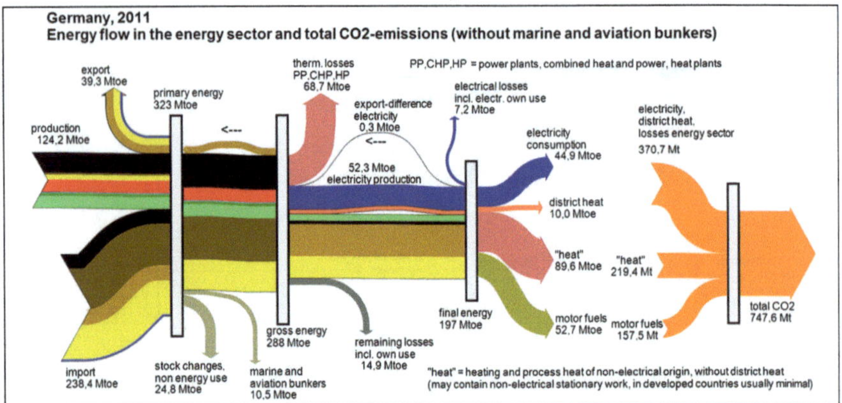


A1. Germany: Energy flow in the energy sector from primary to final energy and total $CO_2$ emissions. Color of the energy carriers as in A2 and A5 (petroleum dark brown, oil products light brown)

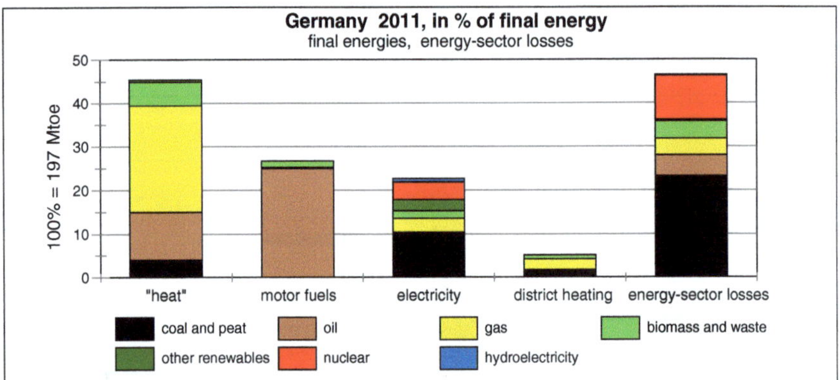


A2. Germany: Shares of energy carriers for obtaining the final energies ("heat", motor fuels, electricity, district heating) and to cover the losses of the energy sector

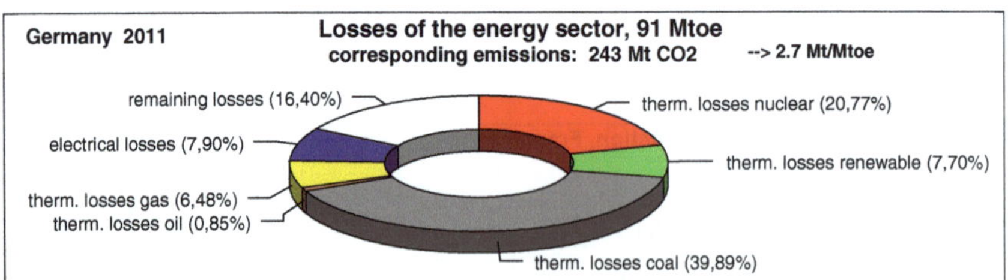


A3. Germany: Percentage distribution of losses in the energy sector; contributors to $CO_2$ emissions include thermal losses of fossil-fuel plants, the electrical losses and the residual losses

## Germany

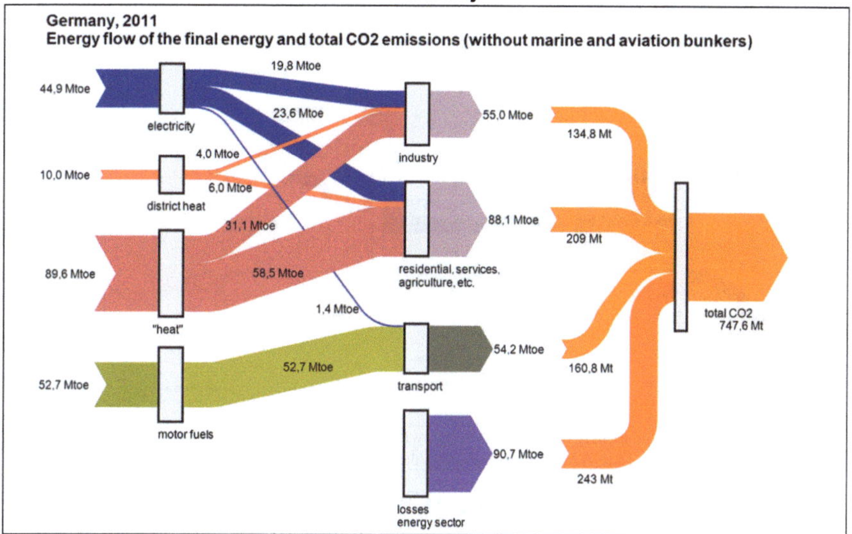


A4. Germany: Flow of final energy to the final users and $CO_2$ emissions of the economy sectors

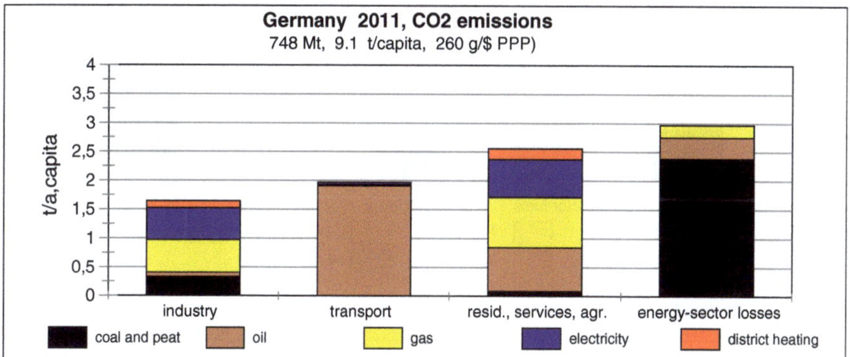


A5. Germany: Energy carriers responsible for the $CO_2$ emissions of sectors of the economy ;
(for electricity see also A2 or A6 and for district heating also A2)

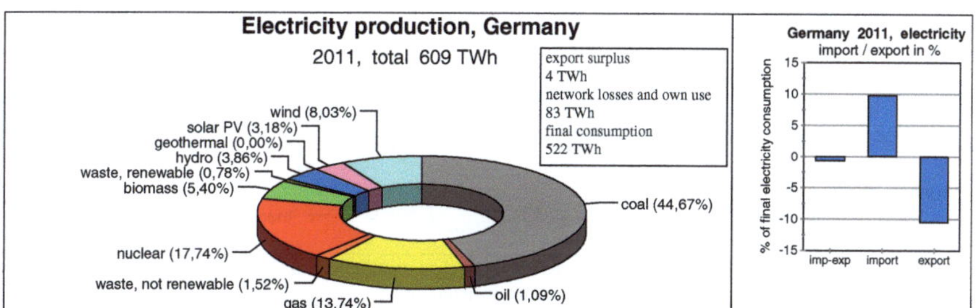


A6. Germany: Generation of electrical energy, final consumption = production + imports - exports - losses

## India

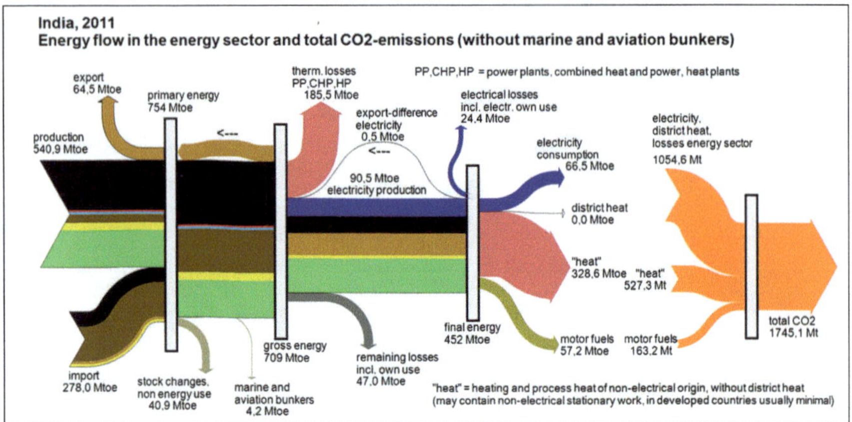


A1. India: Energy flow in the energy sector from primary to final energy and total $CO_2$ emissions. Color of the energy carriers as in A2 and A5 (petroleum dark brown, oil products light brown)

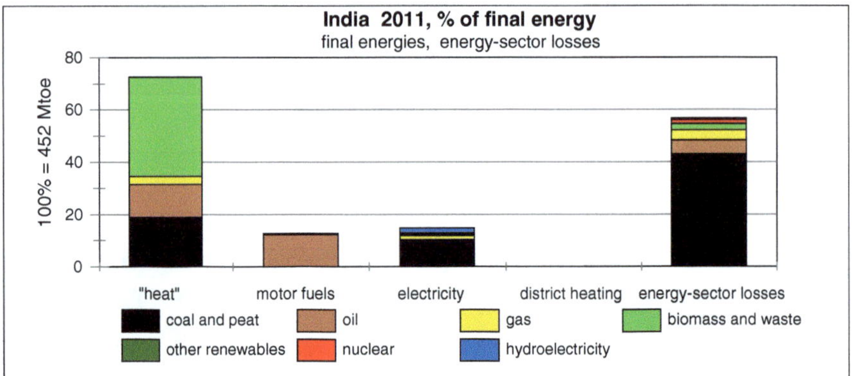


A2. India: Shares of energy carriers for obtaining the final energies ("heat", motor fuels, electricity, district heating) and to cover the losses of the energy sector

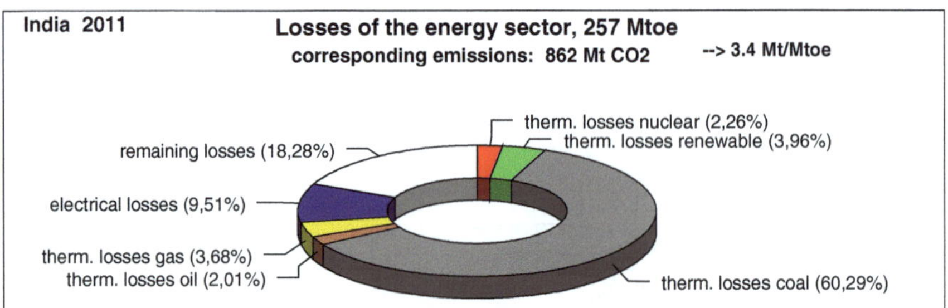


A3. India: Percentage distribution of losses in the energy sector; contributors to $CO_2$ emissions include thermal losses of fossil-fuel plants, the electrical losses and the residual losses

## India

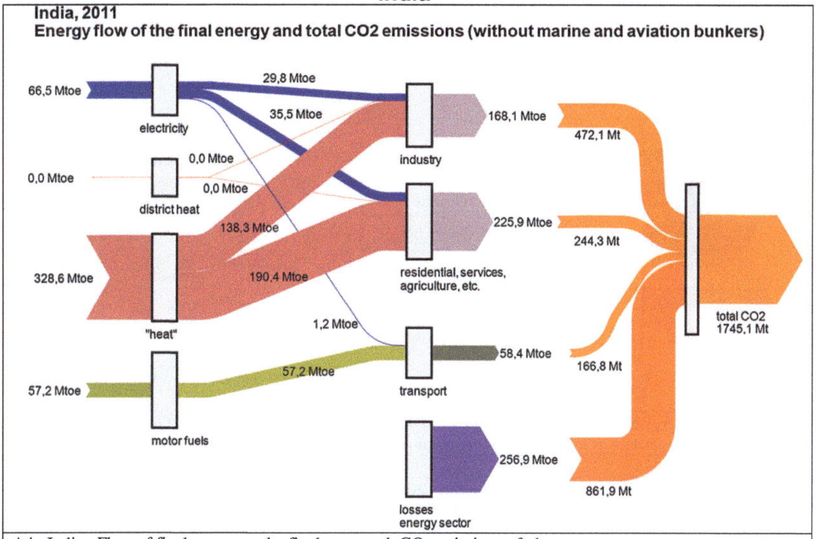


A4. India: Flow of final energy to the final users and $CO_2$ emissions of the economy sectors

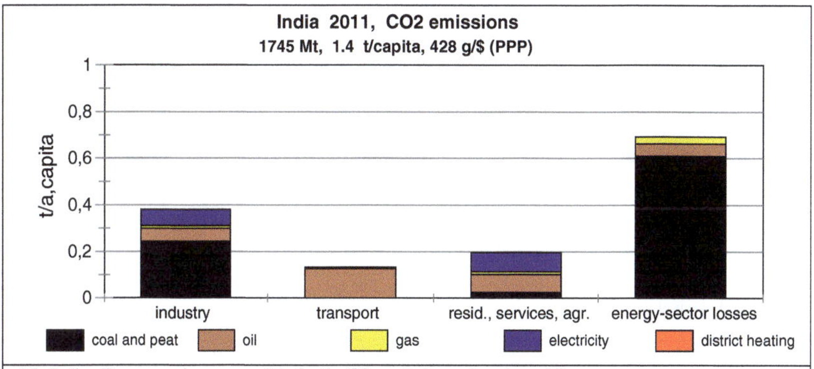


A5. India: Energy carriers responsible for the $CO_2$ emissions of sectors of the economy ;
(for electricity see also A2 or A6 and for district heating also A2)

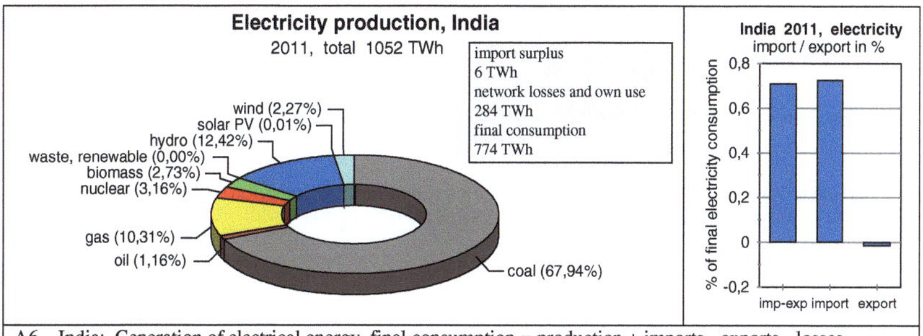


A6. India: Generation of electrical energy, final consumption = production + imports - exports - losses

## Indonesia

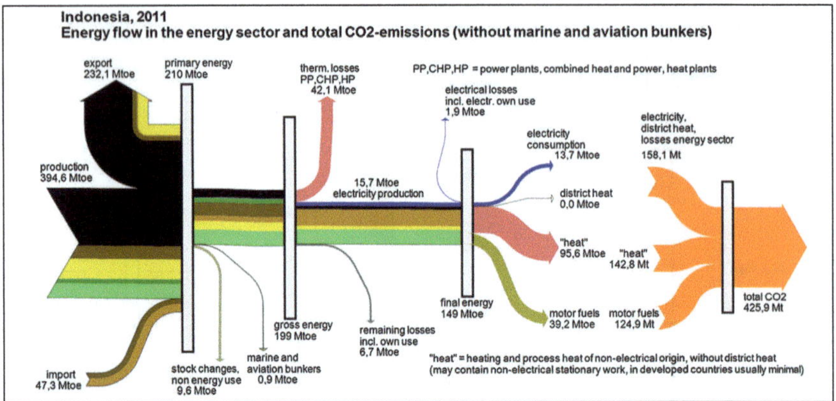


A1. Indonesia: Energy flow in the energy sector from primary to final energy and total $CO_2$ emissions. Color of the energy carriers as in A2 and A5 (petroleum dark brown, oil products light brown)

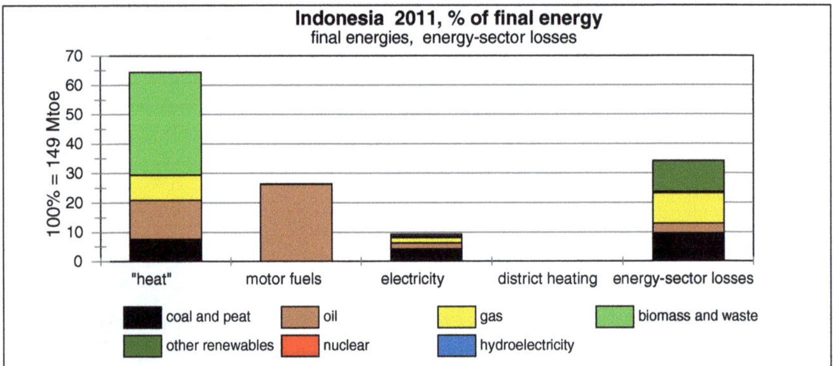


A2. Indonesia: Shares of energy carriers for obtaining the final energies ("heat", motor fuels, electricity, district heating) and to cover the losses of the energy sector

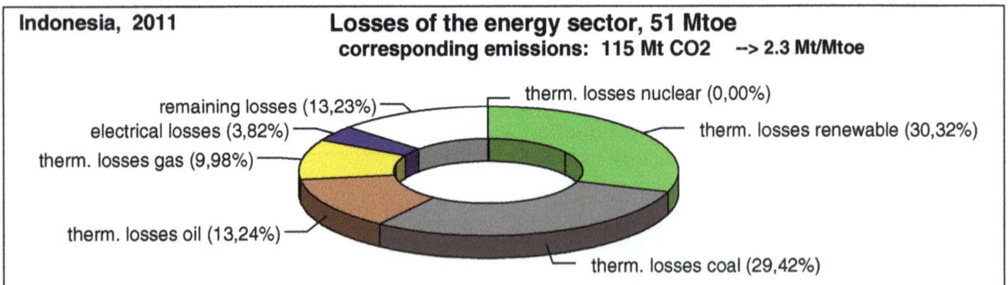


A3. Indonesia: Percentage distribution of losses in the energy sector; contributors to $CO_2$ emissions include thermal losses of fossil-fuel plants, the electrical losses and the residual losses

## Indonesia

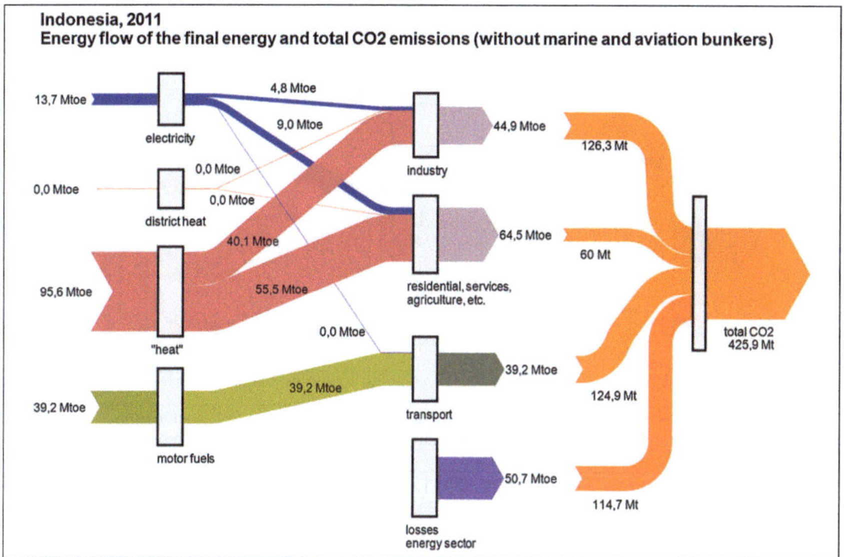


A4. Indonesia: Flow of final energy to the final users and $CO_2$ emissions of the economy sectors

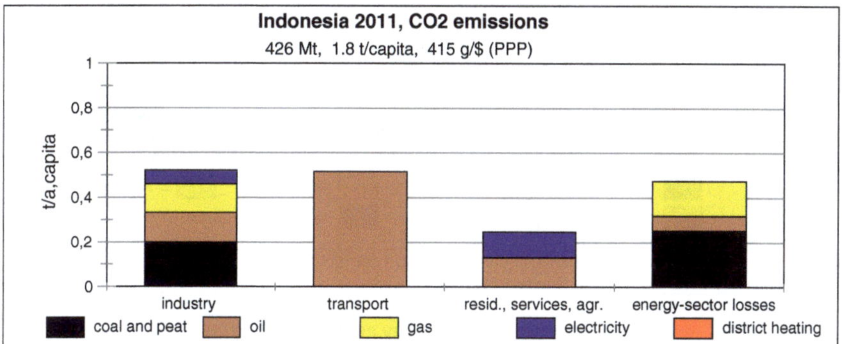


A5. Indonesia: Energy carriers responsible for the $CO_2$ emissions of sectors of the economy ;
(for electricity see also A2 or A6 and for district heating also A2)

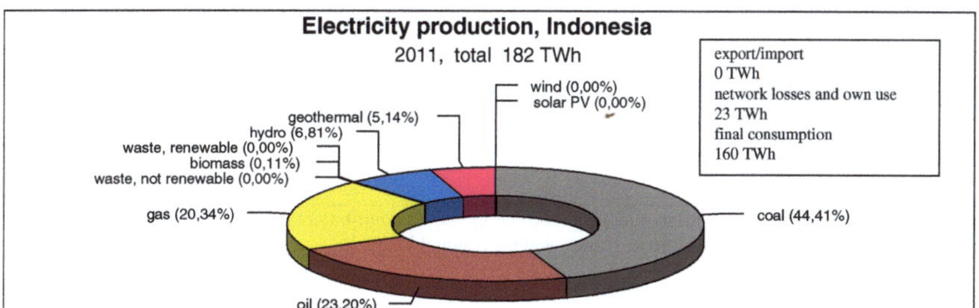


A6. Indonesia: Generation of electrical energy, final consumption = production + imports - exports - losses

## Italy

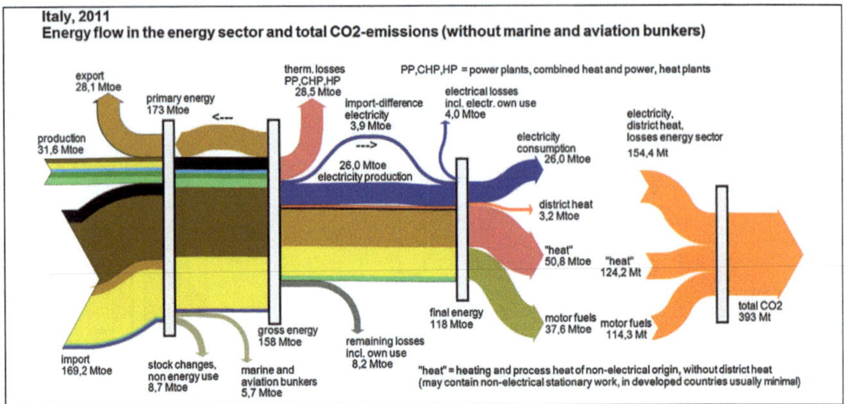


A1. Italy: Energy flow in the energy sector from primary to final energy and total $CO_2$ emissions. Color of the energy carriers as in A2 and A5 (petroleum dark brown, oil products light brown)

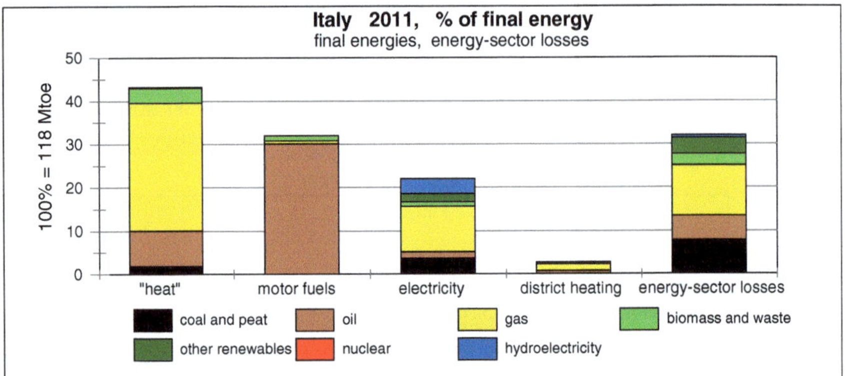


A2. Italy: Shares of energy carriers for obtaining the final energies ("heat", motor fuels, electricity, district heating) and to cover the losses of the energy sector

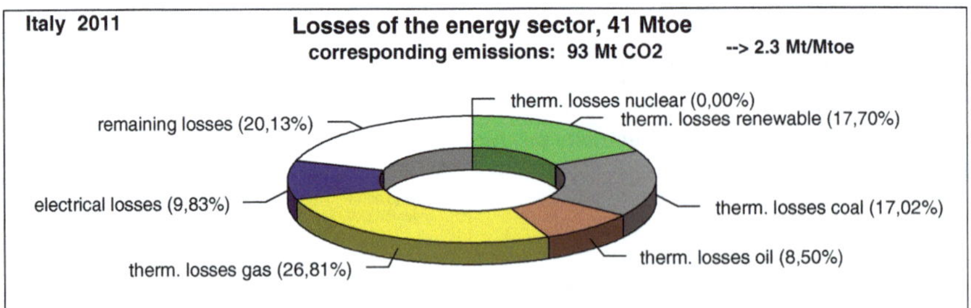


A3. Italy: Percentage distribution of losses in the energy sector; contributors to $CO_2$ emissions include thermal losses of fossil-fuel plants, the electrical losses and the residual losses

## Italy

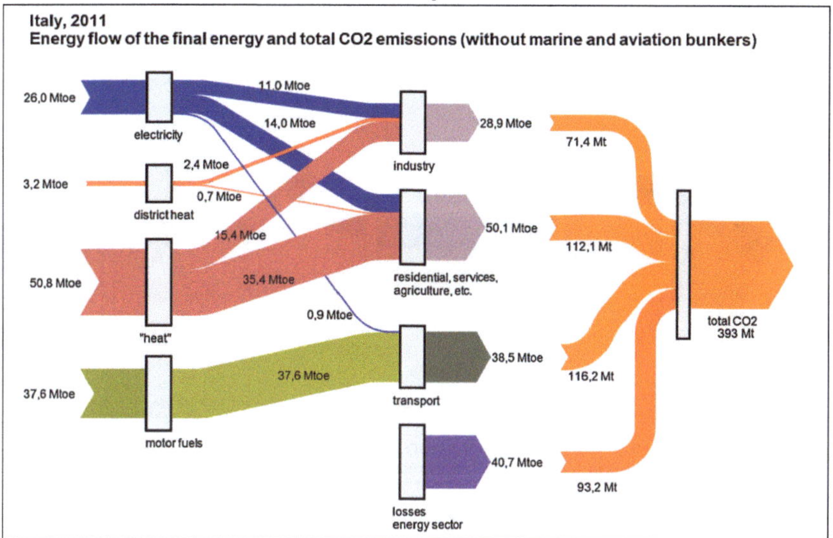


A4. Italy: Flow of final energy to the final users and $CO_2$ emissions of the economy sectors

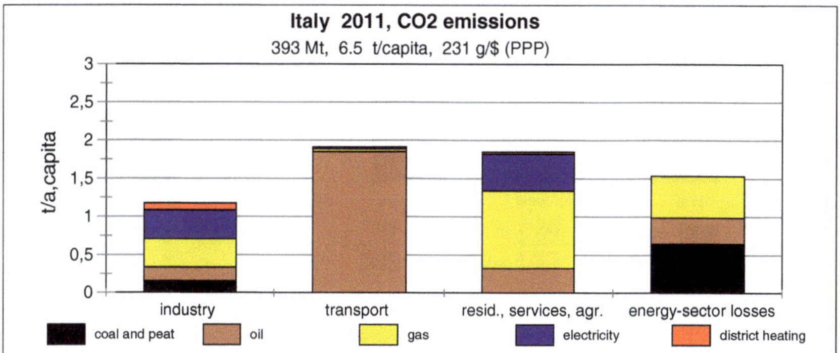


A5. Italy: Energy carriers responsible for the $CO_2$ emissions of sectors of the economy ;
(for electricity see also A2 or A6 and for district heating also A2)

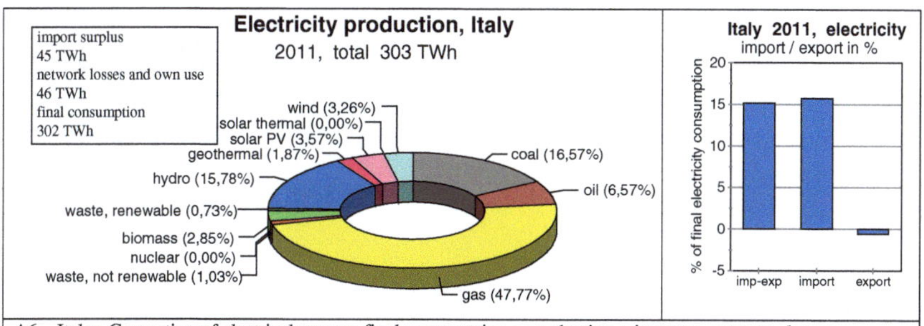


A6. Italy: Generation of electrical energy, final consumption = production + imports - exports - losses

# Japan

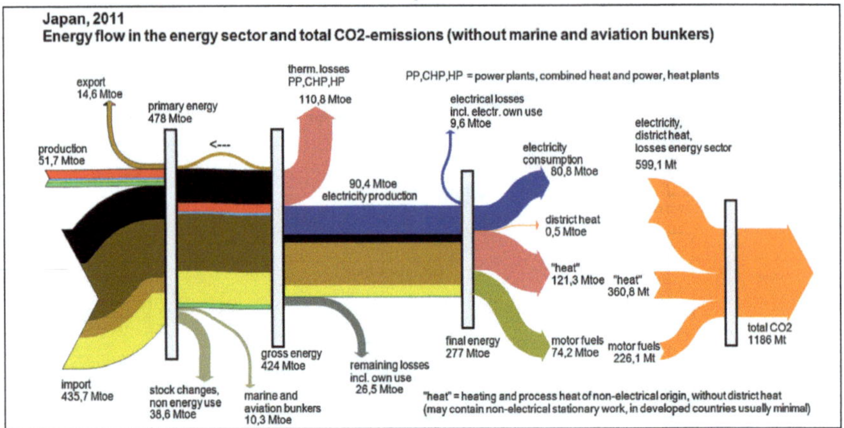


A1. Japan: Energy flow in the energy sector from primary to final energy and total $CO_2$ emissions. Color of the energy carriers as in A2 and A5 (petroleum dark brown, oil products light brown)

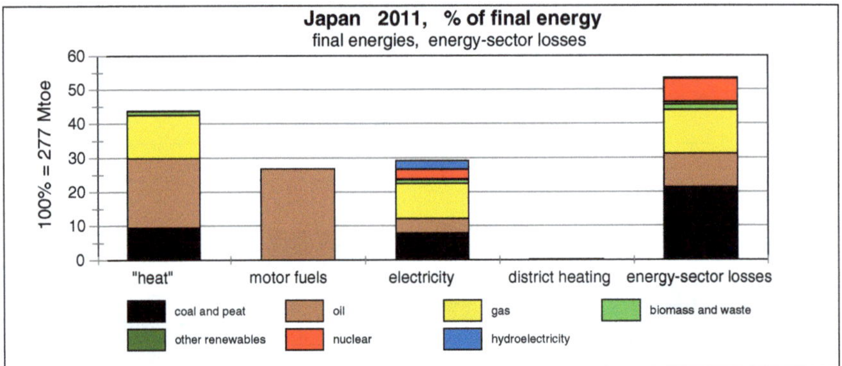


A2. Japan: Shares of energy carriers for obtaining the final energies ("heat", motor fuels, electricity, district heating) and to cover the losses of the energy sector

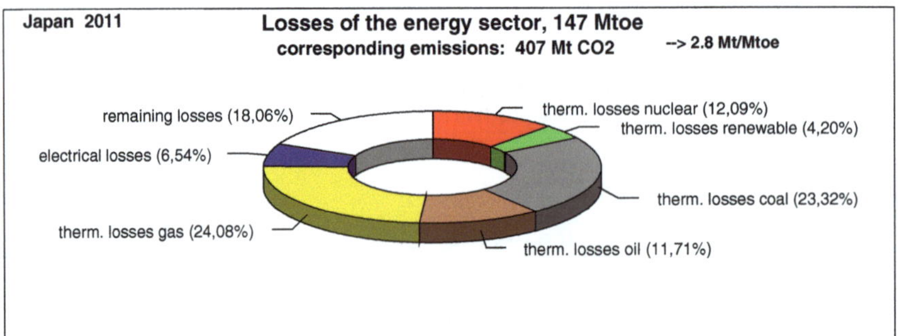


A3. Japan: Percentage distribution of losses in the energy sector; contributors to $CO_2$ emissions include thermal losses of fossil-fuel plants, the electrical losses and the residual losses

## Japan

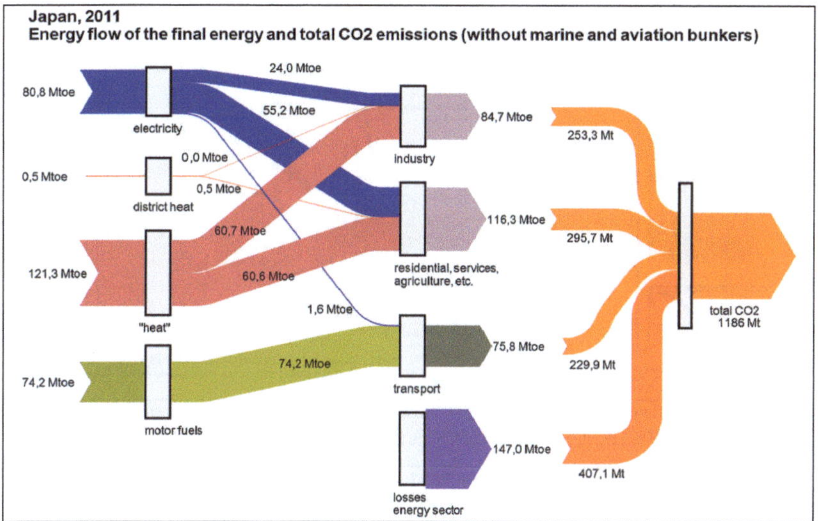


A4. Japan: Flow of final energy to the final users and $CO_2$ emissions of the economy sectors

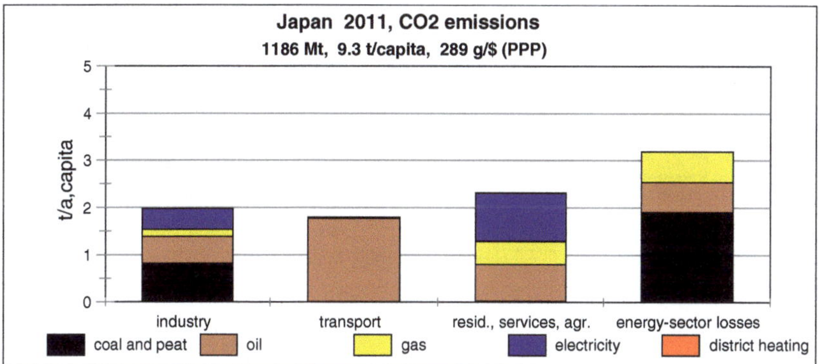


A5. Japan: Energy carriers responsible for the $CO_2$ emissions of sectors of the economy ; (for electricity see also A2 or A6 and for district heating also A2)

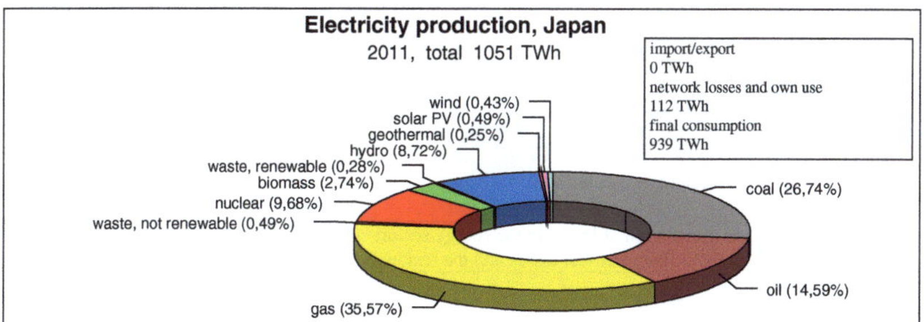


A6. Japan: Generation of electrical energy, final consumption = production + imports - exports - losses

## Mexico

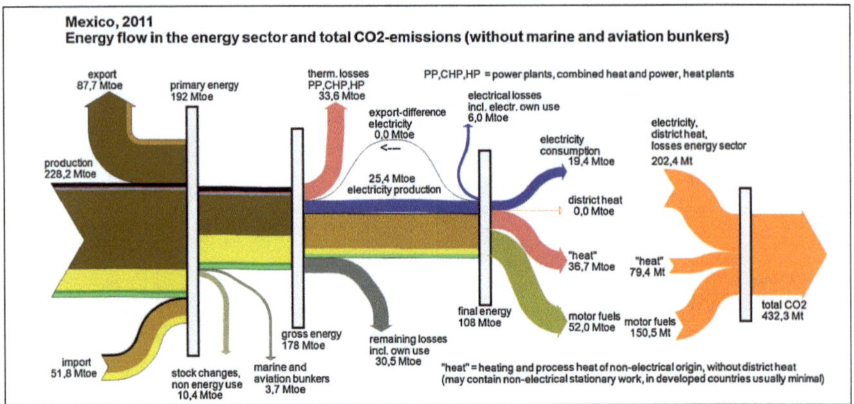


A1. Mexico: Energy flow in the energy sector from primary to final energy and total $CO_2$ emissions. Color of the energy carriers as in A2 and A5 (petroleum dark brown, oil products light brown)

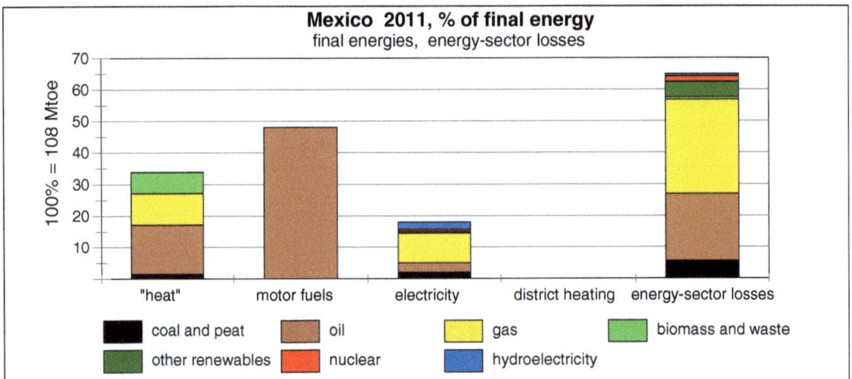


A2. Mexico: Shares of energy carriers for obtaining the final energies ("heat", motor fuels, electricity, district heating) and to cover the losses of the energy sector

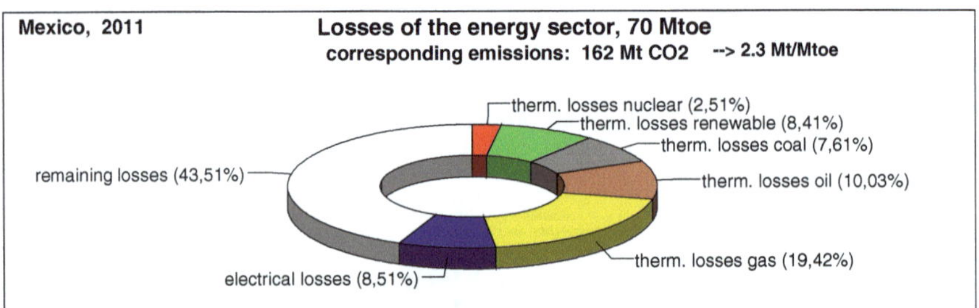


A3. Mexico: Percentage distribution of losses in the energy sector; contributors to $CO_2$ emissions include thermal losses of fossil-fuel plants, the electrical losses and the residual losses

## Mexico

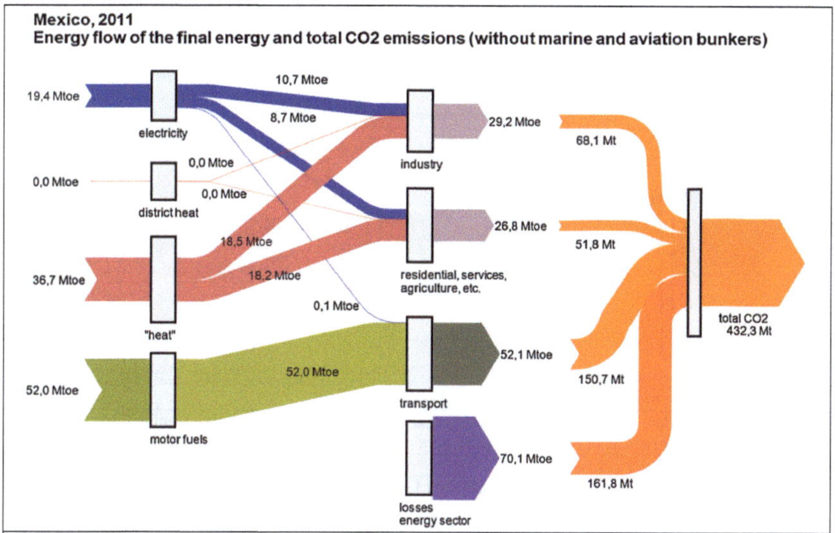


A4. Mexico: Flow of final energy to the final users and $CO_2$ emissions of the economy sectors

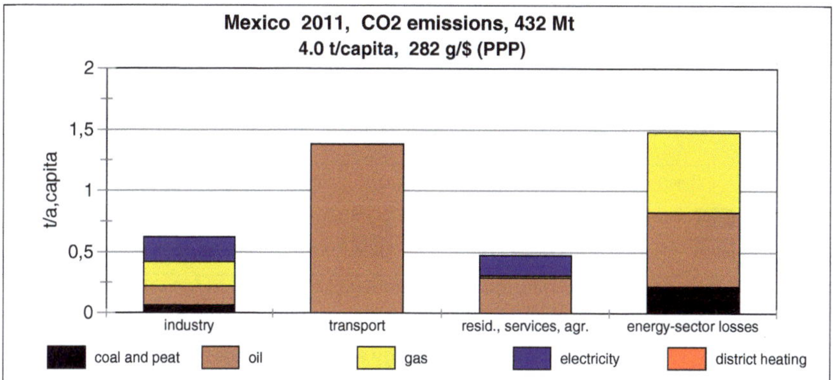


A5. Mexico: Energy carriers responsible for the $CO_2$ emissions of sectors of the economy ;
(for electricity see also A2 or A6 and for district heating also A2)

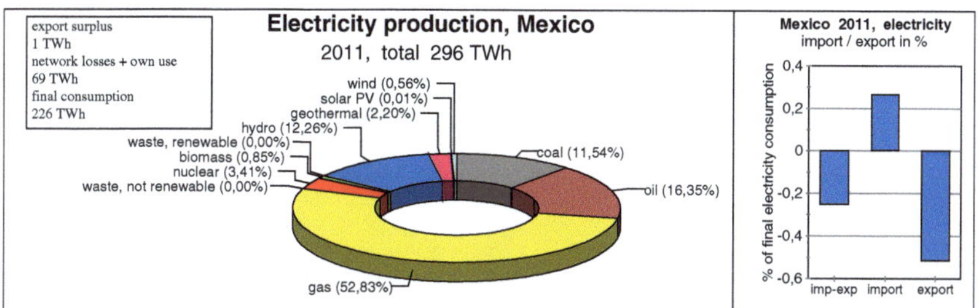


A6. Mexico: Generation of electrical energy, final consumption = production + imports - exports - losses

## Russia

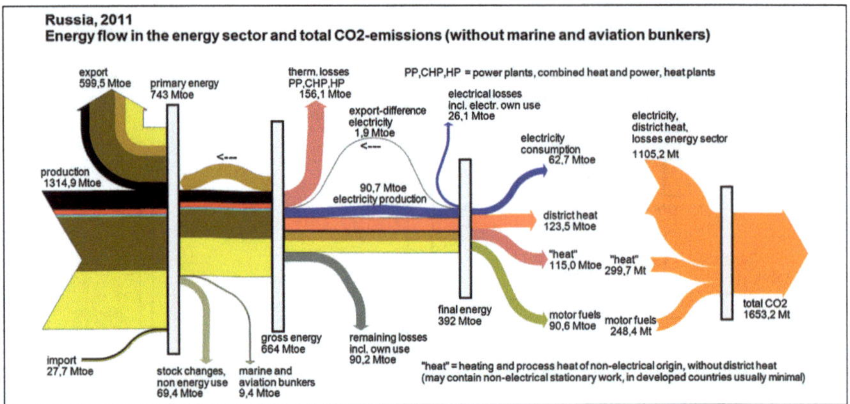


A1. Russia: Energy flow in the energy sector from primary to final energy and total $CO_2$ emissions. Color of the energy carriers as in A2 and A5 (petroleum dark brown, oil products light brown)

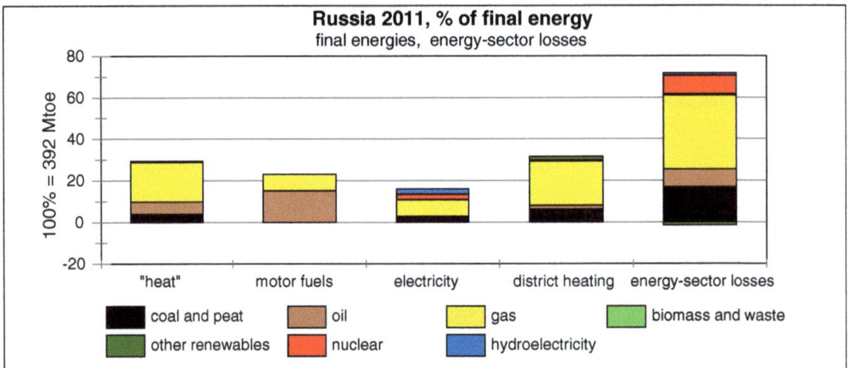


A2. Russia: Shares of energy carriers for obtaining the final energies ("heat", motor fuels, electricity, district heating) and to cover the losses of the energy sector

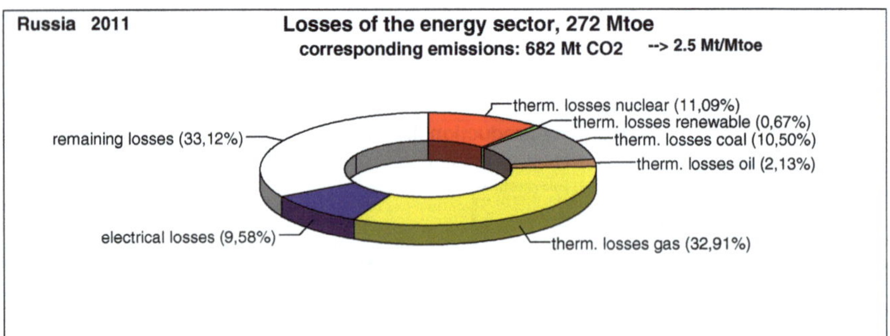


A3. Russia: Percentage distribution of losses in the energy sector; contributors to $CO_2$ emissions include thermal losses of fossil-fuel plants, the electrical losses and the residual losses

## Russia

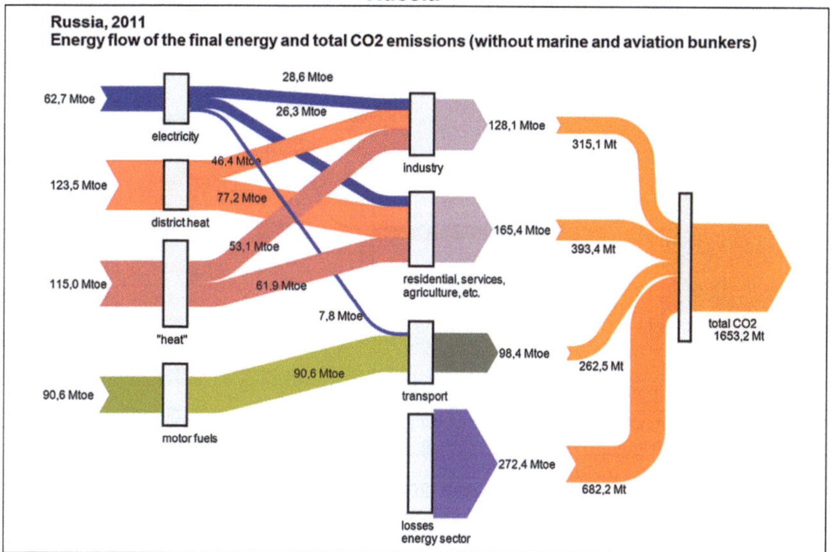


A4. Russia: Flow of final energy to the final users and $CO_2$ emissions of the economy sectors

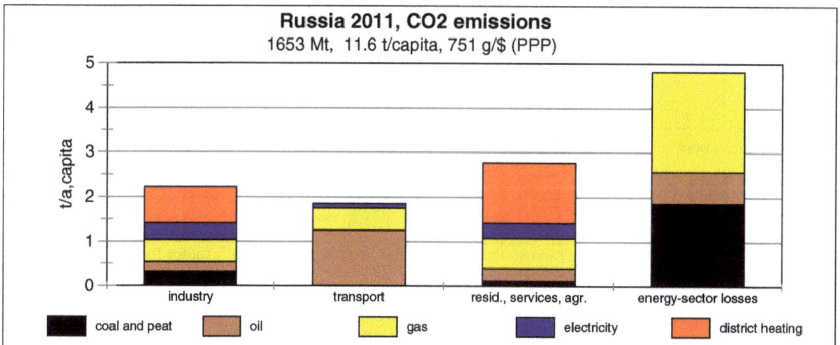


A5. Russia: Energy carriers responsible for the $CO_2$ emissions of sectors of the economy ;
(for electricity see also A2 or A6 and for district heating also A2)

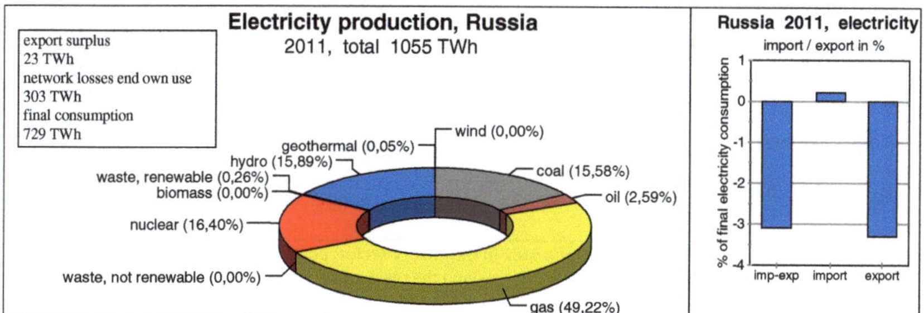


6. Russia: Generation of electrical energy, final consumption = production + imports - exports - losses

## Saudi Arabia

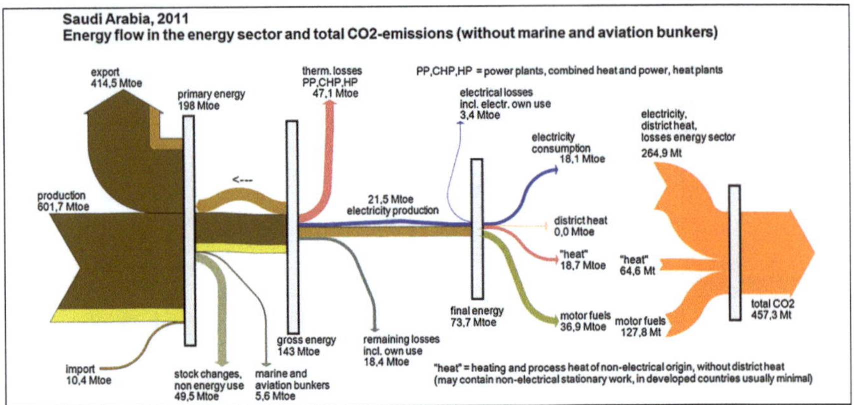


A1. Saudi Arabia: Energy flow in the energy sector from primary to final energy and total $CO_2$ emissions. Color of the energy carriers as in A2 and A5 (petroleum dark brown, oil products light brown)

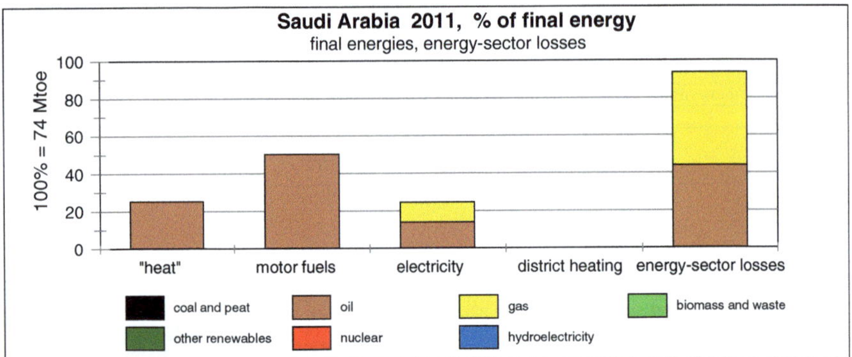


A2. Saudi Arabia: Shares of energy carriers for obtaining the final energies ("heat", motor fuels, electricity, district heating) and to cover the losses of the energy sector

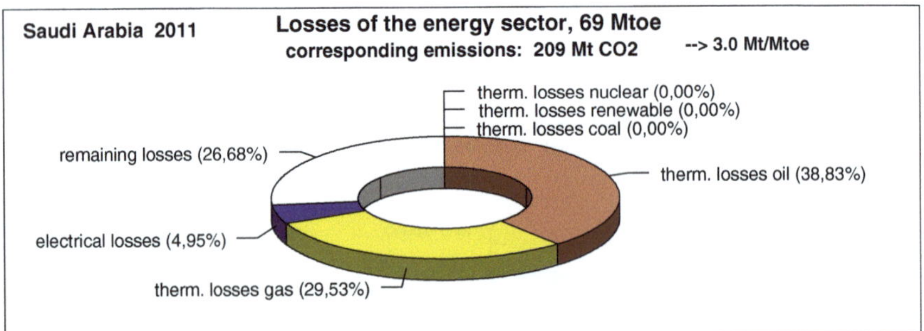


A3. Saudi Arabia: Percentage distribution of losses in the energy sector; contributors to $CO_2$ emissions include thermal losses of fossil-fuel plants, the electrical losses and the residual losses

## Saudi Arabia

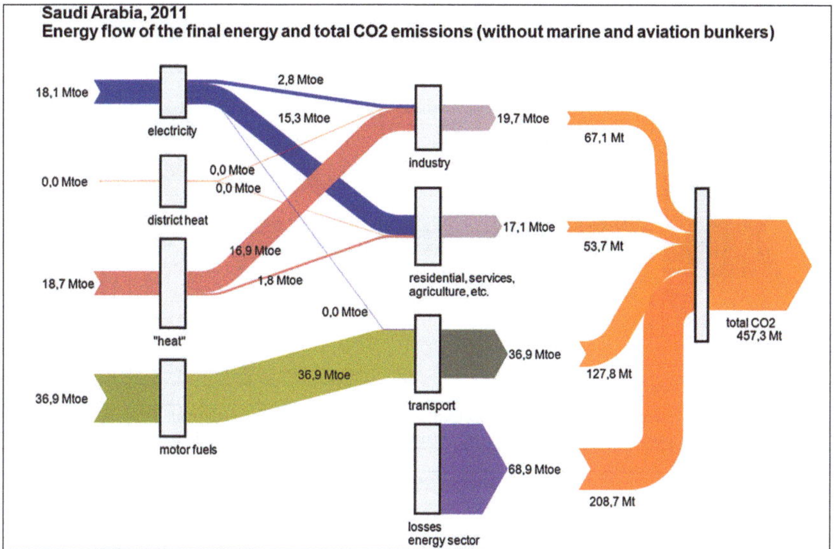


A4. Saudi Arabia: Flow of final energy to the final users and $CO_2$ emissions of the economy sectors

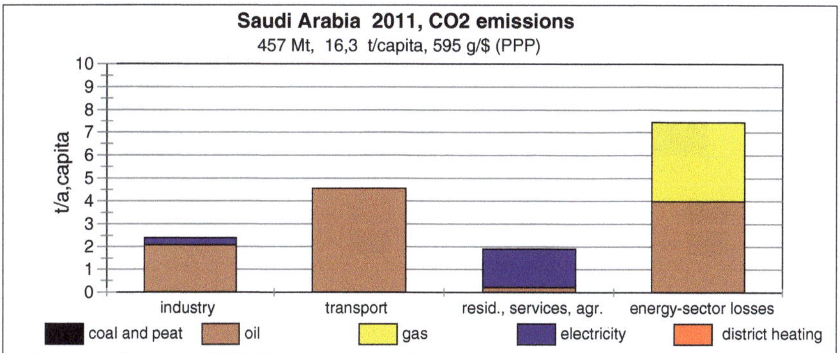


A5. Saudi Arabia: Energy carriers responsible for the $CO_2$ emissions of sectors of the economy ; (for electricity see also A2 or A6 and for district heating also A2)

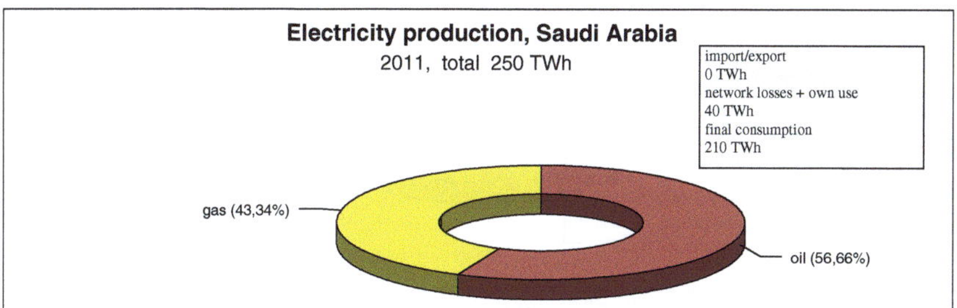


A6. Saudi Arabia: Generation of electrical energy, final consumption = production + imports - exports - losses

## South Africa

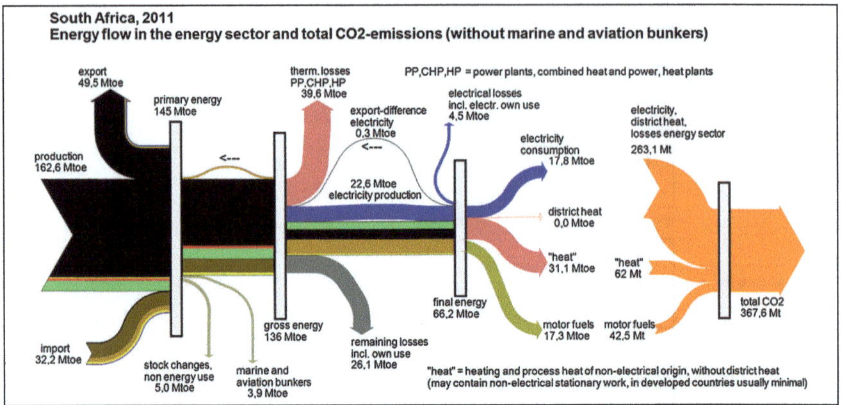


A1. South Africa: Energy flow in the energy sector from primary to final energy and total $CO_2$ emissions. Color of the energy carriers as in A2 and A5 (petroleum dark brown, oil products light brown)

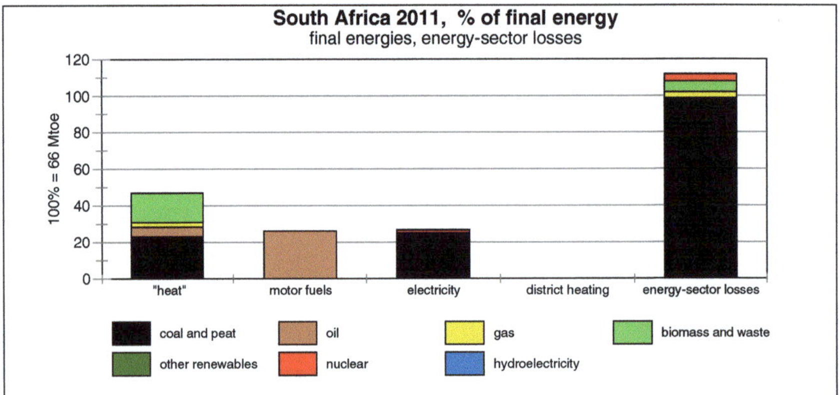


A2. South Africa: Shares of energy carriers for obtaining the final energies ("heat", motor fuels, electricity, district heating) and to cover the losses of the energy sector

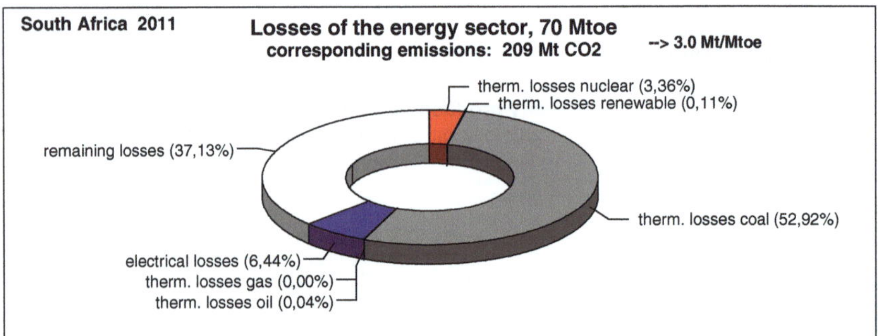


A3. South Africa: Percentage distribution of losses in the energy sector; contributors to $CO_2$ emissions include thermal losses of fossil-fuel plants, the electrical losses and the residual losses

## South Africa

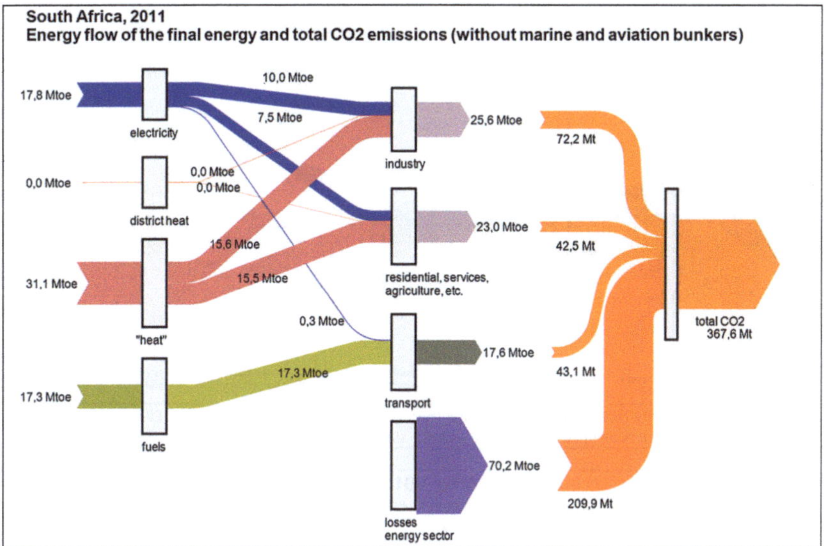


A4. South Africa: Flow of final energy to the final users and $CO_2$ emissions of the economy sectors

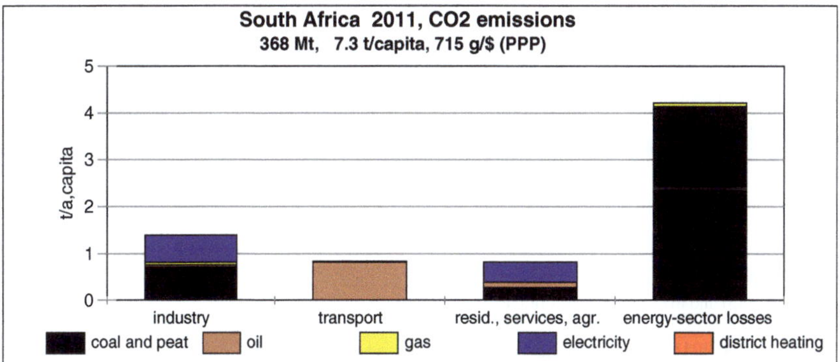


A5. South Africa: Energy carriers responsible for the $CO_2$ emissions of sectors of the economy ; (for electricity see also A2 or A6 and for district heating also A2)

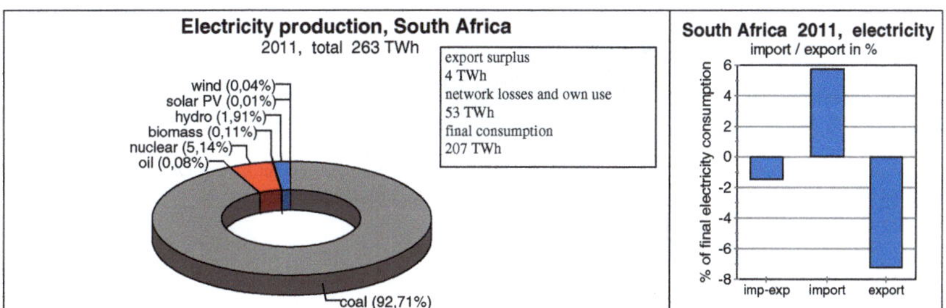


A6. South Africa: Generation of electrical energy, final consumption = production + imports - exports - losses

## South Korea

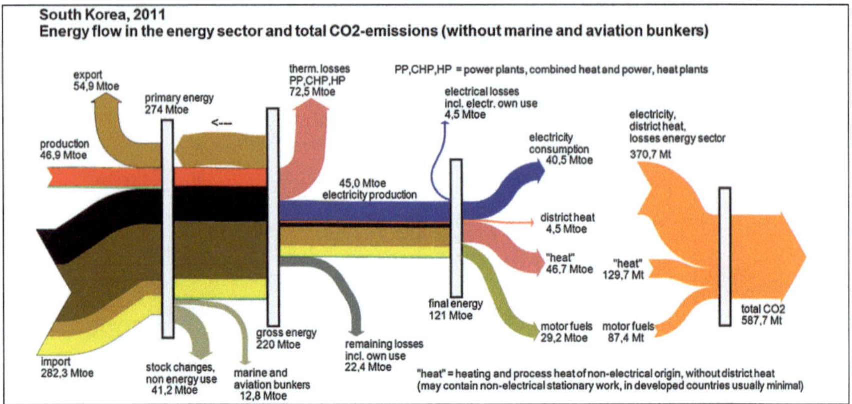


A1. South Korea: Energy flow in the energy sector from primary to final energy and total $CO_2$ emissions. Color of the energy carriers as in A2 and A5 (petroleum dark brown, oil products light brown)

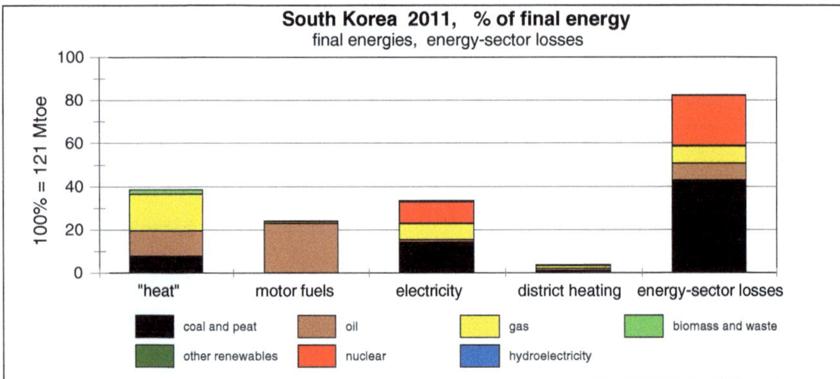


A2. South Korea: Shares of energy carriers for obtaining the final energies ("heat", motor fuels, electricity, district heating) and to cover the losses of the energy sector

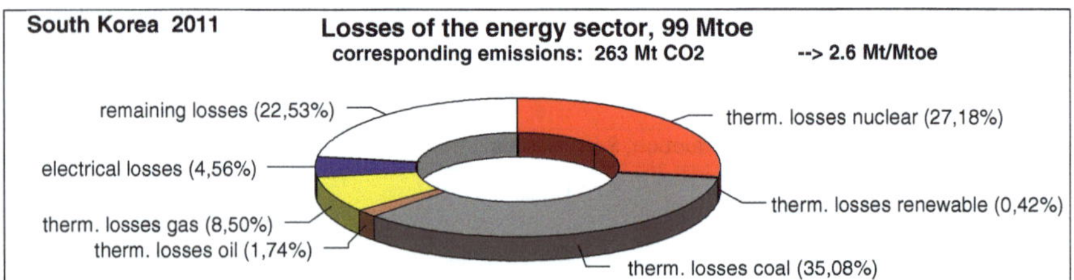


A3. South Korea: Percentage distribution of losses in the energy sector; contributors to $CO_2$ emissions include thermal losses of fossil-fuel plants, the electrical losses and the residual losses

## South Korea

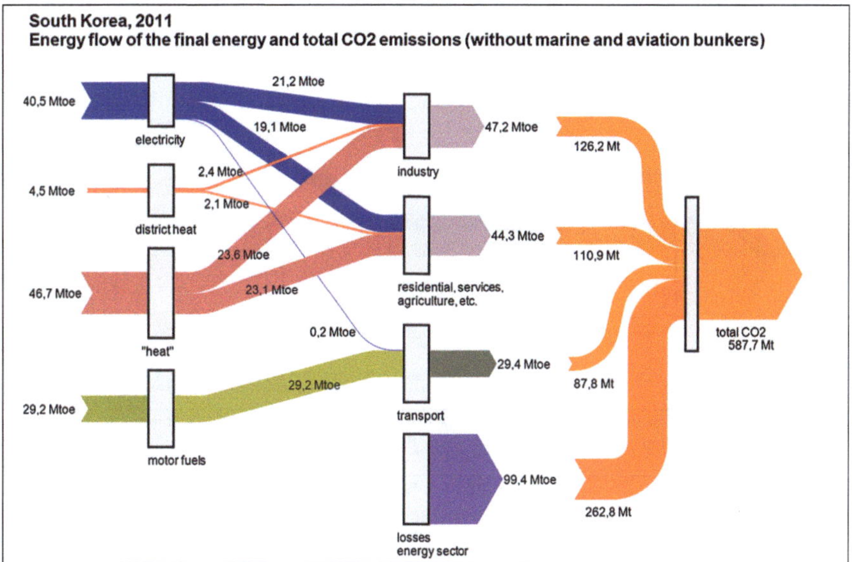


A4. South Korea: Flow of final energy to the final users and $CO_2$ emissions of the economy sectors

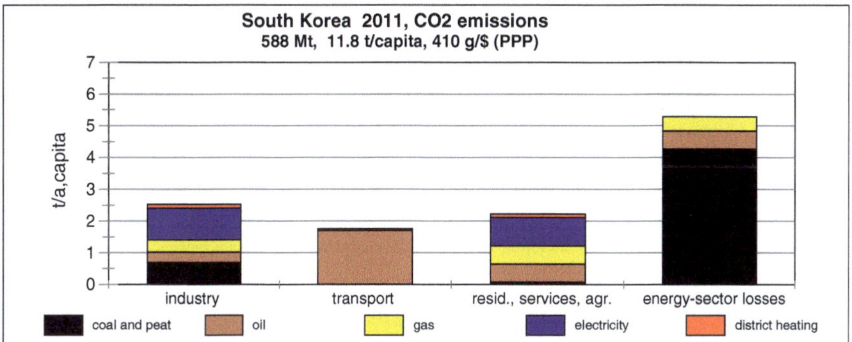


A5. South Korea: Energy carriers responsible for the $CO_2$ emissions of sectors of the economy ; (for electricity see also A2 or A6 and for district heating also A2)

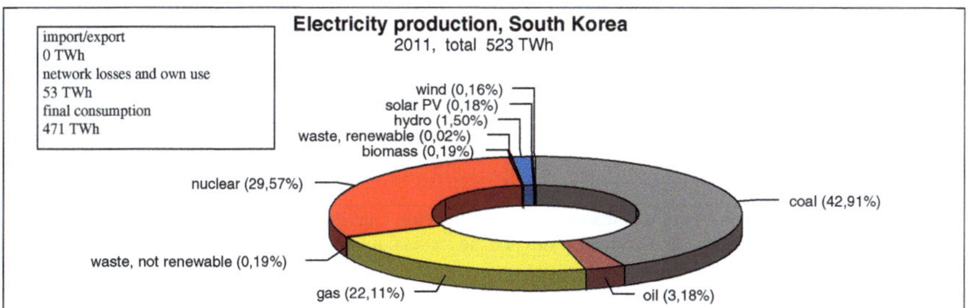


A6. South Korea: Generation of electrical energy, final consumption = production + imports - exports - losses

## Turkey

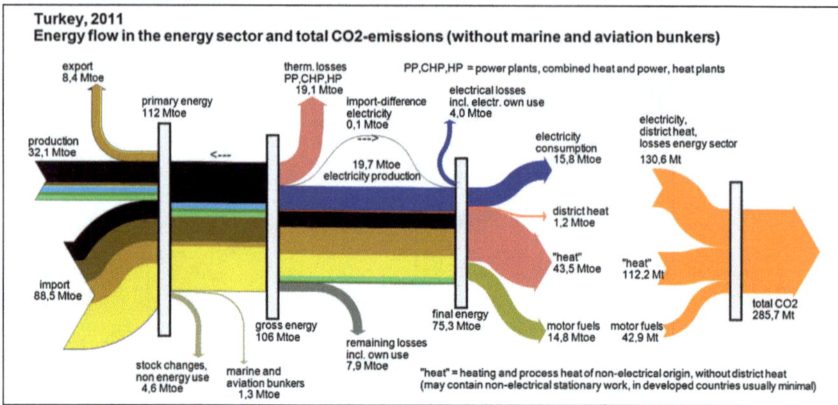


A1. Turkey: Energy flow in the energy sector from primary to final energy and total $CO_2$ emissions. Color of the energy carriers as in A2 and A5 (petroleum dark brown, oil products light brown)

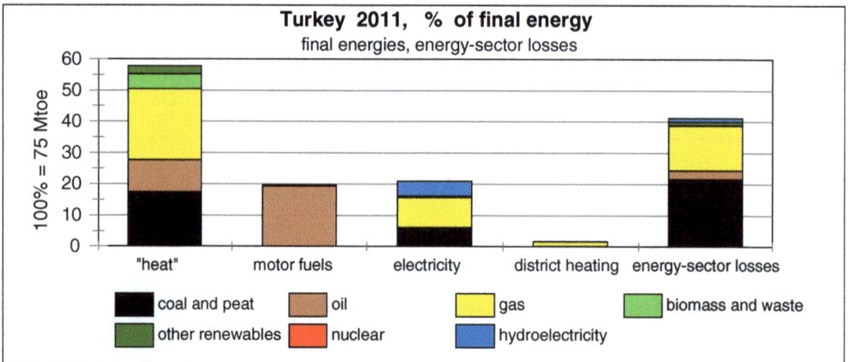


A2. Turkey: Shares of energy carriers for obtaining the final energies ("heat", motor fuels, electricity, district heating) and to cover the losses of the energy sector

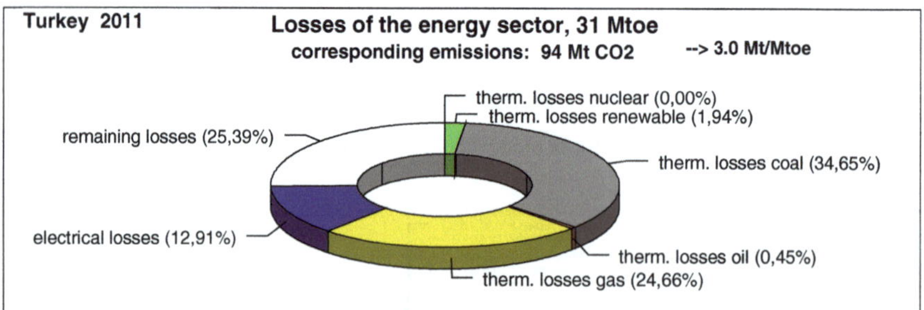


A3. Turkey: Percentage distribution of losses in the energy sector; contributors to $CO_2$ emissions include thermal losses of fossil-fuel plants, the electrical losses and the residual losses

## Turkey

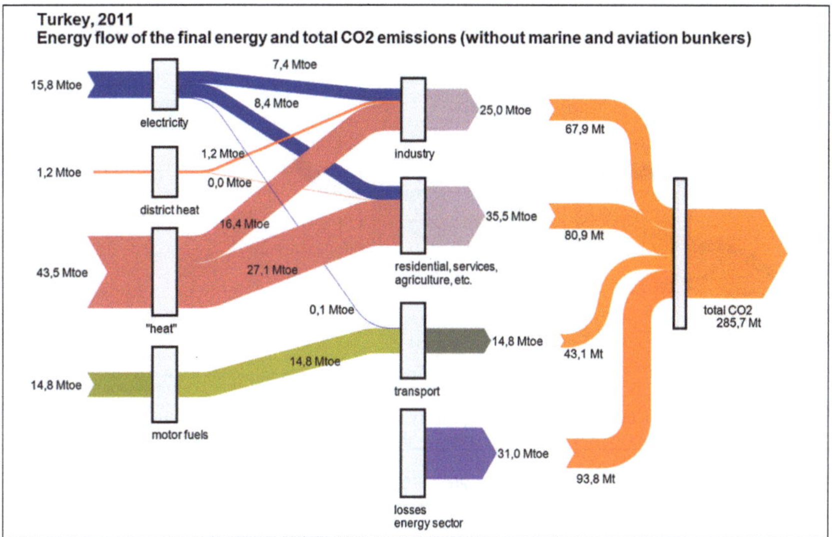


A4. Turkey: Flow of final energy to the final users and $CO_2$ emissions of the economy sectors

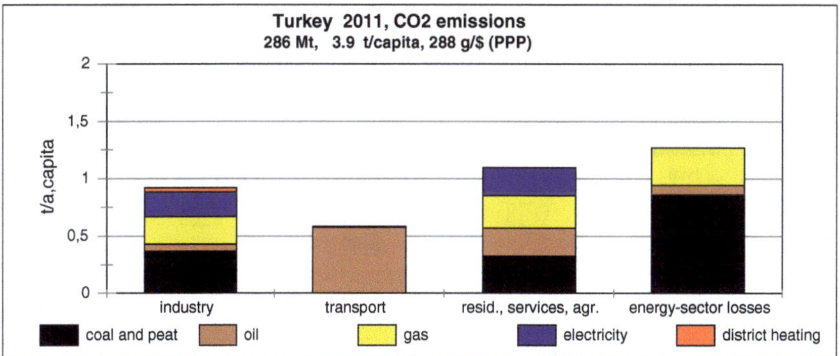


A5. Turkey: Energy carriers responsible for the $CO_2$ emissions of sectors of the economy ; (for electricity see also A2 or A6 and for district heating also A2)

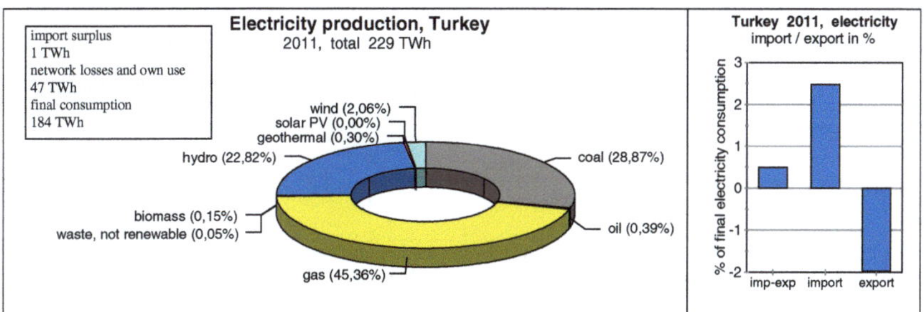


A6. Turkey: Generation of electrical energy, final consumption = production + imports - exports - losses

## United Kingdom

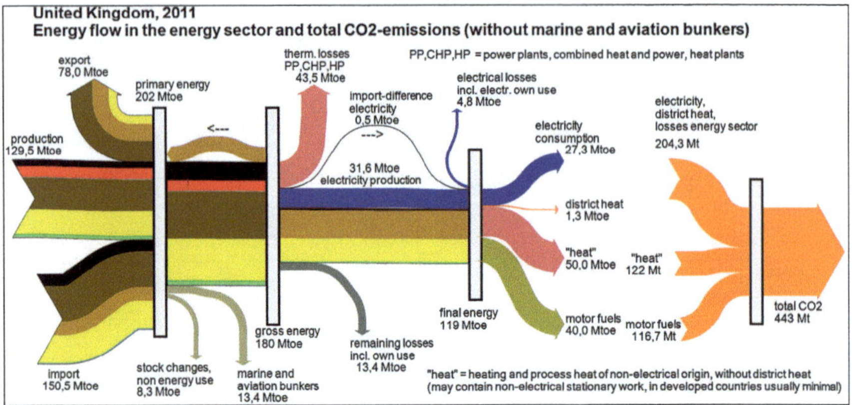


A1. U.K.: Energy flow in the energy sector from primary to final energy and total $CO_2$ emissions. Color of the energy carriers as in A2 and A5 (petroleum dark brown, oil products light brown)

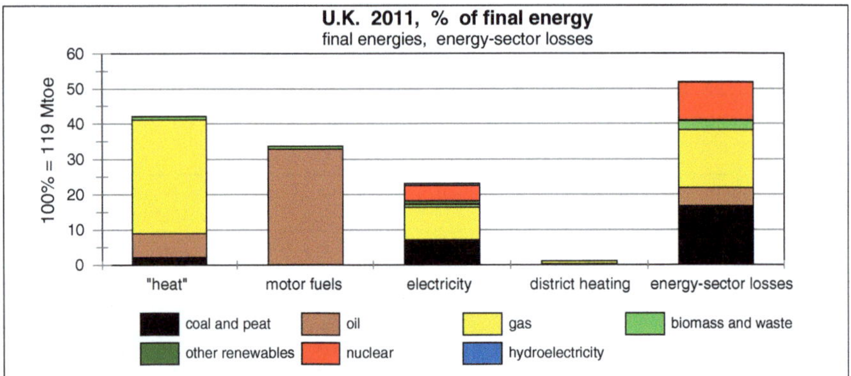


A2. U.K: Shares of energy carriers for obtaining the final energies ("heat", motor fuels, electricity, district heating) and to cover the losses of the energy sector

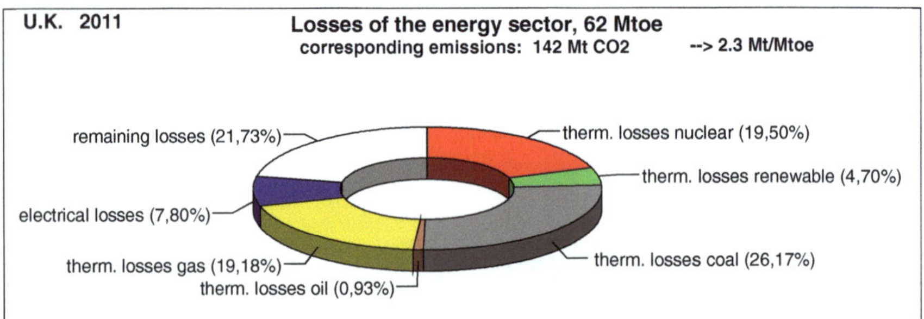


A3. U.K.: Percentage distribution of losses in the energy sector; contributors to $CO_2$ emissions include thermal losses of fossil-fuel plants, the electrical losses and the residual losses

## United Kingdom

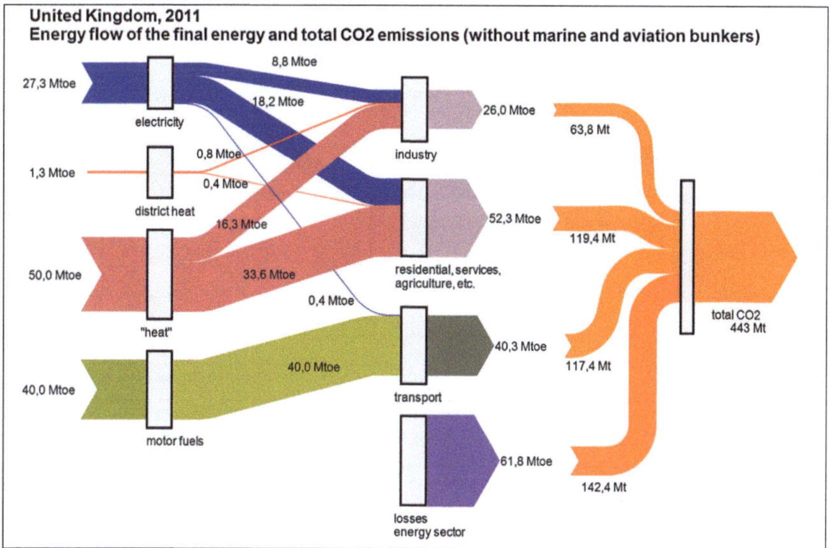


A4. U.K.: Flow of final energy to the final users and $CO_2$ emissions of the economy sectors

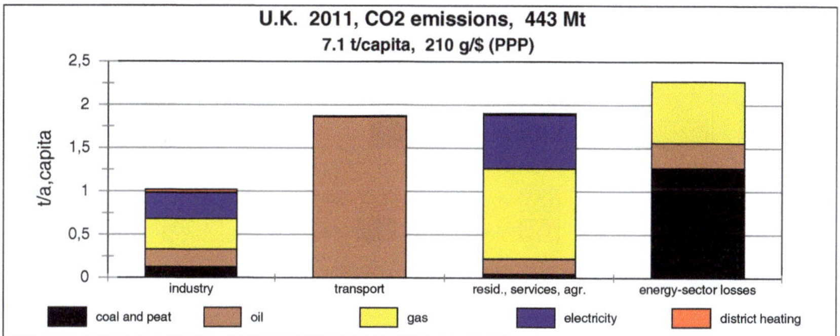


A5. U.K.: Energy carriers responsible for the $CO_2$ emissions of sectors of the economy ;
(for electricity see also A2 or A6 and for district heating also A2)

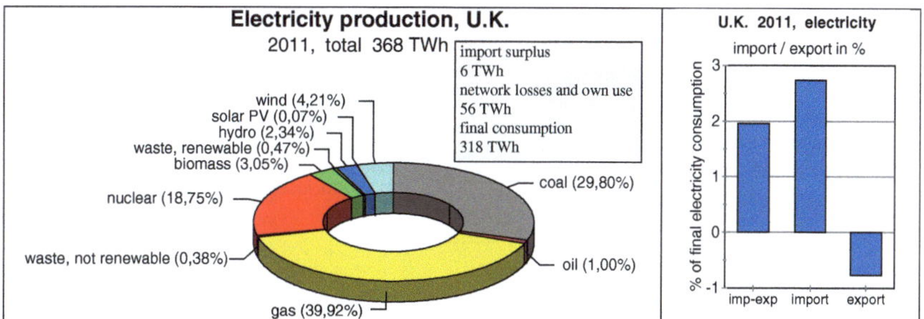


A6. U.K.: Generation of electrical energy, final consumption = production + imports - exports - losses

## United States of America

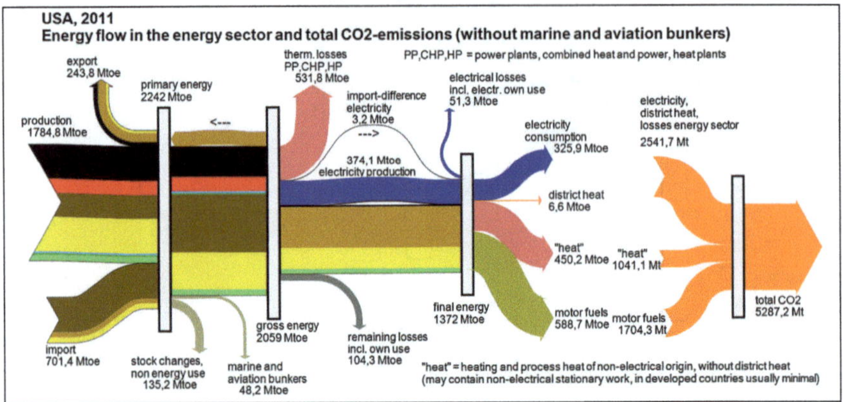


A1. USA: Energy flow in the energy sector from primary to final energy and total $CO_2$ emissions. Color of the energy carriers as in A2 and A5 (petroleum dark brown, oil products light brown)

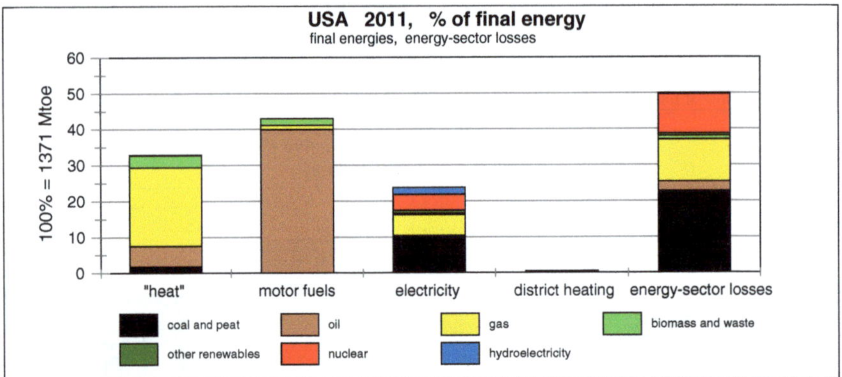


A2. USA: Shares of energy carriers for obtaining the final energies ("heat", motor fuels, electricity, district heating) and to cover the losses of the energy sector

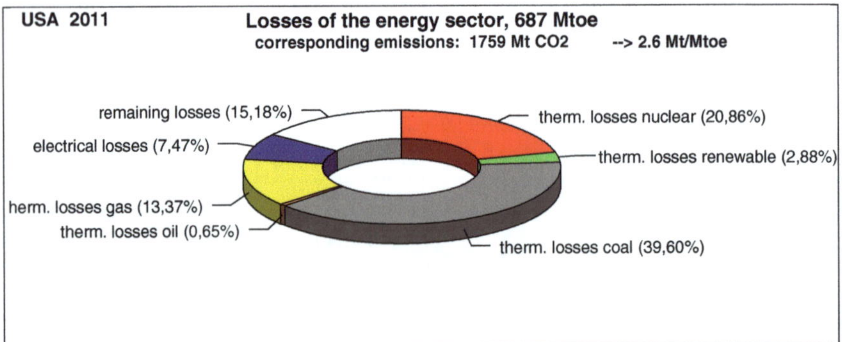


A3. USA: Percentage distribution of losses in the energy sector; contributors to $CO_2$ emissions include thermal losses of fossil-fuel plants, the electrical losses and the residual losses

## United States of America

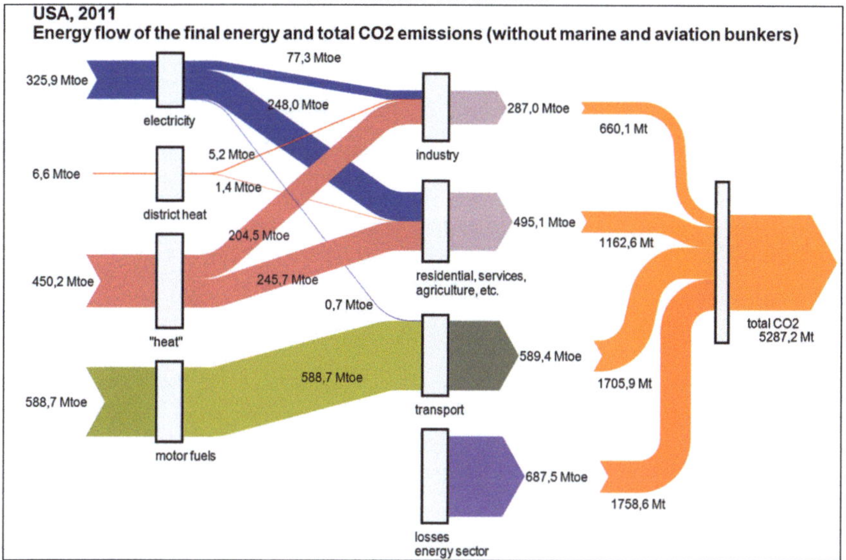


A4. USA: Flow of final energy to the final users and $CO_2$ emissions of the economy sectors

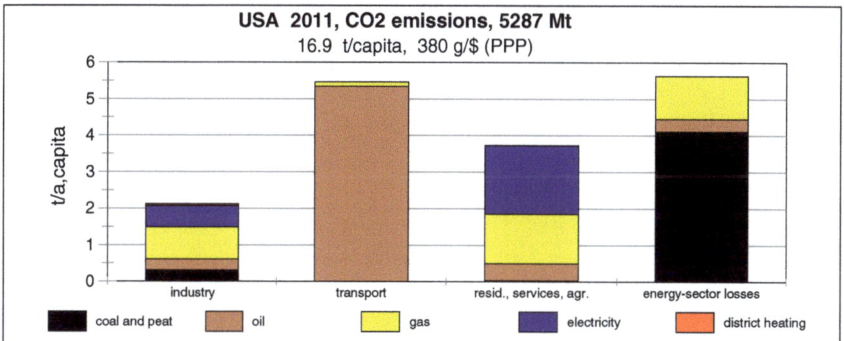


A5. USA: Energy carriers responsible for the $CO_2$ emissions of sectors of the economy ; (for electricity see also A2 or A6 and for district heating also A2)

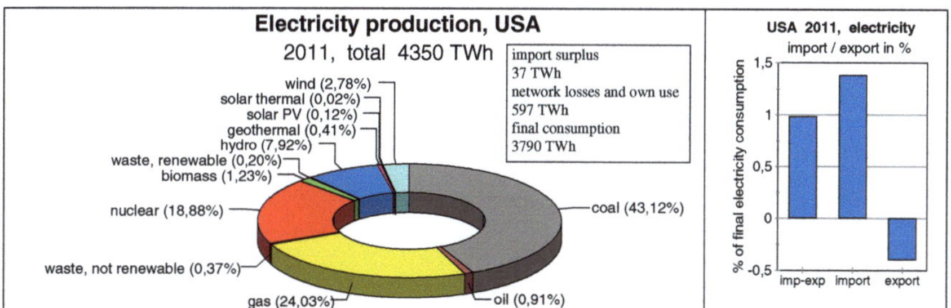


A6. USA: Generation of electrical energy, final consumption = production + imports - exports – losses

# Bibliography

1. IPCC (Intergovernmental Panels on Climate Change) (2013) 5th report, WG I. Sept 2013
2. IPCC (2014) 5th report Working Group II. Mar 2014
3. IPCC (2014) 5th report Working Group III. Apr 2014
4. Steinacher M, Joos F, Stocker TF (2013) Allowable carbon emissions lowered by multiple climate targets. Nature 499:197–201
5. IEA (International Energy Agency) (2014) Statistics & Balances. www.iea.org
6. IEA (2013) World Energy Outlook. www.iea.org
7. Oliver Geden (2012) SWP-Studie, Die Modifikation des 2-Grad-Ziels, Berlin
8. IMF (International Monetary Fund) (2013) World Economic Databases. Oct 2013
9. IEA (2004–2012) World Energy Outlooks.
10. Crastan V (2012) Elektrische Energieversorgung 2, 3rd edn. Springer, Berlin
11. IEA (2011) $CO_2$-Emissions from Fuel Combustion. Highlights 2011
12. IEA (2006–2013) Key World Energy Statistics.
13. IMF (International Monetary Fund) (2014) World Economic Databases. Apr 2014



V. Crastan, *Global Energy Demand and 2-degree Target, Report 2014*,
DOI 10.1007/978-3-319-12310-3

MIX
Papier aus verantwortungsvollen Quellen
Paper from responsible sources
FSC® C105338

If you have any concerns about our products,
you can contact us on
**ProductSafety@springernature.com**

In case Publisher is established outside the EU,
the EU authorized representative is:
**Springer Nature Customer Service Center GmbH**
**Europaplatz 3, 69115 Heidelberg, Germany**

Printed by Libri Plureos GmbH
in Hamburg, Germany